The risks of inclusion

Shifts in governance processes and upgrading opportunities for cocoa farmers in Ghana

Anna Laven

The risks of inclusion
Anna Laven

Royal Tropical Institute (KIT)
KIT Development Policy & Practice
PO Box 95001
1090 HA Amsterdam
The Netherlands
Telephone: +31 (0)20 568 8458
Fax: +31 (0)20 568 8444
Website: www.kit.nl
Email: a.laven@kit.nl

KIT Publishers
Website: www.kitpublishers.nl
Email: publishers@kit.nl

© 2010 KIT Publishers, Amsterdam; Anna Laven

Cover photo: Anna Laven
Other photos: Anna Laven
Cover design: Ronald Boiten, Amersfoort
Printing: Bariet BV, Ruinen

ISBN: 978 94 6022 111 8

All rights reserved. No part of this publication may be reproduced, stored in a retrieval system or transmitted in any form or by any means electronic, mechanical, photo-copying, recording or otherwise without the prior permission of the publishers.

THE RISKS OF INCLUSION

Shifts in governance processes and upgrading opportunities for cocoa farmers in Ghana

ACADEMISCH PROEFSCHRIFT

ter verkrijging van de graad van doctor
aan de Universiteit van Amsterdam
op gezag van de Rector Magnificus
prof. dr. D.C. van den Boom
ten overstaan van een door het college voor promoties
ingestelde commissie,
in het openbaar te verdedigen in de Agnietenkapel
op dinsdag 29 juni, te 10:00 uur

door

Anna Cornelie Laven

geboren te Rotterdam

Promotiecommissie

Promotores: Prof. dr. I.S.A. Baud
                  Prof. dr. G.C.A. Junne

Co-promotores: Dr. J. Post, UHD
                  Dr. A.E. Fernández Jilberto, UHD

Overige leden: Prof. dr. A.J. Dietz
                  Prof. dr. R. Kloosterman
                  Prof. dr. A. Helmsing
                  Prof. dr. R. Ruben
                  Dr. G. van Westen, UHD
                  Dr. F. Ruf, deskundige

Faculteit der Maatschappij- en Gedragswetenschappen

Table of content

List of tables	9
List of figures	10
List of boxes	11
List of pictures	11
Abbreviations	12
Acknowledgements	15

1 Local producers in a global economy — 18
1.1 Introduction — 18
1.2 Rural poverty — 19
    1.2.1 Poor farmers as rural entrepreneurs — 19
    1.2.2 Empowering poor farmers — 20
1.3 Research choices — 21
    1.3.1 A multi-level approach towards upgrading — 21
    1.3.2 Cocoa — 22
    1.3.3 Ghana — 23
1.4 A guide to the study — 23

2 Upgrading for development — 25
2.1 Introduction — 25
2.2 Defining the concept of upgrading — 26
    2.2.1 Inclusive upgrading — 28
2.3 Upgrading and governance — 29
    2.3.1 Global Value Chain analysis — 29
        Chain governance — 32
        Chain coordination — 33
        Limitations of the GVC approach — 34
    2.3.2 The clustering approach — 35
        Collective efficiency, joint action and embeddedness — 37
        Limitations of the cluster approach — 38
    2.3.3 What about the role of the state? — 39
2.4 Linking different governance levels — 40

3 Research questions, methods and respondents — 42
3.1 Introduction — 42
3.2 Research questions — 43

	3.2.1	Research question 1	43
		Data collection	43
	3.2.2	Research question 2	44
		Data collection	45
	3.2.3	Research question 3	46
		Data collection	47
	3.2.4	Research question 4	48
		Data collection	48
3.3	Heterogeneity and agency		49
3.4	Reflections, limitations and the structure of the book		50
	3.4.1	Validating the research findings	50
	3.4.2	Trust and participation	51
	3.4.3	Scope and limitations of the research	52
4	**The risky business of cocoa**		**57**
4.1	Introduction		57
4.2	Governance in the global cocoa chain		58
4.3	Responses to change		63
	4.3.1	Interventions in the cocoa sector in Ghana	66
	4.3.2	Changing relations	68
4.4.	Reflections		69
5	**The role of the state in a liberalised cocoa sector**		**71**
5.1	Introduction		71
5.2	A sector in transition: the experience of West-Africa		72
	5.2.1	Côte d'Ivoire	73
	5.2.2	Cameroon	75
	5.2.3	Nigeria	76
	5.2.4	Responses to reforms	77
5.3	A sector in transition: partial liberalisation in Ghana		78
	5.3.1	From a fairly liberal policy to a more state-controlled economy: 1920-1980	78
	5.3.2	The introduction of gradual reforms: 1980-2008	81
		Internal marketing	82
		External marketing	85
		Institutional reforms	88
		Quality	88
		Extension services	89
		Input distribution and application	90
		Credit	90
		Farmer organisation	91
		Shifts in governance in Ghana	93

5.4	Discussing the impact of reforms: comparisons with fully liberalised countries in West Africa	94
	5.4.1 Comparison 1: price-developments, margins and taxes	95
	5.4.2 Comparison 2: volume of production	95
	5.4.3 Comparison 3: farmer income	96
	5.4.4 Comparison 4: quality and services	97
	5.4.5 Comparison 5: farmer organisation	98
5.5	Public and private interventions in Ghana	99
5.6	Conclusions	102
6	**Who are the cocoa farmers?**	**105**
6.1	Introduction	105
6.2	Different outcomes for different types of producers	106
	6.2.1 Location, domestic migration and landownership	107
	6.2.2 Cocoa production, size of farm and productivity	111
	6.2.3 Position in the community or chain, age and education	112
	6.2.4 Gender and the level of cooperation	116
6.3	When do differences matter?	119
6.4	Reflections	124
7	**The risks of inclusion**	**128**
7.1	Introduction	128
7.2	Upgrading strategies, sub-strategies and interventions in the cocoa sector	132
	7.2.1 Strategy 1: Capturing higher margins for unprocessed cocoa	137
	Sub-strategy 1.1: Capturing higher margins by producing better quality cocoa	137
	Sub-strategy 1.2: Increase in productivity ane higher volumes of production	146
	Sub-strategy 1.3: Producing under more remunerative contracts	160
	7.2.2 Strategy 2: Producing new forms of existing commodities	164
	Sub-strategy 2.1: Producing new forms of existing commodities by producing for niche markets	165
	Sub-strategy 2.2 and Sub-strategy 2.3	173
	7.2.3 Strategy 3: Localising commodity processing and marketing	177
	Sub-strategy 3.1: Processing of cocoa waste	178
	Sub-strategy 3.2: Localising processing of cocoa products	178
	Sub-strategy 3.3: Marketing of cocoa beans	181
7.3	Discussion on more inclusive upgrading for cocoa farmers in Ghana	184
	7.3.1 Upgrading patterns in Strategy 1: Capturing higher margins for unprocessed cocoa	184
	Competitiveness	184
	Remunerative income	185
	Empowerment	185

		7.3.2	Upgrading patterns in Strategy 2: Producing new forms of existing commodities	186
			Competitiveness	186
			Remunerative income	186
			Empowerment	187
		7.3.3	Upgrading patterns in Strategy 3: Localising commodity processing and marketing	187
			Competitiveness	187
			Remunerative income	187
			Empowerment	188
	7.4	Capturing dynamics, thinking in scenarios		188
		7.4.1	Scenario 1: Status quo with passive role of private sector	189
		7.4.2	Scenario 2: Opening up	190
			Inclusive upgrading: moving from scenario 1➤2	191
		7.4.3	Scenario 3: Loosing control	192
			Inclusive upgrading, moving from scenario 1➤3	192
		7.4.4	Scenario 4: Status quo with an active role of the private sector	192
			Inclusive upgrading: moving from scenario 1➤4	193
8	**Conclusions**			**195**
8.1	Introduction			195
8.2	Global Value Chain approach and upgrading			195
8.3	State governance and upgrading			197
8.4	Social structures and upgrading			199
8.5	Interactions between governance structures and upgrading			200
8.6	A framework for more inclusive upgrading			201

Final reflections	202
Notes	205
References	212
Summary	220
Nederlandse samenvatting	232
Appendix 3.1 Overview participants workshops	245
Appendix 6.1 Relations between farmer characteristics	248
Appendix 7.1 Analysing upgrading strategies	249

List of tables

Table 3.1	Cocoa production in Ghana in 2002/03 and 2003/04	54
Table 3.2	Respondents in FS 2003 and FS 2005	55
Table 4.1	The main processing companies and their volumes of production in 2006 and 2008	59
Table 4.2	Location of cocoa grinding activities in season 2005/06	60
Table 4.3	Threats, opportunities and strategies of global cocoa buyers	62
Table 4.4	Actors in the cocoa chain and their representative international organisations	64
Table 5.1	The World Bank 'view' on differences in cocoa marketing and pricing systems in 2001	73
Table 5.2	Regional cocoa purchases by LBCs (in tonnes) for season 2005-06	83
Table 5.3	The composition of the Net FOB price in 2002-03	87
Table 5.4	Objectives of gradual reforms: producer-prices, margins and taxes	94
Table 5.5	The impact of reforms in the major cocoa-producing countries	99
Table 5.6	State interventions in the cocoa sector 2005	100
Table 6.1	Cross-tabulation between position farm and yield; horizontal percentages	110
Table 6.2	Ownership, region and migration	110
Table 6.3	Productivity (in bags per acre) (mean)	111
Table 6.4	Volume of production in 2003/04	112
Table 6.5	Cross-tabulation between position farm and position in community; horizontal percentages	113
Table 6.6	Cross-tabulation between age and position in community; horizontal percentages	114
Table 6.7	Status, age and education	116
Table 6.8	Cross-tabulation between gender respondent and work together with other farmers; horizontal percentages	117
Table 6.9	Gender and working together	118
Table 6.10	Access to quality assistance	120
Table 6.11	Access to a mist blower	122
Table 6.12	Access to a knapsack sprayer	122
Table 6.13	Perceptions on the improvement of farmers' income	123
Table 6.14	Types of farmer organisation: linking upgrading to empowerment	125
Table 7.1	Notions on more inclusive upgrading	128
Table 7.2	Identified interventions affecting cocoa producers in Ghana	134
Table 7.3	Selection of upgrading strategies	136
Table 7.4	Measuring the inclusiveness of the public quality control system	147
Table 7.5	Total bonus payments to farmers 2000-2006	152
Table 7.6	Cross-tabulation between position in community or chain and frequency of mass sprayings; horizontal percentages	158

Table 7.7	Measuring the inclusiveness of the public mass-spraying programme	161
Table 7.8	Growth of the global organic market	166
Table 7.9	Measuring inclusiveness of membership in the Kuapa Kokoo Farmer Union	174
Table 7.10	Main uses of extra land	175
Table 7.11	Cocoa processing in Ghana	179
Table 7.12	Measuring the inclusiveness of cocoa processing in Ghana	182

List of figures

Figure 2.1	Conceptualising clustering and competitiveness	36
Figure 2.2	A comparative framework for labelling public-private interaction in a commodity chain	40
Figure 3.1	Conceptualising shifts in global chain governance and upgrading	44
Figure 3.2	Conceptualising shifts in national governance structures and upgrading	45
Figure 3.3	Conceptualising heterogeneity and inclusive upgrading	47
Figure 3.4	The interaction between global, national and local governance structures and their impact on upgrading	49
Figure 5.1	Organisation of the cocoa sector in Ghana in the 1980s	80
Figure 5.2	The main reasons for farmers to select an LBS in season 2002-03 (n=173)	84
Figure 5.3	Reasons for not working together in the 2003-04 season (n=68)	92
Figure 5.4	Changing governance systems in the Ghanaian cocoa sector	93
Figure 5.5	Margins in the cocoa supply chain 2001-02 main crop	95
Figure 6.1	Patterns of domestic migration	108
Figure 6.2	Cocoa yields for cocoa season 2003/04 in different locations	110
Figure 6.3	Average age of respondents	115
Figure 6.4	Age and education	115
Figure 7.1	Chain empowerment matrix	130
Figure 7.2	Understanding empowerment	130
Figure 7.3	Scenario matrix	131
Figure 7.4	Overview of upgrading strategies, sub-strategies and the involved actors	133
Figure 7.5	Strategy 1: Capturing higher margins for unprocessed cocoa	138
Figure 7.6	Main reasons for the quality decline in 2003/04 – a farmer perspective	140
Figure 7.7	Local quality aspects in the production and control of premium quality cocoa	144
Figure 7.8	Cocoa production and prices in Ghana	149
Figure 7.9	Main reasons for improved yield in cocoa season 2003/2004	150
Figure 7.10	Main reasons for the production decrease in season 2003/4	151
Figure 7.11	Bonuses in seasons 2002/03 and 2003/04	153
Figure 7.12	Frequency of mass spraying in season 2003/04	157

Figure 7.13 Changes in the empowerment matrix due to public interventions 164
Figure 7.14 Strategy 2: Producing new forms of existing commodities 165
Figure 7.15 The organisation of the Kuapa Kokoo Group 170
Figure 7.16 Changes in the empowerment matrix due to a multi-stakeholder initiative 176
Figure 7.17 Strategy 3: Localising commodity processing and marketing 177
Figure 7.18 Changes in the empowerment matrix due to outsourcing of local processing of cocoa to Ghana 183
Figure 7.19 Moving from scenario 1➤2 191
Figure 7.20 Moving from scenario 1➤3 192
Figure 7.21 Moving from scenario 1➤4 193

List of boxes
Box 2.1 The *Filière* tradition 31
Box 6.1 Land-rights for women in Ghana 117
Box 7.1 The Cocoa Abrabo-pa package 153
Box 7.2 Farmer Field Schools 154

List of pictures
Picture 5.1 The Cocoa Research Institute Ghana 81
Picture 5.2 A district buying store in Asankrangwa (Western region) 82
Picture 5.3 A buying station in the Western region 85
Picture 6.1 *left*: Regional Chief Farmer of Western region (Enchi) 113
             right: spiritual leader and large farmer (Asangkrangwa) 113
Picture 6.2 An elder selling cocoa to the Yukwa society (Central Region) 114
Picture 6.3 Knapsack sprayer for sale in an inputshop 121
Picture 6.4 A field visit to farmers that have been trained as trainers in the FFS (2003) 126
Picture 7.1 QCD personnel is checking cocoa quality at a buying depot in Dunkwa (Central Region) 143
Picture 7.2 QCD officials take samples out of cocoa bags 143
Picture 7.3 A Kuapa Kokoo buying station 172
Picture 7.4 Visiting the Cocoa Processing Company in Tema 180

ABBREVIATIONS

ABICAB	Brazilian Chocolate, Cocoa and Confectionary Manufacturers Association
ACRI	American Cocoa Research Institute
ADM	Archer Daniels Midland
AGIDS	the Amsterdam research institute for Global Issues and Development Studies.
ASAP	Agricultural Sector Adjustment Program
ASSR	Amsterdam School for Social Science Research
BBC	British Broadcasting Corporation
Caistab	*Caisse de Stabilisation et de Soutien des Prix de Produits Agricoles*
CAOBISCO	Association of the Chocolate, Biscuit and Confectionary Industries of the European Union
CDC	Cocoa Development Committee
CERES	CEnter for REsource Studies for development
CFC	Common Fund for Commodities
CI	Conservation International
CIRAD	*Centre International en Recherche Agronomique pour le Développement*
CMA	Chocolate Manufacturers Association
CMAA	Cocoa Merchants Association of America
CMC	Cocoa Marketing Company
CODAPEC	National Cocoa Diseases and Pests Control
COPAL	Cocoa Producers Alliance
COS	Convergence of Sciences
CPC	Cocoa Processing Company
CPP	Convention People Party
CRIG	Cocoa Research Institute Ghana
CRP	Cocoa Rehabilitation Project
CSD	Cocoa Services Division
CSR	Corporate Social Responsibility
CSSVD	Cocoa Swollen Shoot Virus Disease
DCC	Divine Chocolate Company
DFID	Department for International Development in the United Kingdom
DPRN	Development Policy Research Network
EBRD	European Bank for Reconstruction and Development
ECA	European Cocoa Association
ECLAC	Economic Commission for Latin America and the Caribbean
EDIAIS	Enterprise Development Impact Assessment Information System
EGFAR	Global Forum on Agricultural Research
EIBE	European Institute for Business Ethics
ERP	Economic Recovery Program
ESI VU	*Economisch en Sociaal Instituut Vrije Universiteit van Amsterdam*

EU	European Union
FAO	Food and Agriculture Organisation
FCC	Federation of Cocoa Commerce
FFS	Farmer Field School
FLO	Fair Trade Labelling Organisation
FMG	*Faculteit der Maatschappij en Gedragswetenschappen*
FOB	Free on Board
FS	Farmer Survey
FT	Fair Trade
GCCSFA	Ghana Cocoa, Coffee and Sheanut Farmers Association
GCFS	Ghana Cocoa Farmers Survey
GCMA	Ghana Marketing Cooperative Association
GDP	Gross Domestic Product
GTZ	*Deutsche Gesellschaft für Technische Zusammenarbeit*
GVC	Global Value Chain
HIVOS	Humanist Institute for Development Cooperation
ICA	International Confectionary Association
ICCO	International Cocoa Organisation
ICPM	Integrated Crop Pest Management Unit
Icco	*Interkerkelijke Organisatie voor Ontwikkelingssamenwerking*
IDRC	International Development Research Centre
IDS	Institute of Development Studies
IFAD	International Fund for Agricultural Development
IFPRI	International Food Policy Research Institute
IITA	Agricultural Research for Development in Africa
ILO	International Labour Organisation
INRA	Institute National de la Recherche Agronomique
IMF	International Monetary Fund
IPM	Integrated Pest Management
ISCOM	Institute for Sustainable Commodities
ISSER	Institute of Statistical Social and Economic Research
IUCN	International Union for Conservation of Nature
KIT	*Koninklijk Instituut voor de Tropen* (Royal Tropical Institute)
KKFT	Kuapa Kokoo Farmers Trust
KKFU	Kuapa Kokoo Farmer Union
KKL	Kuapa Kokoo Limited
LBC	Licensed Buying Company
LIFFE	London International Financial Futures Exchange
MDG	Millennium Development Goal
MMYE	Ministry of Manpower, Youth and Employment
MoFA	Ministry of Food and Agriculture
MOFEP	Ministry of Finance and Economic Planning
NCB	Nigerian Cocoa Board
NCCB	National Cocoa and Coffee Board
NEC	National Excecutive Council

NGO	Non-governmental Organisation
NIDO	*Nationaal Initiatief Duurzame Ontwikkeling*
NPP	New Patriotic Party
NY CSCE	Options Exchange New York Coffee, Sugar and Cocoa Exchange
ODI	Overseas Development Institute
ONCC	*Office National du Café et du Cacao*
ONCP	*Office National de Commercialisation de Produits de Base*
PBC	Produce Buying Company
PC	Purchasing Clerk
POP	Persistent Organic Pollutants
PPP	Public Private Partnership
PPRC	Producer Price Review Committee
PSOM	*Programma Samenwerking Opkomende Markten* (Programme Cooperation Emerging Markets)
PWC	PricewaterhouseCoopers
QCD	Quality Control Division
RIAS	Rabo International Advisory Services
RSCE	Round Table for a Sustainable Cocoa Economy
SAP	Structural Adjustment Program
SME	Small and medium enterprise
SNV	Netherlands Development Organisation
SOMO	*Stichting Onderzoek Mulitinationale Ondernemingen*
STCP	Sustainable Tree Crop Programme
TCC	Tropical Commodity Coalition
ToT	Trainer of Trainers
UK	United Kingdom
UNCTAD	United Nations Conference on Trade and Development
UNIDO	United Nations Industrial Development Organisation
USAID	United States Agency for International Development
USD	United States Dollar
UvA	University of Amsterdam
VU	*Vrije Universiteit Amsterdam*
WAM	West African Mills
WAPCB	West African Produce Control Board
WCF	World Cocoa Foundation
WHO	World Health Organisation

Acknowledgements

Doing a PhD has been a process of constantly finding out how little it is I actually know.

It has taken quite a number of years to finish this book, and there are many people I would like to thank. I am particularly grateful to all the Ghanaian cocoa farmers and their families that welcomed me in their communities. I am thankful to the farmers that have participated in the surveys, in-depth interviews, group discussions and video-sessions. I am particularly grateful to Justice Appiah, Ama Owusu, Charles Adjei, John Busua and Charles K. Boateng. My fieldwork would not have been possible without the support of Conservation International Ghana and Wienco. It was through Yaw Osei-Owusu from Conservation International that I came to know Eric Kofi Doe and David MacMensah Metsowasa. They introduced me to the farmers in cocoa growing communities, assisted me in my field research and became my friends. Marc Kok and Ralph Odei-Tettey from Wienco played a role of similar importance. They made it possible to reach out to a large number of cocoa farmers in more remote areas. I am particularly grateful to Christian Adu-Asare who accompanied me during these field trips. I am also grateful to Ben Appiah Kubi and Benjamin Gyasi who assisted John van Duursen and Derk-Jaap Norde, two Dutch post-graduate students who conducted a part of the farmer survey in 2003. Thanks a lot John and Derk for your enthousiasm, valuable contribution and good company.

In Ghana I spoke to numerous representatives from the Ghanaian Cocobod, Ministry of Food and Agriculture, Ministry of Finance, Department of Cooperatives, farmer groups, World Bank, Licensed Buying Companies, rural banks, NGOs, researchers, etc. Thanks to all of you! A few people I would like to mention here in particular: Peter van Grinsven, Stephan Weise, François Ruf, Kenneth Brew, Toni Fofi, Taco Terheijden, Kwame Amezah, David Preece, Nathan Leibel, Peter van der Wurff, Dick de Bruijn, Nana Appiah, Nana Erhuma Kpanyinly, Nicolas Awartwi, Pieter Frieling and all the people from CRIG and from Kuapa Kokoo. Also thanks to Teun Wolters, Anouk van Heeren and to the Dutch Embassy in Ghana, who introduced me to a number of key players in the Ghanaian cocoa sector.

I am also grateful to Prof. dr. Adarkwa from the University of Kumasi who acted as my local supervisor when I just started my PhD. I am in particular grateful to the staff of the University of Legon/ISSER where I met different cocoa experts. Special thanks to Prof. dr. Nyanteng who became involved in organising a multi-stakeholder meeting in Accra. Also thanks to dr. Asante and dr. Anarfi who introduced me to Henry Anim-Somuah, another cocoaexpert, who I was lucky to work with in the field. Henry brought on board Isaac Asare and Robert Assan-Donkoh, two knowledgeable students who assisted me with the farmer survey and during the conference. I am also particularly grateful to Pedro Arens from SNV and the support he provided to the meeting, together with his colleagues Eric Agyare, Lawrence Attipoe and Maureen Odoi.

In Ghana there are some other people I would like to thank. Piet Bakker, Cecilia and Badi for their friendship and hospitality. Nicolas for helping me to get around the first weeks in Ghana. Ben and Loes for helping us, as agreed, with a house, a car and new friends like Sophie, Alex, Lucienne, Ticho and Suzanne. Thanks also for bringing Okuleh into our lives. Okuleh, you made living in Accra with Jurriaan and Julia a joy. Thanks for being there and being so kind to Julia. Tjalling, it was always a pleasure having you around. Thanks also to my dear friends who came to visit us in Accra: Femke and Mike!

In the Netherlands I would like to thank my PhD supervisory team. My two promotores Isa Baud and Gerd Junne and my co-promotores Johan Post and Alex Fernández Jilberto. Isa, I certainly have a lot to thank you for. You supported and stimulated me to start the PhD process and helped me to bring it to a good end. I admire you for your analytical skills and I have enjoyed your kindness, brightness and hospitality! Gerd, it has been a privilege to have you as my other promotor. I am grateful for your constructive comments, your inspiration and admire your openmindedness and ability to connect. Johan, I am also very grateful to you, never tired to provide me with detailed and constructive comments. I very much enjoyed the discussions we had and the challenging questions you posed. Alex, dear Alex, you are the first that inspired me to do a PhD and I have always felt your support in all the decisions I made. Thanks for this and also for always challenging me to remain true to myself, which I regard as a lesson for life.

Within AMIDSt I would like to thank especially Gert van der Meer, Annemieke van Haastrecht, Clinton Siccama, Guida Morais de Castro, Marianne van Heelsbergen, Puikang Chan and Barbara Lawa for all their support. I am in particular grateful to Sjoerd de Vos.

I would also like to thank Ewald van Engelen, Robert Kloosterman, Ad de Bruijne, Michaela Hordijk, Maarten Bavinck, Fred Zaal, Margriet Poppema, Jan Mansvelt Beck, Jan Markussen, Michiel Wagenaar, Mirjam Ros-Tonen and Len de Klerk. Special thanks to Ton Dietz. Your positive spirit was very encouraging and I am grateful for your support, both as director of AMIDSt and CERES.

Thanks to all my Phd fellows at AMIDSt. Still, whenever I enter the G-building at the Nieuwe Prinsengracht in Amsterdam, I directly feel at ease and at home. There are quite a number of people that contributed to that feeling. First of all my roommates from 'the' room G203: Ellen, Hebe, Perry, Marlie, Robert, Jacob. Plan-B! Other favourite colleagues: Koen, Marloes, Kees, Iris, Amanda, Merijn, Michael, Nadav, Benson, Udan, Inge, Lothar, Mirjam, Niels, Magali, Willemijn, Mendel, Fenne, Edith, Marjolein, Toni, Annika, Rogier, Joran, Thomas, Babak, Arjan, Els and my dear sister in law Edith.

I am grateful for the opportunity I had to supervise several students. I have enjoyed working with you. In particular I would like to mention Rogier van Vuure, Tessa Laan and Richard van Beuningen.

Special thanks to Jean Hellwig and Joshka Wessels for developing the VISTA video course and for teaching me the basics of video-making. Thanks to the CERES graduate school, CERES Board and the PhD Council. I am particularly grateful to their support in setting up the thematic group on value chains, local economic development and social inclusion. Lee Pegler, I would like to mention you here as my partner in crime. Thanks also Peter Knorringa, Bert Helmsing, Sietze Vellema, Roldan Muraldian and Ruerd Ruben.

I would also like to thank Marcel Vernooij from the Dutch Ministry of Agriculture, Nature and Food Quality. I am also grateful to IFPRI and to ODI, especially to Marcella Vigneri.

Financially, the study was made possible by the support of NWO/WOTRO and the University of Amsterdam/AMIDSt. The Royal Tropical Institute (KIT) supported the publication of this thesis by KIT Publishers. Thanks in particular Bart de Steenhuijsen-Piters and Ron Smit. I am also grateful to some of my other new colleagues at KIT, especially Petra Penninkhof and Robert ten Donkelaar for proofreading.

I would also like to thank Nikola Stalevski for editing my thesis.

Sabine Nolens I would like to thank you for reading my thesis and keeping me company in good and bad times. Other dear friends I would like to mention are Larissa, Emilie (vD), Emilie (H), Renet, Eva, Femke, Chris, Bianca, Taco, Kim, Fiona, Lindy, Etienne, Erik and my favourite band Snakwelt. Also I would like to mention Annig, Ard, Elien, Maarten and Jil.

There are three people I would like to mention again. First, Marloes Kraan and Ellen Lammers: my friends and my paranimfs! Second, Koen Kusters, for being more than a friend, even more than a very special friend. I cannot thank you enough.

Finally, living my life would not have been possible without my family. Thanks Cor, Lies, Jeroen, Edith, Rosa and Jonas, Jurriaan, and my *lieve, lieve, lieve* Julia and Ruben.

1
LOCAL PRODUCERS IN A GLOBAL ECONOMY

1.1 Introduction

Proponents of globalisation, consisting of advocates of both neo-liberal free markets and liberal civil society, have long argued that free trade will lead to economic growth and improvements in the livelihoods of all, including poor farmers in developing countries. But since the mid 1990s serious doubts have been raised about the supposed links between economic liberalisation, democracy, growth and equity (Kalb et al., 2004). It became clear that gains from globalisation are not distributed equally. Many of the poor farmers are excluded from integration in the world market and the ones that are included nevertheless suffer. In addition to other factors (such as market failures, failures of governments and NGOs and incompetent farmers), the existing power relations explain the difficulty that small farmers in developing countries face in improving their position in the world market.

Small farmers that produce for the export market are embedded in global value chains with strong multinational 'lead firms' that increasingly exercise power over these chains. Within the country in which the farmers live and work, farmers are embedded in a sector or cluster where power is also being exercised, for example by the government. Within farmer communities and within households, power relations are also present and can enable or hinder farmers to grasp opportunities to improve their position.

In the literature on competitiveness and upgrading for small and medium enterprises it is increasingly emphasised that in order to understand the relationship between power structures, poverty and upgrading it is important to look at the interaction between different governance levels (see Barrientos et al., 2003; Bolwig et al., 2008; Guiliani et al., 2005; Humphrey and Schmitz, 2000). I use this body of literature to look at this interaction and its connection to options for small farmers to improve their position by analysing in detail the cocoa sector in Ghana. In this study I aim to unravel 'upgrading opportunities' for cocoa producers in Ghana and the different outcomes for different types of farmers, with two clear goals in mind. First, I will seek to develop an understanding of the different power relations in which cocoa producers are embedded. Second, I will seek to identify the direction and options for change that favour different small-scale cocoa producers.[1] Such knowledge is important as it adds to the understanding of how individual upgrading strategies and outcomes are linked to long-term trends towards 'collective' inclusion or exclusion. Insights in power relations and in reasons behind different outcomes for different types of producers help policy-makers and NGOs develop more effective interventions that can reach the most vulnerable farmer groups. Moreover, it also helps the private sector to act more strategically in mitigating long-term risks for supplier failure and to aid specific groups in developing sustainable sourcing policies.

In the next sections of this chapter I will explain why I chose the cocoa sector in Ghana as a case-study and present the central question. But first I will present some background information on the concentration of poverty in Sub-Saharan Africa, and the different ideas on how poor farmers could realise change.

1.2   Rural poverty

More than 70 per cent of the world's very poor and food deprived people live in rural areas, with agriculture as their primary source of income. Sub-Saharan Africa is home to the highest portion of the world's poor. In 2002, 50 per cent of the total population (or 300 million people) lived on less than 1 USD per day in this region (World Bank, 2002). Poverty and slow economic growth is linked to the fact that the countries within Sub-Saharan Africa remain highly dependent on primary commodity exports; many of which are agricultural commodities: coffee, tea, cocoa, cotton, bananas, groundnut, rubber, tobacco and sugar. From 1980 to 2002 the prices of a number of such tropical agricultural commodities declined between 50 and 86 per cent (Oxfam, 2004: 3). This has resulted in high losses in export earnings for Sub-Saharan African countries in this period and contributed to the concentration of poverty in rural areas. In recent years many countries in this region enjoyed economic growth that strengthened their balance sheets; however, the recent economic crisis is expected to slow down this growth (IMF,[2] 2009), and pushes commodity prices down again. This will have negative effects on export earnings and the external current account, fiscal revenues, and household incomes and is likely to erode the progress some African countries have achieved in attaining the 'Millennium Development Goals' (MDGs) (IMF, 2009).

The geographical concentration of poverty convinced international policy-makers to (re)focus their poverty reduction strategies on rural areas (e.g. IFAD,[3] 2001; World Bank 2007a). The World Bank stressed the need for the transformation of the agricultural sector 'from considering agricultural activities as simply a way of life to that of a profitable commercial and industrial occupation' (2003: 37-9). It is emphasised that the 'poor themselves have to seize responsibility, as agents of change, for their own development' (IFAD, 2001; interview World Bank Ghana, 2005). This change, emphasising the agency of the poor rather than perceiving the poor as 'passive victims', is linked to a change in perception on the details of poverty. Poverty is not only about income levels, but also about capabilities and vulnerabilities (cf. Verrest, 2007). This change in perception contributed to the recognition of the differences among the poor which have to be taken into account.

1.2.1   Poor farmers as rural entrepreneurs

Small producers in developing countries must improve or upgrade their businesses, if they are to cope with the challenges of globalisation, increased competition and price fluctuations. This may involve acquiring new capabilities that enable them to participate in particular value chains or clusters, or to access new market segments (Humphrey, 2004).

The many actors involved throughout a value chain – producers and traders, processors, manufacturers, (multinational) retailers, supermarkets and consumers – are becoming ever more intimately connected. This is due to market developments, changing consumer preferences and demands, increasing risks for supplier failure and the fact that small and medium enterprises (SMEs) in developing countries are increasingly becoming integrated into the world trading system.

Little is known about the conditions under which rural entrepreneurs in developing countries are inserted in value chains and little is known about the different outcomes for different types of farmers. Upgrading opportunities available for farmers and their outcomes depend (at least) partly on power relations in which they are embedded. For weaker actors within a chain, upgrading does not happen automatically, but can be enabled or hindered by more powerful players, for example governments or existing social structures.

There are different opinions on which strategies small farmers should follow. There are also different ideas on which actors should facilitate and/or support the poor in this process. International institutions, such as the World Bank, emphasise that farmers themselves are responsible for change. In the process of adding value to their product or production process, farmers can learn from 'lead firms' higher up in the value chain (the value chain perspective) or can realise change through joint action and collective efficiency (clustering perspective). Others argue that in less developed countries, where the majority of raw material production is in the hands of small producers, capturing higher margins for unprocessed commodities requires public action (Gibbon, 2001: 352-3; see also Kalb, 2004). Traditionally this 'public action' was in the form of state-supported cooperative systems that combined measures to establish and maintain export quality (input supply, research and extension, price incentives, grading) with the functions of forecasting volumes and forward sales (Gibbon, 2001). However, the introduction of the Structural Adjustment Programs (SAPs) by the World Bank in the 1980s reduced the involvement of the state in the marketing of agricultural export commodities and its provision of services to producers of these commodities. These traditional systems were often removed too fast without providing an adequate replacement.

1.2.2 Empowering poor farmers

In order for small farmers to benefit from participating in global value chains, they need to be empowered to make their own informed decisions about their work and livelihoods. In fact, empowerment can lead to 'self-exclusion'- farmers choosing to remain outside or leave a chain because they foresee too little profit and too many risks (Wennink et al. 2007).

When addressing the role of farmers in value chains there are two key aspects to take into consideration: who does what in the chain (vertical integration), and who determines how things are done (horizontal integration) (KIT et al., 2006). Farmers may be concerned only with production: they prepare the land, plant seeds, apply fertilizer, control pests and weeds, and harvest the crop. But they may also be involved in activities higher up in a chain, including sorting and grading, processing or trading their produce. If farmers are involved in a wide range of activities in addition to production, this contributes to their empowerment. But true 'chain empowerment' requires that these producers gain economic power by becoming involved in *managing* the chain. Farmers can participate in various aspects of management, such as controlling the terms of payment, defining grades and standards, or managing innovation. Important questions include how to obtain this power and what kind of strategies contribute to empowerment and as a consequence insertion under favourable terms or 'self-exclusion'?

1.3 Research choices

1.3.1 *A multi-level approach towards upgrading*

The global value chain (GVC) approach offers an interesting framework for assessing the increased interdependency between global buyers and local suppliers. Within the global value chain literature, the value chain is considered a dynamic open system where producers in developing countries can act as active agents and upgrade their product, process or function in the chain (or apply their competences to a different chain) (Gereffi, 1999; Humphrey and Schmitz, 2002). In theory the identification of upgrading strategies is perceived as a way of changing power relations within a chain. However, in this thesis I will question this 'transformative character' of interventions in value chains by arguing that 'upgrading' often reinforces already existing power relations.

The use of the GVC framework tends to ignore governance structures at the national and local level. The state and other institutions are not perceived as active agents but more as enablers (or hinderers) of economic development. Global value chain analysis is also limited in providing insight into the heterogeneity in outcomes for different types of producers. Therefore, I will complement this global approach with the cluster approach, which focuses on local power structures. Cluster studies focus on local level governance structures, which are viewed as the main facilitators of upgrading and innovation. In order to deal with national governance structures I will make use of a comparative framework for studying public-private interaction in a global value chain (e.g. state involvement versus coordination through market mechanisms) (Griffiths and Zammuto, 2005; Ton et al., 2008). In this framework a link is made between the strategic management theory and political economy.

In this study I focus on the interaction between these global, national and local governance structures. Therefore, the central question of this study is how different governance structures interact in creating opportunities and constraints for more inclusive upgrading among small-scale cocoa farmers in Ghana.

1.3.2 Cocoa

Cocoa production employs around 14 million workers worldwide, and it is estimated that about 3 million smallholders account for the lion's share (90 per cent) of production. The worldwide production of cocoa beans in the season 2007/08 was around 3.7 million tonnes and is concentrated in West Africa. Côte d'Ivoire, Ghana, Nigeria and Cameroon together account for around 70 per cent of the world's cocoa production, and in turn generate substantial export revenues (TCC,[4] 2009).

Cocoa is a primary commodity produced for export, with little added value. Cocoa is traded internationally in the form of beans or as semi-finished products. Looking at the developments in the futures market, high price fluctuations are noticeable. Over the last years the price remained rather stable between 1700 and 2200 USD per tonne, until mid 2007 when it increased significantly (corresponding to price developments in the oil sector). Since the beginning of 2007 the price of conventional cocoa more than doubled, which put traders, processors and manufacturers under financial strains pressure. The drop in cocoa exports from Côte d'Ivoire is a major reason for this price-increase.[5]

The high concentration of cocoa production in West Africa and the low levels of productivity are a risk for global buyers of cocoa. These risks in combination with liberalisation of state marketing systems and the changes in consumer demand for sustainable cocoa have increased the inter-dependency between actors in the chain. Until recently, manufacturers and processors rarely bought directly from producers, unless they had installed processing facilities in the producing countries. Traditionally, the global buyers acquired their beans through a network of trade houses and brokers (Jaeger, 1999: 10). But there is a trend towards building partnerships with development organisations, research institutes, and government agencies, as well as a growing need to develop a 'closer interaction with the cocoa growing regions and with the cocoa farmers' (Helferich, 1999: 2; LMC International and University of Ghana, 2001). Building a sustainable cocoa economy has become *the* concern for key-actors involved in the industry. The common understanding that a sustainable cocoa economy is in the interest of all stakeholders is also reflected in the round table meetings on building a sustainable cocoa economy, which were hosted by the International Cocoa Organisation (ICCO) and took place in Accra, Ghana in 2007 and in March 2009 in Port of Spain, Trinidad & Tobago.

A sustainable cocoa economy refers to the three pillars of sustainability: economic, environmental and social. Often economic sustainability is put forward as a prerequisite for small producers to take up environmental and social challenges. Productivity levels of small-scale producers of cocoa are generally low, because of the prevalence of pests and diseases, soil degradation and the generally old age of farmers, their farms and the trees. The combination of relatively high costs of inputs, weak institutional support and the lack of adequate credit facilities make it difficult for small-scale cocoa farmers to generate a profit. At the same time the environment is deteriorating. The goal of achieving a sustainable cocoa economy makes it both a challenge and necessity to look for win-win opportunities that can accomplish environmental and social objectives while, at the same time, creating economic opportunities for cocoa smallholders.

It is complicated to create the right conditions for sustainable development in cocoa because the developments in the global value chain of cocoa steer in the direction of concentrating production in a small number of countries and vertical integration of upstream activities by a handful of international cocoa processing firms (Daviron and Gibbon, 2005; Losch, 2002; Abbot et al., 2005). This two-fold process of economic concentration affects the bargaining power and price setting between the different chain actors. These concentration processes also determine, together with the taxes and levies applied on various transactions within the chain, the farm-gate price and thus the economic perspectives for producers.

1.3.3 Ghana

Ghana is the world's second largest producer of cocoa. Although its importance is somewhat declining, the cocoa sector makes the highest contribution to the Gross Domestic Product (GDP) (mainly through duties paid on exports) and is considered as the economic backbone of the country (Tiffen et al., 2002: 8; Wayo Seini, 2002: 2). According to recent estimates by Masterfoods,[6] around 30 per cent of Ghana's total earnings come from cocoa exports and around 6.3 million Ghanaians depend on cocoa for their livelihood (representing almost a third of the population). In season 2007/08, Ghana produced almost 700,000 tonnes. In comparison, Côte d'Ivoire, world's largest producer, produced almost 1,400,000 tonnes for the same season.

In addition to being world's second largest producer, Ghana has some particularities that make it an interesting case for assessing the interaction between private and public policies. First, Ghana is the only cocoa producing country in the region that has only *partly* liberalised its marketing and pricing system: the government still plays a governing role in the sector. Second, it is the only country that provides traceable cocoa and the high quality cocoa it produces fetches an additional premium.

A number of important insights will arise from the Ghanaian case. First, it contributes to increased understanding of how producers of agricultural export commodities benefit from being inserted in a global value chain, one which is increasingly driven by multinational cocoa processors and chocolate manufacturers. Second, it contributes to the recent discussion on hybrid governance structures, where both public and private actors play a governing role. Ghana is unique because of this strong role of the state. Examining this strong state will put the discussion of the role of the state in agricultural agenda back on the agenda. Lastly, this study will contribute to understanding how upgrading connects to development aims.

1.4 A guide to the study

After this introduction, *Chapter* 2 will focus on the 'upgrading' debate. In the literature upgrading is presented as a way for small-scale entrepreneurs in developing countries to remain competitive on the world market. The conventional approaches towards upgrading (namely GVC approach and Local Clustering) will be discussed. Chapter 2 will provide insight in the exact meaning of both approaches and their scope. It will show the relevance of the interaction between vertical and horizontal networks in understanding existing opportunities for more inclusive upgrading strategies. Also it will point out the necessity to theorise the role of the state within this process. *Chapter* 3 will elaborate on the research questions and how these are operationalised, thus building a framework for more inclusive upgrading. Furthermore, in this chapter I will share some reflections on the research process and the validity of its results.

Chapters 4 and 5 will demonstrate that upgrading in the cocoa chain is not only driven by local interests, but also by national and global interests. Chapter 4 focuses on the developments in the global cocoa chain. Chapter 5 focuses on developments within the cocoa sector in Ghana. These chapters will show that context is an integral part of the analysis that helps to explain the dynamics of current governance structures. Three developments are particularly relevant: (1) the processes of liberalisation and privatisation in the cocoa sector; (2) the growing importance of non-price competition, together with increasing concerns for sustainable development issues; and (3) the increasing risk of supplier failure that (international) lead firms face. In order to validate the shifts in governance that are taking place and show how they affect producers of cocoa, I will make some comparisons over time (before and after the reforms) and some comparisons with the current situation in other cocoa producing countries in West-Africa. *Chapter* 6 will explore the heterogeneity among cocoa producers. This is an important exercise as it will contribute to increasing the understanding of how upgrading strategies are linked to development goals. Although not presented as such, upgrading is a selective process. In order to evaluate the relative strength of upgrading strategies, it is necessary to take a look at issues of social in- and exclusion. 'Social position' is an important intermediate variable that helps explain why, under equal conditions, some producers benefit more from upgrading strategies than others. The aim is not only to explain how structures determine the behaviour/actions of agents, but by understanding this relationship, to seek improvements and successful interventions by the competent agents.

In *Chapter* 7, I will analyse different upgrading strategies by looking at sub-strategies and their interventions. I will propose a more inclusive way of looking at upgrading: focusing the analysis on the goal of interventions, their impact and who is targeted. In addition, I will consider economic, environmental and social trade-offs. In this chapter I link individual upgrading strategies to collective upgrading by looking at possible future scenarios. *Chapter* 8 is the concluding chapter. In this chapter I will come back to the questions posed in this study and propose a model for upgrading that acknowledges the contribution (positive or negative) of hybrid governance structures to upgrading. In this chapter, I will also reflect on the transformative character of governance structures, and how power relations on different scale levels can be altered in order to produce more favourable outcomes for small producers. Finally, I will reflect on current policies and practices.

2
UPGRADING FOR DEVELOPMENT

2.1 Introduction

The discussions on globalisation and its impact on development emphasise the challenges faced by entrepreneurs in developing countries, i.e. intensified competition and price-fluctuations. Since the late nineties, the focus has been especially on small and medium enterprises (SMEs) (Humphrey and Schmitz, 2000; Lambooy, 2002; Maskell and Malmberg, 1999; UNCTAD, 2001[7]). Recently, with agriculture back on the development agenda as the 'engine for economic growth', the emphasis has shifted from SMEs to small-scale agricultural producers (EGFAR, 2003[8]; World Bank, 2007[9]). At the same time, these small-scale producers are increasingly viewed as independent entrepreneurs. In order to support sustainable growth and reduce poverty, these agricultural 'firms' have to improve their competitiveness (Wenner, 2006; World Bank, 2007).

In the literature on competitiveness, the concept of upgrading highlights the options available to producers for obtaining better returns. Historically this concept is linked to the process of shifting from 'Fordism' to 'post-Fordism'. This change includes a shift in the understanding of the process of change, moving away from the idea that state driven interventions build-up capital and technological innovation towards the idea that upgrading is the outcome of organisational learning and inter-firm networking. This paradigm, which finds its origin in political economy and industrial economics, has been applied in various bodies of literature, from cluster studies to the value chain approach (Gibbon and Ponte, 2005: 87-8). Cluster studies focus on local level governance structures, which are viewed as the main facilitators of upgrading and innovation (Helmsing, 2002; Humphrey and Schmitz, 2000; Westen, 2002). Alternatively, the value chain approach concentrates on 'global chain governance', which views inter-firm co-operation within the chain as a competitive advantage (Gerrefi, 1999; Humphrey and Schmitz, 2000: 14; Vargas, 2001). Essentially the value chain approach proposes that inserting entrepreneurs in value chains offers the possibility to engage in learning processes and to acquire new knowledge from external buyers (Humphrey and Schmitz, 2000; Vargas, 2001: 5).

Several authors have highlighted the limitations of both approaches and the necessity to combine the 'horizontal networks' with the 'vertical networks' (see Giuliani et al., 2005; Humphrey and Schmitz, 2000; Lambooy, 2002; Palpacuer, 2000; Westen, 2002).

In addition to being complimentary, both approaches have some important similarities. For example, both approaches assign a limited role to public sector actors. The public sector is not perceived as active intervener but rather more as an enabler (or hinderer) of economic development. Recently the role of the state has been re-examined (cf. Griffiths and Zammotto, 2005; Lall, 2005), emphasising that national governments and institutions did and continue to play an important role in fostering the competitive advantage of industries/firms. An increased focus on the state also follows from the recent trend to directly link upgrading to development goals. The link of upgrading to development originated from the recognition that globalisation knows 'winners' and 'losers' – the gains of globalisation are not distributed equally. In the agricultural sector the small farmers are seen as the losers; only those locked into larger farm production have a chance of making a profit (Kaplinsky, 2001: 127). This recognition of varied success in reaping the gains from globalisation opened another discussion on social 'inclusion' and 'exclusion' and 'inclusive growth' (Giuliani et al., 2005: 6; IDRC, 2006[10]) and on the reasons behind unequal benefits that the different actors receive. The notion that 'global processes produce different outcomes in different settings' also opened space to discuss ideas on 'structure and agency'[11] (Post et al, 2002: 1-4).

It seems worthwhile to investigate how different governance levels separately (and in interaction) contribute to creating the upgrading opportunities and to influence the upgrading outcomes. Therefore, in this chapter I will discuss these different governance levels and how they relate to the concept of upgrading and development. I will start by introducing the concept of upgrading, its different typologies and its relevance for agricultural commodities. In this section I will elaborate on the concept of 'inclusive upgrading'. Next, I will discuss the potentials and shortcomings of the two conventional approaches for upgrading, the global value chain approach and the local clustering approach. I will continue with a discussion on the changing role of the state in development and its theoretical underpinnings by making use of the concept of 'state governance'. Lastly, I will discuss the need to study the interactions between different governance levels to identify more inclusive upgrading.

2.2 Defining the concept of upgrading

In the discussions on how small and medium enterprises in developing countries cope with globalisation there is an apparent agreement and a clear preferred course of action. In order to remain competitive these firms should upgrade by learning and acquiring new knowledge (Humphrey and Schmitz, 2000; Gereffi, 1999, Schmitz, 1999; Vargas, 2001). There are differences in opinion on the exact definition, typology and use of the concept of upgrading. For example, there are different types of upgrading with particular categories, such as: product, process, functional and inter-sectoral (or inter-chain) upgrading (Gereffi, 1999; Humphrey and Schmitz, 2002). 'Product upgrading' refers to moving into more sophisticated product lines, with increased product value. 'Process upgrading' is defined as transforming inputs into outputs more efficiently by re-organizing the production system or introducing superior technology. 'Functional upgrading' can be understood as acquiring new superior functions in the chain, such as design or marketing, or as abandoning the existing low-added value functions in favour of higher value added activities. 'Inter-sectoral upgrading' refers to applying the acquired competences in order to move into a new sector (Humphrey and Schmitz, 2002).[12]

Although recognised as helpful, this classification of upgrading has been criticised for several reasons. For example, Meyer-Stamer (2002) opposed this typology because it does not provide information about the direction upgrading takes (for example 'downgrading')[13] and neglects the idea that the direction can be perceived differently.[14] Smakman (2003) argued that the distinction between 'upgrading as a process' and 'upgrading as an outcome' is not always clear. Other doubts, raised by Gibbon and Ponte (2005: 88), criticised the difficulty of distinguishing between product and process upgrading. For example, in the case of agricultural commodities the introduction of organic processes of production generates a new category of products (for example organic coffee). Should this be considered product or process upgrading? Gibbon and Ponte also had a critique on the emphasis placed on functional upgrading, which implied that this type was more optimal than the other available upgrading options. Excessive emphasis on this option also overlooks key findings of various authors (e.g. Schmitz and Knorringa, 1999; Gereffi, 1999), namely that global firms can make it very difficult for local producers to progress in functional upgrading ('lock-in').

Because earlier studies on upgrading mainly concentrated on small-scale industries in developing countries, some adaptation is necessary in order to adequately apply the upgrading concept and its terminology to producers of agricultural commodities. Therefore, looking at upgrading issues for producers of this type of commodities, Gibbon (2001) proposed an alternative (provisional) classification: 1) capturing higher margins for unprocessed commodities, for example through higher levels of productivity; 2) producing new forms of existing commodities; and 3) localising commodity processing (2001: 352-4).

More recent work provides alternatives to unpacking relations between upgrading and governance in GVCs, which intentionally avoids using terms such as process, product, functional and inter-sectoral. Gibbon and Ponte (2005: 91) proposed a detailed empirical analysis (on a chain-by-chain basis) 'that identifies concrete roles that offer suppliers higher and more stable returns, as well as the routes that they typically use for arriving at them'. Such an approach would make it possible to identify the returns that actors below the level of the 'lead agent' accrue. Studying the global coffee trade, Daviron and Ponte (2005) tried to avoid the vocabulary of upgrading and chose instead to focus on its components: 'the ability of producers to create and control the value' (2005: 30).

In my study the different views on the exact definition, categorisation and adequate use of the concept are not perceived as contradictory but rather as complementary to each other. I link the different types of upgrading *(what?)* to the process of upgrading *(how?)*, to its main driving forces *(by whom?)* and its outcomes for producers *(for whom?)* (see Chapter 7). I question the validity of the identified 'goals' (competitiveness) and 'means' (learning) for upgrading for producers of agricultural commodities, despite the existing agreement in the upgrading debate. These notions seem to be grounded in two basic assumptions: 1) the presence of an open market system and 2) the absence of a limit to upgrading, as long as there is access to new knowledge and technologies. But these conditions are not always 'fully' present for the producers. In some large sectors of several commodity producing countries, there are no truly 'free markets'. Also, with respect to access to new knowledge and technologies, small producers often face serious constraints to apply the already existing knowledge/technologies let alone seek out and adapt new innovations. Nevertheless, I recognise the importance of learning, especially in light of the increasing demand for more sophisticated commodities (for example more sustainably produced commodities) that is likely to make learning more urgent for these types of producers. In this context, it is important to know the gatekeepers of knowledge and the channels for transferring this knowledge. It also requires the acknowledgement of other 'non-economic' indicators of success, for example improvements in process quality, and increasing concerns about safety, health, environmental and labour standards (cf. Barrientos et al., 2001; Laven, 2007).

The idea that upgrading is not only based on better pricing and improved quality of products but that it is also important whether there is sustainable production within the chain is increasingly gaining ground (Abbot et al., 2005; Daviron and Ponte, 2005). This provides impetus and input for discussions on 'inclusive upgrading'.

2.2.1 Inclusive upgrading

In most debates on inclusive upgrading the focus is still on insertion of the poor (in a chain or cluster). Insertion in a value chain or cluster does not automatically result in upgrading. Especially for weaker actors, upgrading can be enabled or hindered by more powerful players (including governments and NGOs) and by existing social structures. Also the gains that result from upgrading are often unequally distributed (Giuliani et al., 2005; Tiffen, no date). For example, fair trade organisations aim to pay poor farmers a fair price for their produce, but membership in such schemes is selective. Many development initiatives attempt to secure benefits for small producers by making the value chain shorter or by reconfiguring a chain. In practice this results in the exclusion of middlemen, who may also be poor and may have difficulty finding alternative employment.

As used by policymakers the concepts of social exclusion refers to a lack of material resources but also to a lack of rights. Inclusion was defined as a policy aim and a desired situation (Hospes and Clancy, 2009). Various scholars embraced the concepts of social inclusion and exclusion. In these studies it is increasingly recognised that 'inclusion' is not always a desirable aim and is not always wanted by the 'excluded' (Blowfield, 2003; Hospes and Clancy, 2009; Wennink et al., 2007). 'Self-exclusion' can be preferred by for example farmers that foresee too little profit and too many risks in the chain. But for 'self-exclusion', or inclusion under favourable terms, small producers need to be empowered to make their own informed decisions about their work and livelihoods (Wennink et al., 2007).

In value chain and cluster literature various authors took up the discussion on social inclusion and exclusion (e.g. Altenburg, 2006; Cortright, 2006; Gibbon and Ponte, 2005; Knorringa and Pegler, 2006; KIT et al., 2006; Nadvi and Barrientos,

2004).[15] In this study I make a distinction between a more general notion observed by Altenburg (2006) that 'value chains become more exclusive as small-scale producers fail to meet rising scale and standard requirements', and the idea of 'inclusive upgrading', whereby I refer to the outcomes and impact of interventions on weaker actors within a chain. In Chapter 7, I will further define and operationalise the concept of inclusive upgrading and highlight some of the different notions on its exact meaning.

Further in the next section, I will discuss different governance structures and will link these to the upgrading concept. The main question of interest is to discover to what extent dominant governance structures support or hinder 'more inclusive' upgrading.

2.3 Upgrading and governance

There are different possibilities of theorising governance. 'Governance' instead of 'government' implies a reduced role for the state in development. It also implies an increased role for other actors, and the configurations in the relationship between the state and these other actors (Nuijten, 2004). Although some authors asserted that the state continues to be one of the central actors in governance (e.g. Gibbon, 2001; Lall, 2005; Nuijten, 2004), others stressed that states primarily act as the facilitators of international capital instead of being the principal caretaker of social equity and well-being (Post et al, 2002: 2). This last observation, which views the state mainly as an enabler of economic development, dominates the discussion on upgrading approaches. The GVC approach focuses on global chain governance, where private actors are the main drivers in the chain. The cluster approach concentrates on the level of local governance structures, where firms and institutions compete and cooperate. While the government's influence on the 'determinants of regional advantage' is recognised (for example in Porter's Diamond Model, introduced in 1998) (Neven and Dröge, no date), nevertheless, the government is not regarded as a steering actor within a cluster. Recent debates on upgrading and development partially repositioned the role of the government, thus opening a debate on the role of the government in chain development and cluster development.

In this section I will discuss these different levels of governance and link them to the upgrading approach. I will look at governance from a wide perspective, including not only formal arrangements but 'any form of institutionalised practice' (Nuijten, 2004: 104-5), whereby governance will refer to 'both steering processes themselves and the results of these processes' (see also Kooiman, 1993; Rhodes, 2000).

2.3.1 Global Value Chain analysis

Global value chain (GVC) analysis is an analytical tool used in a variety of domains. GVC analysis has its roots in world systems theory and dependency theory. The GVC approach first appeared in the literature under the term global commodity chain analysis, as introduced by Hopkins and Wallerstein (1986; 1994) and further developed by Gereffi and Korzeniewicz (1994). Originally, 'global commodity chain analysis' was concerned with agricultural products, but Gereffi was mainly concerned with industrial commodity chains and the development of a unified theoretical framework that would make it possible to identify upgrading strategies for firms and thus change existing power relations within a chain (DFID[16], 2004). A commodity chain refers to 'the whole range of activities involved in the design, production, and marketing of a product' (Gereffi, 1999: 38). The terms value chain and commodity chain are often used interchangeably. Using the term value chain reflects the understanding that value is added at each point of the chain (Smakman, 2003; Vermeulen et al, 2008: 14). Essentially the primary returns – economic rents – accrue to parties who are able to protect themselves from competition by creating entry barriers (Kaplinsky and Morris, 2003).

Gereffi distinguished between four dimensions in the value chain: 1) their input-output structure, or the sequence of interrelated value-adding activities (including product design and engineering, manufacturing, logistics, marketing and sales); 2) the geographical coverage, which refers to the spatial dispersion or concentration of activities within and across locations; 3) the global chain governance, which is defined as 'authority and power relationships that determine how financial, material, and human resources are allocated and flow within the chain' (Gereffi 1994: 9); and 4) the institutional framework that defines the local, national and international conditions and policies that in turn shape the environment where firms operate (Gereffi, 1994; Gibbon, 2001; Smakman, 2003). Because of the discussions on upgrading and its link to development goals, the institutional dimension has increasingly gained in interest (see Daviron and Ponte, 2005; Gibbon 2001; Tiffen, no date; Westen, 2002). It is essential to include the institutional framework in value chain analysis as it recognises that chains are not 'closed systems'. They receive external inputs in terms of knowledge management (technical research institutes, extension services) and they are influenced by: *advocacy movements* (trade unions, NGOs) that work on environmental or social issues; *policy priorities* set by national governments or international organisations (e.g. World Trade Organisation, World Bank, or United Nations agencies), and by *social structures* (e.g. on the organisational level of producers or traditional hierarchical relations). Furthermore, the institutional framework is important because it either provides effective channels through which quality standards can be introduced as part of upgrading or it creates barriers that block this exchange. For example, sanctioning may also take place outside the chain (Kaplinsky and Morris, 2003: 16-8).[17]

Although the importance of the institutional dimension is recognised in GVC analysis, institutions are not regarded as active actors with governing power. In the value chain literature there is a different 'approach' that looks more at 'the totality of structures and relations around specific commodities, including relations of power', namely the Francophone filière tradition (see Box 2.1). This approach was mainly used to look at the upgrading patterns in primary export commodities.

Despite the benefits of the *filière* approach and methodological similarities between this tradition and my study, I prefer the GVC approach. The main reason is that the

> **BOX 2.1** **THE FILIÈRE TRADITION**
>
> The *filière* tradition, influenced by studies on agriculture in the the US in the 1950s and 1960s, was developed by French researchers at the Institute National de la Recherche Agronomique (INRA) and the Centre Internationale en Recherche Agronomique pour le Developpement (CIRAD) as an analytical tool for empirical agricultural research. In contrast to the GVC analysis (which aims at developing a unified framework), the *filière* approach includes 'several different schools of thought', with the 'common characteristic that they use the *filière* (or chain) of activities and exchanges as a tool to delimit the scope of their analysis'. The *filiére* approach is more a 'meso-field of analysis' than a real theory. It is seen as a 'neutral, value-free technique applied to analysing existing marketing chains for agricultural commodities'.
>
> The *filière* approach started in the 1960s by studying contract farming and vertical integration in French agriculture. It was soon applied to the analysis of the production and marketing chains for selected export commodities that were produced in France's (African) colonies, such as: rubber, cotton, coffee and cocoa. Initially the studies mainly dealt with local production systems and consumption patterns, because state institutions controlled all trade and processing. The main focus was on the way in which public institutions affected local production systems. It has been argued that this type of analysis was used to justify the maintenance of interventionist systems (such as the price stabilisation funds) because the French research showed negative consequences of market liberalisation on developing countries. Recently, the *filière* approach has started to focus more directly on issues of trade and marketing.
>
> One of the main traditions within the *filière* approach is its 'empirical research tradition', with its main objective 'to map out actual commodity flows and to identify agents and activities within a filière, which is viewed as a physical flow-chart of commodities and transformations'.
>
> Sources: DFID, 2004 and Raikes et al., 2000.

analyses of the *filière* tradition attach more importance to the technical side of the material flow than to the role of social actors. It has also been criticised for an excessively strong 'quantitative tradition' and its rather static analyses (Raikes et al, 2000).

The majority of studies that use a GVC perspective focus on labour-intensive manufacturing (cf. Humphrey, Knorringa, Morris, Schmitz). Cramer (1999) was the first to highlight the necessity for broadening the focus of value chain analysis to also include primary commodities. He stressed the particular importance that upgrading agricultural export commodities has for developing countries; it is their main link to the global economy. A number of authors have taken up this notion and analysed several agricultural commodity chains, primarily focusing on cocoa and coffee (see Gibbon, 2001, 2003; Fold, 2002, 2004; Gibbon and Ponte, 2005; Kaplinsky, 2004; Ponte, 2002; Daviron and Ponte, 2005) but also looking at cotton (Larsen, 2003) and tobacco (Vargas, 2001). Studies on these agro-based commodities tend to be more normative than studies on labour-intensive manufacturing, which seldom go beyond 'observing' differences in power. They reflect the initial thinking on global commodity chains, which has its roots in dependency theory.[18]

Development agencies and policymakers have also adopted GVC analysis, by employing this perspective as a point of departure for drafting international agricultural development strategies.[19]

So far GVC analyses have focused on global chain governance. However, the value chain literature can seem confusing as the concepts of 'chain governance' and 'chain coordination' are often used interchangeably. Because they are a central focus of this study, I will further elaborate on these concepts in separate paragraphs.

Chain governance

Value chain governance helps unravel the determinants of income distribution and opportunities for adding value, by highlighting the factors that determine the nature of the insertion of different producers into the global division of labour (Kaplinsky, 2001: 124-9). In discussions on chain governance, generally a distinction is made between two types of chains where the producers have different positions (Gereffi, 1999). Producer-driven chains are found in capital and technology intensive sectors. Technical knowledge and high levels of capital prevent new producers from entering these sectors. Multinationals are the central players in such chains, which are complex and multi-layered, often marked by international subcontracting of the more labour-intensive parts of the process. In contrast, buyer-driven chains are found in more labour-intensive sectors, where design and marketing are centrally controlled (such as garments and footwear) and in agro-commodity chains (such as cocoa and coffee).

Both producer-driven and buyer-driven chains are seen as vertical networks. In buyer-driven chains the so-called 'lead firms' include large retailers, branded marketers, and branded manufacturers. They act as strategic brokers who link producers and markets; their privileged knowledge of strategic research, marketing and financial services grants them this privileged position (Gibbon, 2001; Gereffi, 1999). Lead firms (as a group) control certain functions that allow them to dictate the terms of participation by other actors in different functional positions in the value chain. Lead firms use entry barriers to generate different kinds of 'rents' (Kaplinsky and Morris, 2003). Immediately upstream of lead firms, there are other powerful agents who do most, or at least a large share, of the day-to-day work of chain coordination. These firms are defined as first-tier suppliers and have their own suppliers, so-called second-tier suppliers, which can have their own suppliers, and so on (Gibbon and Ponte, 2005: 99-104).

This distinction between producer and buyer driven chains has been challenged from different perspectives, which questioned its relevance in analysing agricultural commodities. For example, it has been argued that this classification does not reflect governance patterns in agricultural commodity chains. According to Gibbon (2001), international traders govern a growing number of agricultural commodity chains (e.g. coffee, cocoa, cashews), for which he proposes to use the term 'international trader-driven chain'. Fold opposes the distinction between buyer and producer driven chains from another point of view. According to Fold (2002: 230) this:

> crude dichotomy (...) fails to acknowledge the more complicated patterns of power relations between lead firms in global chains – or, at least those for agro-industries. (...) This distinction does not help specify the dynamics of 'drivenness' in certain global chains.

This idea of 'degrees of governance' is very relevant for a growing number of agricultural value chains that are becoming 'increasingly buyer/trader-driven' (Fold, 2002; Humphrey and Schmitz, 2000). The concept of chain coordination was introduced in this context; it focused on 'inter-firm relationships and institutional mechanisms through which non-market coordination of activities in the chain is achieved' (Humphrey and Schmitz, 2002: 7). Gibbon and Ponte (2005: 163-4) made a clear distinction between governance and chain coordination, arguing that 'a GVC may be characterized by different forms of coordination in various segments, yet a relatively coherent form of overall governance'. In other words, a lead firm may control the value chain without controlling each segment of the chain.

Chain coordination

Humphrey and Schmitz (2000) distinguished between three types of coordination: 1) 'network relationships' based on co-operation between equals; 2) 'quasi-hierarchy', or 'captive relationships', combining cooperation with asymmetrical power relationships in which buyers dominate over suppliers; and 3) 'hierarchy', associated with vertical integration, where the buyer takes direct ownership of the operations. When buyer and supplier do not need to collaborate in defining the product, because either the product is standard or the supplier defines it without reference to particular customers, the term 'arm's-length market relationships' is used. In the literature, the arm's-length market relationships are often considered a fourth type of chain coordination.[20]

According to Humphrey and Schmitz (2000: 15), the buyers' (traders) risks for losses from the supply chain failure is a factor that determines the type of relationships between local producers and external buyers. According to them due to 'the increasing importance of non-price competition based on such factors as quality, response time and reliability of delivery, together with increasing concerns about safety and standards', buyers have become more vulnerable to shortcomings in the performance of their suppliers. Different studies in developing countries have demonstrated that in response to the upgrading challenge, the relationships between local producers and their external buyers (traders) change (e.g. Gereffi, 1999; Nadvi, 1999). In addition, there are also changes in the relationships between other actors, for example between manufacturers and retailers (Gibbon and Ponte, 2005). However, the reasons for the types of shifts differ greatly. Humphrey and Schmitz (2000) expected that increased risks of supplier failure would result in a shift from arm's-length relations to more active forms of cooperation between buyers/traders and suppliers, such as network and quasi-hierarchical relations. But, according to Gibbon and Ponte (2005: 163), this shift towards so-called 'hands-on' forms of coordination does not necessarily occur. They argue that 'if economic actors are able to embed complex information about quality in standards, labels and certification procedures, they may still be able to operate with more hands-off forms of coordination' (ibid). Also, it is not clear how the specifics of the value chain ('tight or loose' organisation) are linked to a particular outcome for producers (see also Kaplinsky, 2000; Kaplinsky and Morris, 2003). These uncertainties imply that there are power relations in a chain that have to be unravelled, as poor producers run up against them (see also Kaplinsky, 2001: 140). They also indicate that besides risks for buyers there are other risks involved, namely for producers, who are generally not the main drivers in a chain.

In recent debates on upgrading and social in- and exclusion an important question was raised: how can 'the poor' enter global chains? This emphasised the risks for small-scale producers to remain (or become) excluded from these types of chains. However (as already argued in Section 2.2) including the poor in value chains does not automatically result in upgrading. Inequalities that exist in a society – endorsed by social and political structures at local, national and global levels – largely determine who will benefit from inclusion. In this study I chose to focus more on the changing conditions under which producers are inserted in global value chains, what I call 'the risk of inclusion'.

Limitations of the GVC approach

Notwithstanding its potential use, the GVC approach has some limitations. Despite its institutional dimension, it has the tendency to ignore the importance of local governance structures, the role of the government and international regulation (Gibbon, 2001; Humphrey and Schmitz, 2000; Smakman, 2003; Vargas, 2001). Although the influence of public regulation and trade policy instruments is acknowledged (e.g. by Gibbon and Ponte, 2005) it does not recognise that institutions can play a steering role in value chain development and upgrading issues. Another observation is that in studies that apply a value chain approach authors who incorporate the institutional dimension tend to focus only on formal arrangements (e.g. Daviron and Ponte, 2005). They leave out the influence of informal institutions, such as the existing social structures in which local producers are embedded.

The tendency to focus on 'lead firms' is another limitation in the discussion on GVCs' impact on development and on their potential to become more inclusive in the context of liberalisation. As lead firms are the main coordinators of agricultural value chains, their decision-making patterns have been extensively reviewed (cf. Altenburg, 2006). But, in order to obtain a full understanding of GVCs and their links to development, in addition to lead firm also attention should be paid to the suppliers/producers in the sector (i.e. agricultural export commodity sector) and the observed heterogeneity among them. I think that upgrading strategies should intentionally address different types of producers. Not only do these suppliers form the large majority of all the actors who are involved in the production of agricultural export commodities but they also are the weakest 'link' in the chain; they do not control their own upgrading agenda. Leaving them out makes the case that there is an inherent bias in existing studies that use GVC analysis. They focus on parts of the chain where upgrading is more manifest and ignore the actors/parts of the chain where upgrading is marginal. I assert that in order to link upgrading to development it is important to focus on the heterogeneity among suppliers (see Chapter 6 for more details).

Regarding the differences among producers, generally a distinction is made between large and small primary commodity producers (Gibbon, 2001; Kaplinsky,

2004), which neglects non-economic factors (e.g. social position, gender, regional backgrounds). This excessive focus on vertical relationships in a chain constrains the analysis further as it does not capture more local inter-firm relations, for example joint actions among producers. The clustering approach addresses this shortcoming.

2.3.2 The clustering approach

In contrast to the value chain literature, the cluster literature views inter-firm cooperation within a single geographic area (rather than within the chain) as the source of competitive advantage (Humphrey and Schmitz, 2000: 14). This body of literature suggests that local level governance – by networks of public and private sector institutions – facilitates upgrading strategies (Helmsing, 2002; Humphrey and Schmitz, 2000). Having relationships in a cluster facilitate the creation of new products and services.

Despite a growing body of literature on clustering the concept is still rather vague offering multiple versions of a comprehensive definition (Boschma and Kloosterman, 2005; Cumbers and Mackinnon, 2007: 959-60; Guiliani et al, 2005). Simply defined, clusters are 'agglomerations of firms operating in the same or in interconnected industries, within a spatially bounded area' (Pietrobelli and Rabelloti, 2005). Other scholars consider the presence of institutions and of linkages between institutions and firms as a minimal requirement for defining an economic locality as a cluster (cf. Porter, 1990). Porter, who is regarded as one of the most influential geographical economists (Cumbers and MacKinnon, 2007), introduced the cluster approach defining clusters as,

> 'geographical concentrations of interconnected companies, specialised suppliers, service providers, firms in related industries, and associated institutions (for example universities, standard agencies and trade associations) that compete but also co-operate'. (Porter, 1998)

The cluster approach has been widely adopted in developed countries, focusing on small-scale industries in Europe. Building upon these experiences (in particular from Italy), an agenda was set for research in developing countries (Knorringa, 1999; Schmitz, 1989; Schmitz and Nadvi, 1999: 1503-4). When looking at clusters in developing countries, the emphasis tends to be on 'industrial upgrading' and the rise of specialisation (Ceglie and Dini, 1999; UNIDO[21], 2004; McCormick, 1999; Humphrey and Schmitz, 2000; Lambooy, 2002; Maskell and Malmberg, 1999; UNCTAD[22], 2001[23]). A number of case studies, looking at clustering in developing countries, concluded that in these countries clustering: favours incremental innovation, does not appear spontaneously (as was the case with Italian clusters) and often depends on external interventions (Knorringa, 2002; Cegli and Dini, 1999). There are some interesting examples of successful interventions aimed at fostering co-operative relations within SME clusters drawn from the experiences of Brazil, Mexico, India (by Nadvi, 1995) and Chile (by Humphrey and Schmitz, 1995).

Two frameworks are widely applied when studying clusters in developing countries: *flexible specialisation* (by Piore and Sabel in 1984) and *collective efficiency* (by

Schmitz in 1995). Both models have been criticised for their shortcomings. The framework of flexible specialisation could not be applied to many developing countries, because it made unrealistic assumptions on the availability of machinery, skills and trust in clusters. The framework of collective efficiency was criticised because it missed critical elements like external linkages (Neven and Dröge, no date: 5-9). In 1998, Porter introduced a new framework, the 'Diamond Model', which alleviated some of the limitations of the existing frameworks. This model proposed four interrelated factors, each representing a determinant of regional advantage: 1) firm strategy, structure and rivalry; 2) demand conditions; 3) factor conditions; and 4) related and supporting industries. Two additional factors, 'chance' and 'government', complete the model. These two factors are not determinants but influence the first four factors. Together these six factors form a system that differs from location to location; thus it can explaining why some firms (or industries) succeed in a particular location while others fail (adapted from Neven and Dröge, n.d.: 4). In recent work, Porter (2008) conceptualised the relationship between clustering and competitiveness by taking into account the role of the macroeconomic-context (see Figure 2.1).

Figure 2.1 Conceptualising clustering and competitiveness

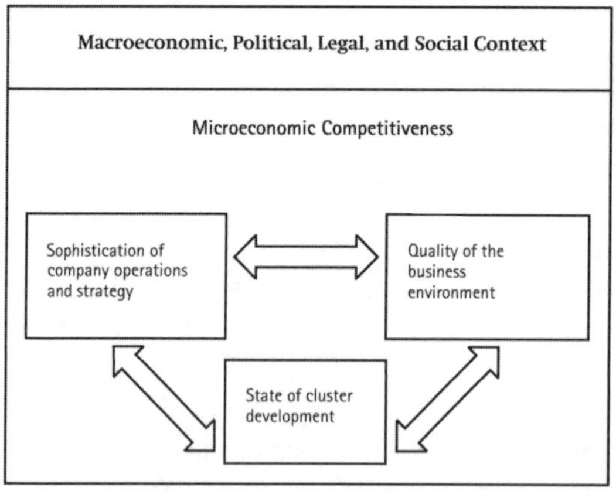

Source: Adapted from Porter, 2008.

Over the years various scholars have emphasised the importance that external linkages to the cluster have for enhancing competitiveness and reducing the possibility of negative lock-in (e.g. Bell and Albu, 1999; Guerrieri et al., 2001; Humphrey and Schmitz, 2001; Neven and Dröge, no date; Schmitz and Nadvi, 1999). Nevertheless, these linkages are still largely ignored and weakly theorised. This limitation does not pose a problem for my study, as I intend to combine this horizontal approach with the global value chain approach. The consideration of the interactions between these approaches incorporates the importance of external linkages within the analysis. The clustering approach complements the GVC approach and generates some new insights regarding the local governance structures and the facilitators of competitive advantage.

An associate problem with the clustering approach is its over-emphasis on industrialisation and SMEs in development debates, while paying little attention to other ways of adding value and to other actors. This focus has unfortunately shifted the attention away from agricultural sectors, where poverty still is concentrated. In the cluster literature the recent shift back to agricultural development is reflected in the growing attention paid to agro-food clusters, for example fishery, tobacco, diary, and others (Porter, 2008; Visser, 2004). Viewing these types of 'clusters' mainly from a value chain perspective has produced limited effects. It chiefly examines inter-firm cooperation within the chain as the source of competitive advantages while neglecting the added value of the spatial approach and local-level governance processes.

In this study, I will explore the added value of the cluster approach in analysing upgrading strategies for small-scale producers of agricultural export commodities in developing countries. I will concentrate on some of the central concepts used in the cluster literature: 'collective efficiency', 'joint action' and 'embeddedness'.

Collective efficiency, joint action and embeddedness
A central hypothesis is that 'upgrading [which is] necessary to respond to the new pressures[,]requires a greater joint action by local firms', which can be facilitated through strategic intervention in areas such as technological development or environmental upgrading (Kennedy, 1999). Joint action is perceived as a way to overcome problems of size, dependence on buyers and lack of knowledge and capital. 'Collective efficiency' and 'joint action' as a more 'deliberate force at work', introduced by Schmitz in 1995, have become central concepts in the cluster literature. 'Collective efficiency' is defined by Schmitz and Nadvi (1999) as the 'competitive advantage derived from external economies and joint action'. External economies (positive or negative externalities) are seen as a passive component of collective efficiency, while the intended effects resulting from joint action are perceived as the active component (Neven and Dröge, no date: 5). Nevertheless, joint action and collective efficiency do not automatically take place. Different studies have shown that specific conditions are required, such as the existence of effective sanctions and trust (both within clusters and within their trading connections). In this context, clearly the existence of a shared language, culture and norms are important factors. Trade networks are another key precondition for collective efficiency (Schmitz and Nadvi, 1999: 1506-7).[24] For joint action to take place it is important that there are incentives to work together, as joint action can involve opportunity costs (see also Deven and Dröge, no date: 5).

Joint action as a potential way for upgrading is a relevant strategy for small producers of primary export commodities. Individually, these suppliers have little to no bargaining power. This disadvantage is especially worrying for producers who have to negotiate directly with big buying companies. Also for the buyers joint action is important and they even may exert pressure on their producers to cooperate, for example for quality reasons or for requested volumes. In a recent trend, buying from producer groups has become a kind of marketing strategy for buyers to demonstrate that they source their cocoa in a socially responsible way ('Fair Trade' products are a clear example). The farmers can achieve economies of scale by cooperating and working together. It can also contribute to time-efficiency and learning. Before promoting joint action, it is important to consider the existing extent of joint action among producers, in case it is absent, why it is lacking. If joint action is desirable, the relevant questions are who can facilitate such a process and what type of organisation, if any, is required?

The cluster literature uses the concept of 'cluster governance' to refer to the intended, collective actions of cluster actors aimed at upgrading a cluster (Gilsing, 2000). Gilsing warned that clusters can have very cohesive and integrated structures but may not be very inclined to adapt when circumstances change. As a consequence, firms can lose their competitive advantage because of emerging weaknesses in their environment. Westen (2002: 51) was one of the authors who pointed out that clustering does not automatically favour innovation or help local firms to compete globally. Advantages from clusters usually derive from an 'optimal mix between cooperation and competition among its members'. When this balance is disturbed clustering can jeopardize competition. Understanding the conditions under which such an optimal balance can occur requires insight in the level of 'embeddedness' of economic activity. In this study, I will follow the version of Mark Granovetter, who reintroduced the concept of 'embeddedness' in 1985, emphasising the importance of social groups in which people are embedded. Many of these social relationships are geographically localised. People are not simply workers or managers; they are also consumers, citizens, church-goers, kin, and community members. Such social interactions that enhance economic efficiency are also known as social capital (Westen, 2002: 229).

Limitations of the cluster approach

One of the main limitations of the cluster literature is its lack of theorising external linkages. The cluster literature overemphasises the need to improve co-operation and local governance. Even the resources for product and functional upgrading are seen as mainly deriving from within localities themselves (Humphrey and Schmitz, 2001). The dynamics of change within clusters themselves is another area that has been neglected; clusters are often treated as static. Especially clusters in developing countries need to be able to deal with radical changes in their environment (Gilsing, 2000; Halder, 2002; Knorringa, 1999). Another limitation is that the path a cluster should follow is not always clear. Some authors have argued that a focus on 'key turning points' can facilitate our understanding of the trajectories of clusters, such as the introduction of marketing reforms in developing countries. Another already mentioned observation is the assumption that upgrading occurs on the cluster level. This seems to overlook the idea that clustering does not benefit all of its firms equally; some groups of firms may even be completely left out (this point was also raised in earlier work of Knorringa, 1999; Nadvi and Schmitz, 1994). Recent studies emphasised that 'exlcusion' does not only exist among firms, but also among groups within a firm. For example, there are clear signs that particular categories of workers, especially women and unskilled workers, lose out when a cluster upgrades. So far only few cluster studies have explicitly addressed these poverty concerns (Cortright, 2006; Knorringa and Pegler, 2006; Nadvi and Barrientos, 2004: v).

Similar to the GVC approach, the cluster approach pays little attention to the governing role of the state. With all these limitations noted, nevertheless concentrated assessment of the local governance structures (focusing very much on networks between institutions and firms) does provide more room for examining hybrid types of governance, where institutions and firms jointly govern a cluster.

2.3.3 What about the role of the state?

Linking upgrading to development reopened the discussion on the role of the state in both 'chain development' and 'local economic development'. In the 80s, the idea that the state was the governor of trade relations and competitiveness was widely accepted. The success of industrial policy of the 'Asian Tigers' and its failure elsewhere, let some authors to conclude that it is not about *whether* a government should intervene but more on *how* it should intervene (cf. Lall, 2005: 58). From a political economy perspective, the role of institutions and their interactions with the private sector play a prominent role in fostering the competitive advantage of industries/firms (cf. Griffiths and Zammuto, 2005). Griffiths and Zammuto proposed an integrative framework that drew on both the strategic management and political economy literature to explain variations in national industrial competitiveness. Their study gave an important input for the development of a more dynamic comparative framework for labelling public-private interaction in a GVC. This new framework included two fairly independent dimensions that influence economic governance and decision-making, and thus define the conditions for economic transitions within a particular sector: 1) the fragmentation of the integration of the value chain (McGahan and Porter, 1997); and 2) the level of state involvement versus coordination through market mechanisms (North, 1981; Hall and Soskice, 2001). The framework points out four types of interaction in a commodity chain: 'state governance' (a situation where transactions are coordinated through state involvement and the value chain is coordinated through market forces), 'joint governance' (a situation where transactions are coordinated through state involvement and the value chain is coordinated through chain integration), 'market governance' (a situation where transactions and the value chain are both coordinated through market forces) and 'corporate governance' (a situation where transactions are coordinated through state involvement and the value chain is coordinated through chain integration) (figure 2.2).

This framework makes it possible to identify the direction of change over time. The concept of 'state governance', originating from the political economy approach, traditionally paid more explicit attention to the role of the state. In this school of thought, the state is regarded as the governor of trade relations and competitiveness.

It is often argued that trade liberalisation and structural adjustment have reduced the mandate and the ability of the public sector to make specific pro-poor interventions. However, even within liberalised markets governments have a significant role to play. For instance the public sector plays an important role in advocating sustainability and the fair distribution of power among actors in value

Figure 2.2   A comparative framework for labelling public-private interaction in a commodity chain

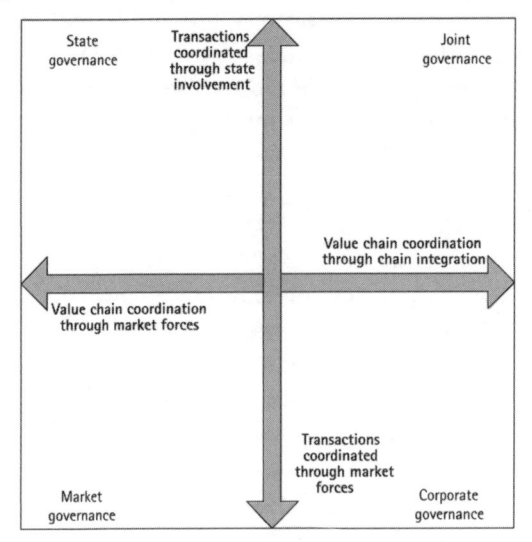

```
              Transactions
State         coordinated          Joint
governance    through state        governance
              involvement
                    ↑
                    |
                    |    Value chain coordination
                    |    through chain integration
  ←─────────────────┼─────────────────→
  Value chain coordination
  through market forces
                    |
                    |
                    |    Transactions
                    |    coordinated
                    |    through market
  Market            ↓    forces       Corporate
  governance                          governance
```

Source: Ton et al, 2008: 5, based on Griffiths and Zammuto 2005.

chains. It can also play a role in creating more favourable conditions for agricultural sector development (see also Joosten and Eaton, 2007: 1[25]). In fact, government led interventions in the agricultural sector have been crucial in most countries with successful agricultural sectors (Berdegue et al., 2008).

Although it is important to reconsider the role of the public sector, I have some doubts that the public sector is truly an 'enabler' and also assert that this does not reflects the reality in the economies of developing countries. I believe governments in developing countries cannot be considered neutral players in their economies, the lessening of their role in agricultural development is still a fairly recent phenomenon in most countries. Governments (still) represent the interests of certain economic sectors and groups within society, with some powerful personal interests also playing a role. Moreover, the government of the developing country has a very limited capacity, especially in an African setting, to manage successful interventions geared at pro-poor agricultural development. The same holds true for other actors, such as producer organisations, private companies and non-governmental organisations. These actors also need to be strengthened in order to enable them to overcome pervasive market failures and to secure desirable social outcomes in their countries (World Bank, 2007). To ignore the role of state in policy designs is to ignore the fact that in some countries or sectors the government still plays a steering role. This oversight severely limits the analysis and it ignores the opportunities and constraints for action.

2.4   Linking different governance levels

Both the value chain approach and the cluster approach focus on processes of economic governance, where the interactions between private sector actors are the potential catalysts for change. The GVC approach fails to integrate the importance of local-firm cooperation and local governance structures. On the other hand, most studies that utilise a cluster approach fail to integrate global governance processes, with some notable exceptions (see Guiliani et al., 2005;[26] Humphrey and Schmitz, 2000; Nadvi and Halder, 2002:). Both approaches also pay little attention to the role of the state. Several authors emphasised the interaction between different governance levels as a good avenue for understanding upgrading, both the processes and their outcomes on competitiveness and poverty. In addition to approaches that link global chain governance (GVC analysis) to local level governance (cluster literature) (cf. Humphrey and Schmitz, 2000; Westen, 2002; Guiliani et al., 2005) there are other possible combinations. Earlier I introduced approaches that link strategic management theory to political economy (Griffiths and Zammuto, 2005; Ton et al., 2008). This can provide a more dynamic comparative framework for studying public-private interaction (e.g. state involvement versus coordination through market mechanisms) in a global value chain.[28]

In this study I use both combinations and I look at the interaction between different governance level in creating opportunities and constraints for more inclusive upgrading among small-scale cocoa farmers in Ghana. The idea of 'rational choice' dominates the upgrading discussion and emphasises the leading role of 'modern' actors. For the African setting, it is much more appropriate to consider the interaction between governance structures in the global value chain, national governance structures and social structures, because there the functioning of the producers is very much affected by traditional structures, the role of the state and the capacity of the state to play this role.

3
RESEARCH QUESTIONS, METHODS AND RESPONDENTS

3.1 Introduction

The central question in this study is how different governance structures interact in creating opportunities and constraints for more inclusive upgrading among small-scale cocoa farmers in Ghana. I distinguish between three levels of governance: first, the global chain governance (referring to power relations in the global cocoa chain); second, state governance (referring to the level of state involvement in the Ghanaian cocoa sector); and third, the social structures (in which cocoa farmers are embedded locally). The different dimensions contained in this question demand for a combination of different research tools and concepts. The value chain approach is used as a tool to identify upgrading opportunities and constraints for cocoa farmers in Ghana by considering the existing power relations in the global chain and by looking at changes in these relations. At the national level, the introduction of reforms in Ghana is taken as a 'key-turning point' to understand local upgrading opportunities and constraints, and how these have changed overtime. The changing role of the state, the entrance of new players and changes in the farmer's enabling environment are the main factors that determine the conditions under which cocoa farmers produce their cocoa. At the local level, I seek to explain the different impact that shifts in governance structures and upgrading opportunities (along with the constraints that result from these changes) have on farmers, resulting in unequal benefits. Central concepts of the cluster literature, such as 'embeddedness' and 'joint action', are used to identify social structures which constrain or facilitate upgrading strategies of individual cocoa farmers locally and thus affect the way they benefit from these strategies.

An overall assumption is that analysing opportunities and constraints for upgrading among producers of primary export commodities requires a multilevel (i.e. disentangling and analysing the different processes that operate at different spatial scales)[29] and dynamic perspective, not only in terms of developments over time but also in terms of interactions between agents and institutions.[30] By looking at the shifts in different levels of governance and the interaction between (changing) global, national and local power structures I attempt to develop a framework for more inclusive ways of upgrading, by making an explicit link between upgrading and development goals. In this process, I build upon some of the key features from the literature on 'new economic geographies'. This school of thought views social relations as power relations. Another key feature is that the role of the context 'in shaping and understanding economic behaviour in time and space' is an integral part (not external) of the subjects/objects under investigation (Yeung,

2003: 444-5). In analysing the crucial role of context and power relations, I build upon the notion that structures tend to be reproduced by agents (Sewell, 1992).

In this chapter, I will present the four research questions that together comprise the conceptual framework. For each research question I will explain the use of different research methods and data collection. Next, I will discuss the validity of the research findings. Finally, I will highlight briefly some common features and differences among Ghanaian cocoa producers, which potentially have a defining influence on their participation in upgrading strategies and the ways in which they benefit from interventions.

3.2 Research questions

3.2.1 Research question 1

> *How are the main interests of global actors who currently govern the cocoa chain being manifested locally, both through their involvement in local upgrading strategies in Ghana as elsewhere, and through their establishment of more direct relations with cocoa suppliers and the formation of new public-private partnerships?*

Global Value Chain (GVC) analysis of the cocoa chain yields crucial information on the following themes: first, the way added value and power is distributed along the cocoa chain; second, insight in shifts in governance and the changing role of institutions; third, the understanding of the main interests of the main drivers of the cocoa chain and their perception on opportunities and threats in the sector; fourth, information on changing relationships with other suppliers, new public-private partnerships and alliances; and finally, the identification of interventions affecting the Ghanaian cocoa farmers' opportunities and constraints for upgrading. In order to understand the significance of buyer-driven strategies in Ghana, I made some comparisons with comparable strategies that were implemented in other cocoa-producing countries in the region. Figure 3.1 conceptualises the shifts in global chain governance and how they affect choices for buyer-driven upgrading strategies and relationships between the Ghanaian government and the Ghanaian cocoa suppliers.

Data collection
Several studies have already examined the cocoa chain (cf. CREM, 2002; Fold, 2002; Gibbon and Ponte, 2005; Kaplinsky, 2004; Norde and Duursen, 2003), uncovering essential knowledge on the composition of the cocoa chain, its main driving actors/mechanism and recent developments. In this study, I will take the discussion one step further and analyse the implications for producers at the beginning of the chain. To answer the first research question, I administered a (small-scale) survey among cocoa processors, chocolate manufacturers and some of the institutions that represent their interests. The questions covered the opportunities and threats they faced in addition to the types of strategies and strategic interventions they used to

Figure 3.1   Conceptualising shifts in global chain governance and upgrading

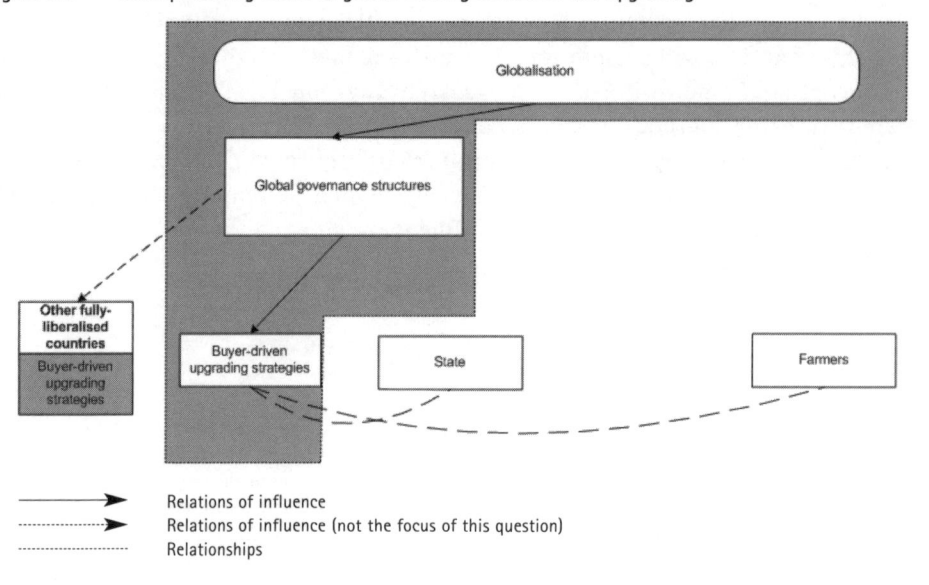

[Diagram: Globalisation → Global governance structures; Other fully-liberalised countries (Buyer-driven upgrading strategies); Buyer-driven upgrading strategies; State; Farmers]

→ Relations of influence
⇢ Relations of influence (not the focus of this question)
⋯ Relationships

respond to these challenges. The respondents included two major chocolate manufacturers, one major cocoa processor and several institutions representing industry.[31] I complemented this data by making use of already existing case studies, reports and literature. In addition, I gathered information through a number of informal discussions with the world's major cocoa buyers. Usually these meetings took place at conferences where the industry discussed relevant developments (for example the round table meetings for a Sustainable Cocoa Economy in 2007 and in 2009[32], and the World Cocoa Foundation partnership meeting in 2007). In addition, I organised two multi-stakeholder workshops, one in the Netherlands (2003) and one in Ghana (2005), with key representatives of industry and other actors (see attendance lists in Appendix 3.1). These workshops provided additional insights in the perspectives and position of the industry players, making it possible to reflect on the provisional outcomes of the study. Also, these participatory multi-stakeholder meetings contributed to a process of trust building. As a final information gathering tool, I conducted in-depth interviews with actors closely involved with the industry.

3.2.2   Research question 2

> How does the changing role of the state, of the other actors and of the institutions in the Ghanaian cocoa sector affect the conditions under which cocoa farmers operate and in turn define their opportunities and constraints for upgrading?

The concept of 'state governance' is used to assess the level of state involvement in contrast to coordination through market mechanisms; it influences the decision-making processes and determines the conditions under which economic transactions take place in the sector (Ton et al, 2008). Like many other sectors in developing countries, the cocoa sector in Ghana is a sector in transition, with marketing and institutional reforms being gradually introduced. As a result the role of the state is changing and new actors have entered the sector. In contrast to other cocoa-producing countries in the region, the Ghanaian government has remained the main coordinator of the cocoa supply chain. In order to understand the impact that the current organisation of the cocoa sector in Ghana has on the position of cocoa farmers and in order to evaluate the developments in the sector, I conducted some comparisons over time. Specifically, I compared the current state of affairs in Ghana with the conditions in Ghana prior to the reforms and with the current situation in other major cocoa-producing countries in the region. For these comparisons, I build on a study of the World Bank (Akiyama et al., 2001) that assessed the impact of structural adjustment programmes in cocoa producing countries. This analysis focused principally on the developments in price levels, taxes, marketing costs and the volume of production. I complemented these economic variables with other types of indicators that reflect relevant institutional changes and opportunities/constraints for upgrading.

Figure 3.2 conceptualises the shifts in national governance structures and the effect they have on the enabling environment and upgrading strategies of cocoa producing farmers.

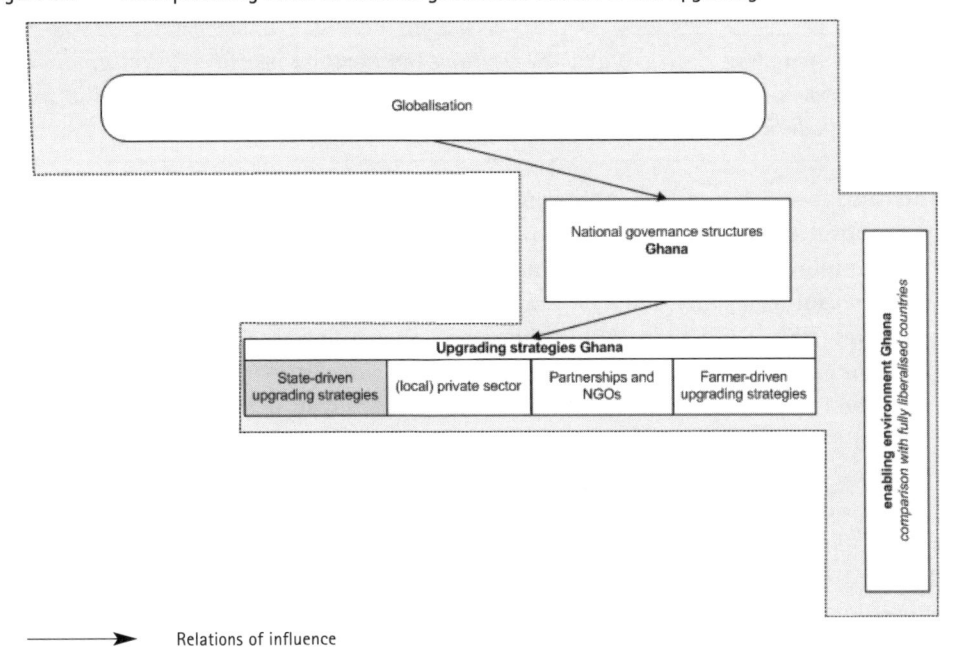

Figure 3.2   Conceptualising shifts in national governance structures and upgrading

Data collection
The shifts in the national governance structures and their impact on the institutional environment of cocoa farmers are analysed mainly through the in-depth

interviews with Ghanaian actors active in the supply chain and by using secondary quantitative and qualitative data. In addition, a number of reports and case-studies are used, in particular for the comparison with neighbouring cocoa producing countries. In addition to the collection of qualitative data, I conducted two farmer surveys (explained in greater detail in Section 3.2.3) in order to analyse the direct impact of the reforms on the level of the individual farmer.

Because of the changing role of the state and the emergence of 'new' public, private and civil society actors who attempt to fill the 'vacuum', I interviewed a number of representatives of governmental bodies, licensed buying companies (LBCs), non-governmental organisations (NGOs), international institutes, banks, input providers, research institutes, farmer's organisations and others. Including the perception of other actors on developments within the sector provides a better understanding of the underlying dynamics and tensions of the partially liberalised system in which the sector is embedded. As many of the interviewees were promoted[33] to different posts during the research period, (in these cases) I was forced to interview their successors. It helped that I interviewed most actors twice; coming back for a second round was appreciated and it contributed to a process of trust building and the creation of a valuable network. A number of respondents allowed me to record the interview on video and participated in the workshop organised in Accra in 2005.

3.2.3 Research question 3

> *How do upgrading strategies and the resulting interventions benefit different groups of cocoa farmers, and to what extent do these groups participate actively in the activities and initiate their 'own' strategies?*

The third research question builds on the concept of 'embeddedness' (as put forward by Granovetter in 1985, who reconstructed earlier work of Polanyi [1944]). The notion that economic behaviour of firms, markets or economic institutions is embedded in wider social relations implies that in order to identify 'inclusive upgrading strategies' first it is necessary to understand the differences in the farmers' social relations. Central is to examine the different identities among cocoa farmers (explained in greater detail in Section 3.3); it will make it possible to explain the economic activities and the responses to interventions and to come up with a framework for more inclusive upgrading strategies in terms of development.

But, constructing a framework for more inclusive upgrading requires more than simply identifying the upgrading strategies and the responses to interventions for different groups of farmers; it requires an analysis of their impacts. Therefore, in this study I will begin by analysing the interventions' impact on their outreach, the number and types of farmers that they reach. (More specifically, the questions posed are: how many farmers are reached; what types of farmers are more at risk of being excluded or are unable to benefit from the interventions; and what kind of impact can be observed?). Consequently, I will continue by identifying the mechanisms behind the interventions (the way in which the intervention takes place, for example through learning). Finally, I will integrate the identification of constraints and the potential trade-offs.

Figure 3.3   Conceptualising heterogeneity and inclusive upgrading

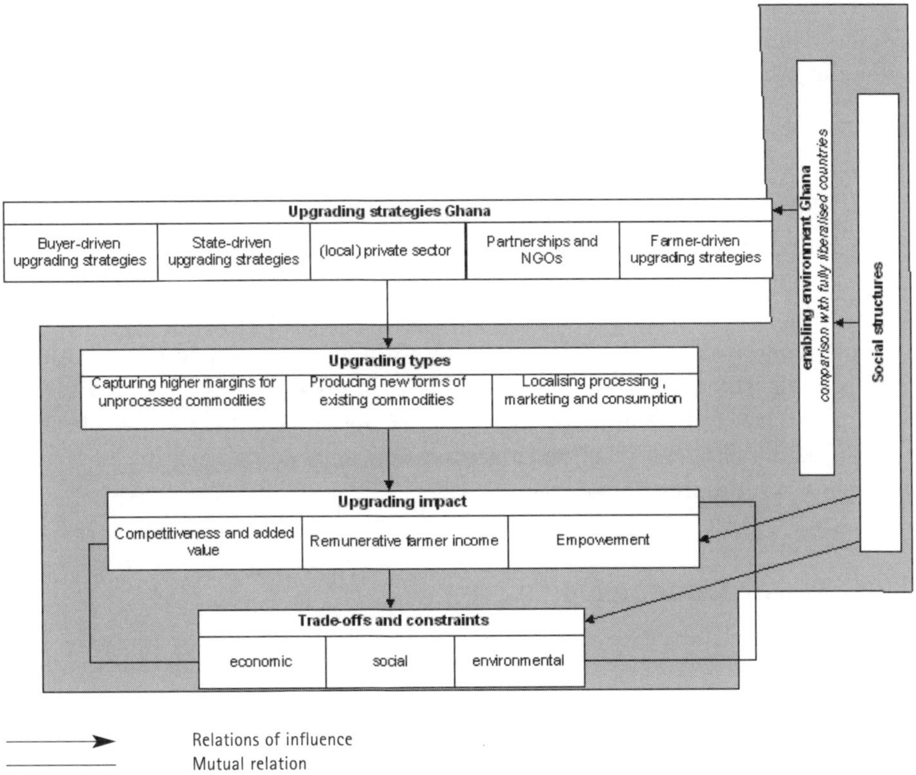

Upgrading strategies Ghana				
Buyer-driven upgrading strategies	State-driven upgrading strategies	(local) private sector	Partnerships and NGOs	Farmer-driven upgrading strategies

Upgrading types		
Capturing higher margins for unprocessed commodities	Producing new forms of existing commodities	Localising processing, marketing and consumption

Upgrading impact		
Competitiveness and added value	Remunerative farmer income	Empowerment

Trade-offs and constraints		
economic	social	environmental

Sidebar: enabling environment Ghana — comparison with fully liberalised countries; Social structures

⟶ Relations of influence
—— Mutual relation

Data collection

To provide answers to this third research question I combined quantitative with qualitative data. The two farmer surveys (2003 and 2005) were conducted in 17 districts in 4 cocoa-growing regions of Ghana (Western region, Brong Ahafo, Ashanti and Central region). These regions are part of Ghana's cocoa belt, which lies in the south of the country. Ashanti is the traditional cocoa-growing area. The south of Brong Ahafo represents the zone where cocoa belt is moving into (by extension of the dry savannah areas of Northen regions) (Vigneri, 2007). Cocoa production is currently concentrated in the Western region, which is the relatively new area of production for cocoa (Gockowski, 2007). The Central region is a somewhat smaller in terms of cocoa production. Other cocoa-growing regions, which were not selected for data collection, are the Volta region and the Eastern region (MMYE,[34] 2007) (for more information on the scope and limitations of the farmer surveys see Section 3.4.3). I combined the two surveys with 30 in-depth interviews with a selection of the farmers, which I call 'farmer profiles'. These qualitative profiles complemented the data gathered in the two surveys and made it possible to check the outcomes of the surveys, ask for clarifications and to check inconsistencies. In addition, I held group discussions in around one third of the communities that I visited. Group discussions helped me identify the main challenges that farmers faced and indicated whether these were individual problems or shared by the majority of the group.[35] It also gave insight into some of the existing power relations among farmers; chief farmers and cocoa buyers participated more actively in discussions than for example female farmers. Sometimes group discussions helped me to identify new farmers for subsequent interviews.

Video

In 2005, I recorded the farmers (upon expressed prior consent by the farmers) during the group discussions and during in-depth interviews on video, in order to create output to be used for educational purposes. The use of the camera was no obstacle; on the contrary, it generally contributed positively to the discussions. It seemed that the farmers felt that by recording their discussions they were taken seriously and given an official voice. Some of the participants dressed up nicely for the occasion. The use of the camera also had an effect on me; it forced me to ask open questions and to interrupt as little as possible. It helped me observe the farmers' comments and reaction in greater depth, by going back to the recordings and re-interpreting the data.[36]

3.2.4 Research question 4

> *How do different governance structures interact in creating opportunities and constraints for upgrading among different groups of farmers?*

In the literature it is recognised that in order to understand upgrading opportunities and constraints it is important to look at the level of interactions and the changing conditions under which upgrading takes place. Even though, the importance of studying interactions between vertical and horizontal relationships has been emphasised by several authors (see Bolwig et al., 2008; Humphrey and Schmitz, 2000; Lambooy, 2002; Palpacuer, 2000; Westen, 2002) it has not been sufficiently examined. I decided to look at the interaction between different governance levels (global, national and local) by zooming in on some of the interventions. By looking at the actions, one can grasp the different interests/intentions involved (who is behind it and who benefits). This approach also sheds more light on who exactly is targeted by interventions. Although it I believe that upgrading is per definition a selective process, from a development point of view it is important to understand why some benefit more than others. It would enable the creation of better targeted development policy, thus contributing to the effectiveness of pro-poor development strategies.

Data collection

This fourth research question corresponds to the central question of this study. The central question did not require the collection of additional field data as it is largely based on the interpretation of information already gathered.

Figure 3.4   The interaction between global, national and local governance structures and their impact on upgrading

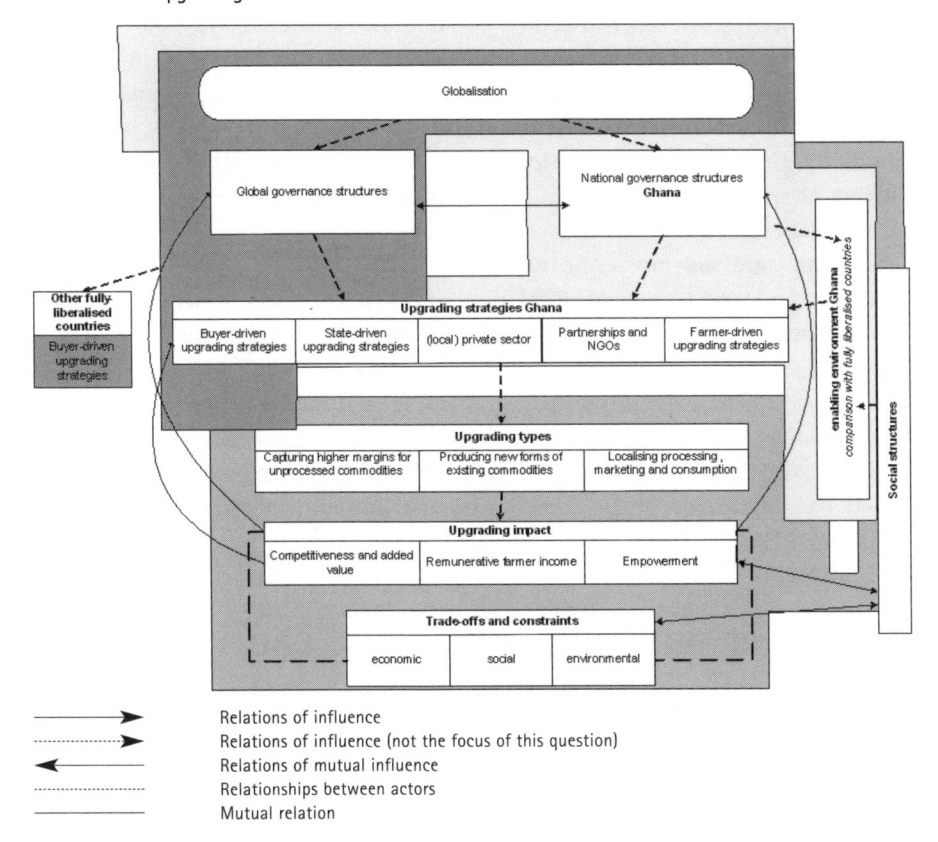

[Figure 3.4: Diagram showing the interaction between Globalisation, Global governance structures, National governance structures (Ghana), Upgrading strategies Ghana (Buyer-driven, State-driven, (local) private sector, Partnerships and NGOs, Farmer-driven upgrading strategies), Upgrading types (Capturing higher margins for unprocessed commodities, Producing new forms of existing commodities, Localising processing, marketing and consumption), Upgrading impact (Competitiveness and added value, Remunerative farmer income, Empowerment), and Trade-offs and constraints (economic, social, environmental). Other fully-liberalised countries with Buyer-driven upgrading strategies are shown, along with enabling environment Ghana comparison with fully liberalised countries, and Social structures.]

Legend:
- Relations of influence
- Relations of influence (not the focus of this question)
- Relations of mutual influence
- Relationships between actors
- Mutual relation

In order to capture changes over time I participated actively in global and local meetings, such as the round tables (RSCE1 and RSCE2) and the IFPRI-ODI[37] Cocoa Workshop in Ghana[38] (2007). On these occasions I shared my data and discussed my findings, both with the international research community (e.g. CIRAD, ODI, STCP[39]) as well as with representatives of the public sector, the private sector and members of civil society.

In this study I will outline potential avenues for development in the future by describing a number of scenarios built around two dimensions crucial for the development of Ghana's cocoa sector: 1) the process of liberalisation; and 2) the shift in demand from 'product quality' to 'process quality'. This exercise helps to understand the relation between individual upgrading strategies (and outcomes) and long-term trends towards 'collective' inclusion or exclusion.

3.3 Heterogeneity and agency

To avoid determining relevant differences among farmers *a priori*, I integrated a range of questions on respondent characteristics in the first farmer survey. This allowed me in the second survey to analyse what types of characteristics and social relations constrain or facilitate upgrading. Building on lessons drawn from the GVC literature and cluster literature, I questioned farmers on several issues spanning: land-ownership, volume of production, gender, age, social network, and other. Even though this was not a comprehensive selection (for example leaving out variables such as ethnicity, membership of a political party, or being a producer of mono crops versus mixed crops) it still helps me to identify differences between the respondents. This knowledge clarifies the extent to which social relations, economic features and spatial characteristics influence the respondents' decision-making in economic choices, their responses to interventions and the extent to which they benefit from interventions. Another part of the analysis, based only on data obtained in the FS 2005, is looking for significant correlations between the different variables.[40] Insight in social relations helps me to understand even better what kind of characteristics (or a combination of characteristics) define the farmers' decision-making processes and the possibility of farmers to upgrade. In anticipation of the analysis I have clustered the characteristics around four themes: 1) Location, migration and farm-ownership; 2) Cocoa production, size of farm and productivity; 3) Position in the community, age and education; and 4) Gender and joint action.

3.4 Reflections, limitations and the structure of the book

3.4.1 Validating the research findings

Triangulation is a good method to assure that the research process and instruments indeed explain what they are designed to explain. There are different types of triangulation (data triangulation, investigator triangulation, theory triangulation, methodological triangulation and environmental triangulation) (Guion, 2002: 1-3). Most researchers who work with qualitative data use all of the types or a combination of several; I opted for data triangulation, theory triangulation, methodological triangulation and environmental triangulation.[41]

'Data triangulation' involves the use of different sources of data/information and is an essential component for analysing the value chain. It provides insights into the perception of different actors (both the ones involved in the change and those outside of it) and the recent developments in the specific chain. Outcomes that are agreed upon by different actors can be an indicator for the validity of the findings. Also, the inclusion of a variety of actors as respondents can produce contradicting outcomes. It is obvious that contradicting outcomes can also be the result of differences in perception.

'Theory triangulation', which 'involves the use of multiple professional perspectives to interpret a single set of data/information' (Guion, 2002: 2), is another method I applied. By combining the GVC approach with the concepts of 'state governance' and 'embeddedness' it is possible to get a better understanding of the constraints and opportunities available to the farmers.

'Methodological triangulation' involves the use of multiple qualitative and/or quantitative data (Guion, 2002: 2). I used both qualitative and quantitative methods. The two farmer surveys enabled me to do a statistical analysis and to identify significant correlations. The in-depth interviews, conducted with around thirty of the farmers, made it possible to check the answers for inconsistencies and obtain additional clarifications. For a qualitative analysis I used the qualitative data analysis programme *Atlas-ti*. The group-discussions with farmers yielded additional information and provided valuable insight as I was able to observe first hand the group-dynamics. By using video and relying on an interpreter to facilitate communication with the farmers I could focus on being an observer and limit my engagement as a participant.[42] In gathering information from other stakeholders, especially from members of industry, semi-structured interviews and informal discussions were an important source of data. The combination of quantitative and qualitative data is especially useful when measuring the level of inclusiveness of interventions. The quantitative data made it possible to obtain statistical information on the type of economic activities that the respondents of the survey are involved in and shed some light on the way different types of farmers respond to interventions. Furthermore, it yielded information about precisely which respondents are reached by interventions. The qualitative data helped to explain why some strategies have no effect or have unexpected side effects. Moreover, it helped to understand the role of power dynamics and contextual features. In this respect, the use of different methods is not only of value for measuring validity, but also a way of complementing and enriching my data and it also contributes to a better understanding of the complex (and always changing) reality.

Finally, 'environmental triangulation' involves the use of different locations. I chose one research location (Ghana) and used other cocoa producing countries serving as context setting comparisons to explain Ghanaian developments. These comparisons are mainly based on literature studies. On the farmer level, farmers were interviewed in four different cocoa growing regions, making it possible to verify for my respondents if and when 'location' matters.

To some extent the reliability of the data is safeguarded by triangulation; however, I doubt that other researchers would obtain exactly the same research results if they used the same methods. During the two periods of fieldwork, I became aware of the constant fluctuations in the farmers' environment and that the respondents were most sensitised to the problems which were most acute at the moment of conducting the interview.[43] For the other actors involved in the cocoa sector the environment and mentality over the years also changed. For example, the involvement of global buyers in social and environmental programmes is no longer only 'window dressing'. A continuous dialogue with (new) public and private partners, including members of civil society, contributed to the recognition that a sustainable cocoa economy requires the active participation of different actors and some level of mutual cooperation.

3.4.2 *Trust and participation*
In qualitative research the role of the researcher and the relationships that he/she builds with the respondents are crucial. I used different strategies to build trust

during my field research. First of all, Ghanaian extension officers and researchers active in the areas introduced me to the village chiefs, to the local purchasers of cocoa and to the farmers, with whom they already had built up relationships. Second, after a first round of interviews with the farmers I tried to trace them back for a second survey. Two years later, the farmers seemed to appreciate my return. Most of them were willing to participate in another round of interviews, including an in-depth interview that I was allowed to record on video. Over the years I also built a relationship of trust with other actors involved in the cocoa sector. After the Dutch consultancy firm CREM introduced me initially to a number of key-informants, the process of finding additional respondents progressed quite naturally. The two multi-actor workshops that I organised during my research were also helpful; there I disseminated my findings and organised participative sessions. These workshops contributed to a process of trust building among a wider variety of actors. In 2003, during the first workshop in Amsterdam, the over fifty participants represented a broad cross-section of stakeholders: industry (chocolate manufacturers, processors, traders, warehousing), international institutions, the Dutch government, NGOs, researchers, and others. In 2005, at the end of the second period of fieldwork, I organised another get-together that was supported by the Dutch NGO SNV and professor Nyanteng, a cocoa expert from the University of Legon (ISSER)[44] (Laven, 2005). The participants in this second event included members of: the private sector, the public sector (representing the Ghanaian government), civil society (including the research community) and farmers. I was particularly proud that around a quarter of the total number of participants were indeed farmers. In this workshop held in Ghana, I presented some of my provisional research findings and gave input for the discussion in a plenary session. After the plenary, I divided the different participants in working-groups. In contrast to the workshop in Amsterdam, I organised the participants in homogeneous groups, mainly due to language considerations and the fear that farmers would feel intimidated or uncomfortable in the presence of officials and international business men. The dynamic of discussion during the workshop seemed to confirm that I made the right decision.[45]

Trust-building was also an important factor in developing good relationships with research assistants. I was lucky to work together with three Ghanaian cocoa experts, who had extensive knowledge of the cocoa sector and agricultural development, spoke the local language and were already extensively engaged with the local cocoa producers. Frequently, we went together on fieldtrips, during which we reflected on the day and shared the main observations over a good meal. The same holds true for two Dutch post-graduate students who were involved in the first farmer survey held in 2003.

3.4.3 Scope and limitations of the research

I interviewed farmers in four different cocoa growing regions in Ghana, covering 34 communities concentrated around 5 locations (see Map 3.1 and Table 3.2). The data collected is especially relevant for the respondents, the communities where they live and work, and for Ghana as a whole. Because Ghana is regarded as quite an exceptional case, I did include (secondary) data from other researchers who studied similar developments in Ghana and in other cocoa growing countries.

Map 3.1   Location of surveys in 2003 and 2005[46]

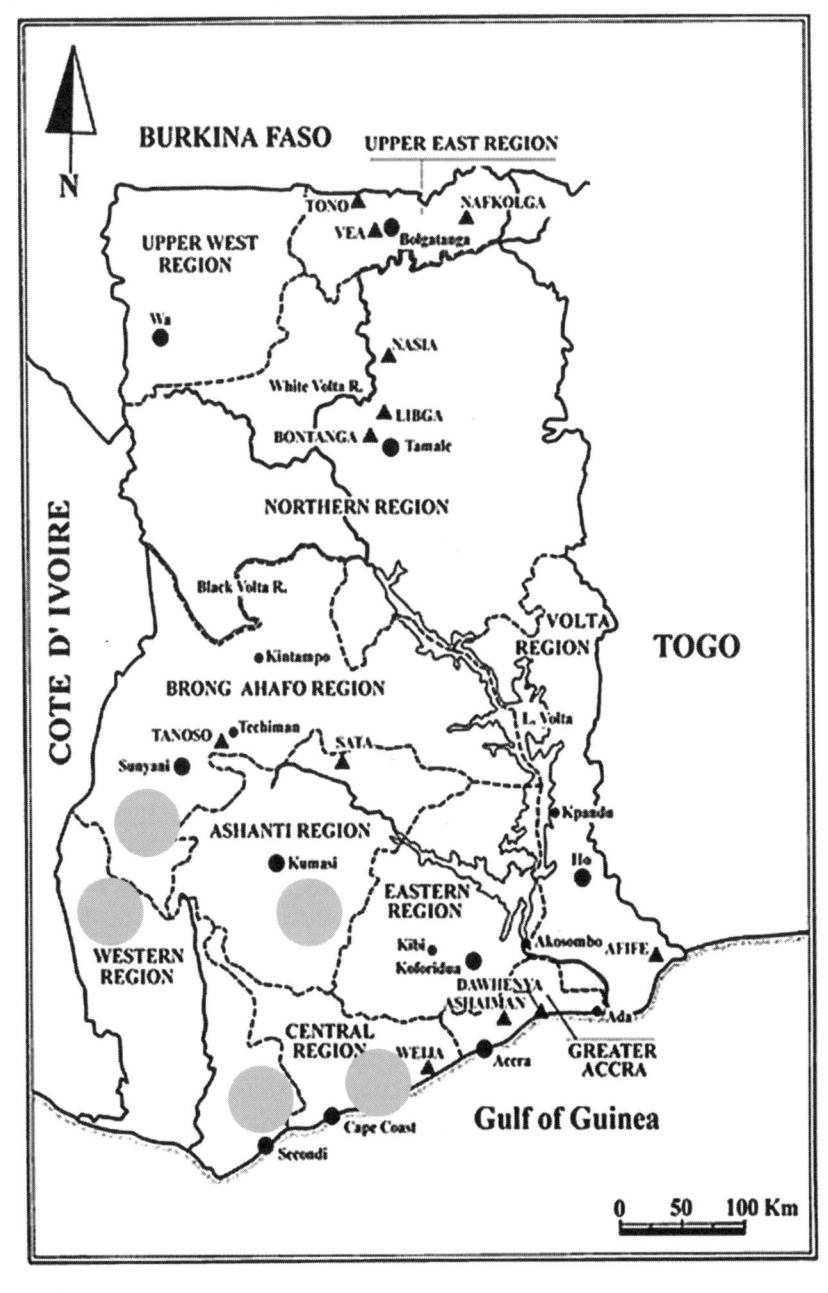

The limited number of farmers who participated in the study is clearly a restricting factor; they represent only a very small proportion of all cocoa producers in Ghana. However, I have used a stratified sampling procedure in order to provide a greater degree of representativeness in the sample. It is estimated that the cocoa sector in Ghana employs around 800,000 smallholder families (Anim-Kwapong and Frimpong, 2004). Together these families produced almost 500,000 tonnes of cocoa beans in 2002/03. In 2003/04, total production was about 637,000 tonnes (Anim-Kwapong and Frimpong, 2004).[47] In total I interviewed 280 farmers (see table 3.2).[48]

Cocoa is mainly produced in five cocoa growing regions (see table 3.1).[49] The Western region is of major importance. In cocoa season 2002/2003 around 55 per cent of total cocoa production is produced in this region. Therefore, the choice was made to select a similar percentage of the total number of respondents from this region (60 per cent). The selection of the farmers from the other regions, was based on available support and logistics. I selected a number of farmers from the Brong Ahafo, Central and Ashanti region. Cooperation with private providers of extension services (Wienco), the Non-Governmental Organization (NGO) Conservation International (CI) Ghana and several research assistants helped in getting access to these cocoa growing locations as well as to cocoa farmers. The result was that 173 farmers in 27 villages were interviewed in 2003.

Table 3.1   Cocoa production in Ghana in 2002/03 and 2003/04

Year	Total	Western region	Ashanti region	Brong Ahafo region	Central region	Eastern region
2002/03	496,846	276,586	82,445	45,309	39,989	51,604
2003/04	636,957	419,710	121,233	69,688	56,631	67,904

Source: Anim-Kwampong and Frimpong, 2004.

This first farmer survey provided me with considerable insight into the sector. However, it also made me realize that some important categories of farmers were underrepresented. Therefore, in 2005, I chose to re-interview the same farmers, but also to select a number of caretakers and additional female farmers. In this second farmer survey I was able to locate and interview around 60 per cent of the farmers that participated in the previous farmer survey. Although the survey conducted in 2005 reflects the diversity among farmers better than in 2003, I still cannot assume that my second sample is completely representative because I was not able to get insight in estimated population proportions for some of the categories I distinguished. In processing and analyzing data I can therefore only give outcomes for my respondents and not for cocoa farmers in general. Another restriction is that I only give outcomes on changes over time (comparing the two samples) for the respondents that participated in both surveys (n = 103). To provide further understanding, I complemented my samples with qualitative data (for example obtained in group discussions and in in-depth interviews). I also made comparisons with national data (where available) and quantitative data from other researchers (for example Teal et al, 2004; Vigneri, 2007; Zeitlin, 2006; Ruf, 2007a/b).

Table 3.2   Respondents in FS 2003 and FS 2005

Region	District	Village	No of farmers in 2005	No of farmers in 2003	No of farmers interviewed in both years
Central region	Twifo Hemang	Sumnyamekudu	1	5	1
		Abeka	2		2
		Afiaso	4	5	4
	Upper Denkyira East	Asikuma	14	7	7
		Ayanfuri	4	3	2
	Fosu	Kruwa	5	4	3
	missing	Missing	1		
	Other			3	
Total Central region			31 (14,8 of the total of farmers interviewed in 2005)	27 (15,6 of the total of farmers interviewed in 2003)	19(18,4 of the total of farmers interviewed in both years)
Western region	Bibiabi Akwiaso Bekwai	Dansokom	2		
	Sameraboi	Kokoase	3	1	1
		Ohiamatuo	6	4	3
	Amenfi West	Bonsie	2	7	2
	Enchi	Enchi	5	9	4
	Aowin District	Enchi	2		
	Bia	Elluokrom	16	8	8
		Asuantaa	10	5	2
		Cashierkrom	3	7	3
		Attakrom	2	7	1
		Asantiman	4	8	3
	Bibiani Anmiso Bekwai	Attannyamekrom	2		2
		Adeambrah	6	8	3
		Ampenkrom	8	8	8
		Soroano	9	12	9
		Donkorkrom	4		4
		Dansokrom	8	10	2
	Bibiani Anhwisa	Adeambrah	1		
	Akotom	Wassa West	1	1	
	Mpohor Wassa East	Mpohor	5		
		Sekyese Krobo	14		
		Sekyese Nsuta	3		
	Other			7	
Total Western region			116 (55,2%)	102 (58,9%)	55(53,3)
Ashanti region	Ahafo Ano South	Abesewa	5	6	5
		Nyamebekyere	13	8	6
Total Ashanti region			18 (8,6%)	14 (8,0%)	11(10,8%)
Brong Ahafo region	Dormaa	Asikasu	9	6	5
		Nsuhia	8	6	2
		Kokurasna	5	4	2
	Tano (8)	Buokuokwa	5	4	3
		Boaso	2	2	2
		Missing	1		
	Asutifi (14)	Asemapnaye	4		2
		Atwedie	9		2
		Afwedia	1	4	
	Other			4	
Total Brong Ahafo region			44 (21 %)	30 (17,3%)	18 (17,5%)
Missing			1 (0,5% of the total of farmers)		
Total			210	173	103

FS 2003 and FS 2005.

Another restriction was my inability to communicate directly with the farmers in their local language. I made use of several interpreters, all knowledgeable in cocoa production and experienced in working with farmers. Although the interpreters did a very good job, not being able to intervene directly was clear disadvantage. Sometimes it was also difficult for the interpreter to give an accurate translation. In order to overcome this limitation, I recorded the in-depth interviews with the farmers and some of the group discussions on video and had them translated from Twi into English. This enabled me to verify the given summaries and also to revisit material for additional clarity.

Table 3.2 provides a detailed overview of respondents of two surveys. These surveys are referred to as Farmer Survey FS 2003 and FS 2005 (table 3.2). In one-third of the communities also focus group discussions were held.

4
THE RISKY BUSINESS OF COCOA

4.1 Introduction

The integration of small-scale producers in developing countries into global markets is a fact. For producers of a number of agricultural commodities (such as coffee, cocoa and cotton) their insertion in 'global value chains' is nothing new; however, the manner of organising and governing these chains is new. In addition, the risks and opportunities for producers and other actors in the chain also changed.

In terms of organisation the main shift has been from the use of forced labour on large plantations (beginning of the nineteenth century) to a smallholder system. This also entailed a shift from a centrally organised production and marketing to open competition. New traders entered, new standards emerged, and a future market developed.

In terms of governance, an important shift was the transfer from 'colonial planters' (as central actor) to a number of private traders who had to negotiate with organised smallholders on price formation. Collective action of producers was a powerful force to reckon with. Later, the state became the main actor in the management of agricultural commodities (Daviron and Ponte, 2005: 11). The state fulfilled the role as intermediary, which made it possible for international buyers to buy tropical products without establishing any direct relationship with their suppliers. The introduction of reforms in producing countries again reduced the role of the state. With the diminishing role of the government in the provision of marketing channels and services international manufacturers, traders and processors became the main driver of agricultural commodity chains. These multinationals concentrated their operations and economic power and became strong entities in global value chains (Kaplinsky, 2004; Oxfam, 2004).

In terms of risks and opportunities for multinational buyers of primary commodities a number of changes occurred. The introduction of market reforms in producing countries made these buyers more vulnerable to the performance flaws of their suppliers. Another source of this vulnerability were the structural adjustment programmes, imposed by the World Bank, which resulted in the abandonment of public quality systems. This occurred in an era when quality is becoming increasingly important as one of the parameters for competitiveness. Quality no longer simply refers to product characteristics but encompasses a wider variety of 'process' criteria, including environmental and labour conditions. This is partly a response to the rising public concern and pressure from civil society (such as consumer organisations and NGOs). They call upon processors to take more responsibility for social and environmental issues, not only in own factories but in the whole value chain (Boomsma, 2008; Dyllick and Hockerts, 2002; Helferich, 1999; Heslin and Ochoa, 2008). In addition to pressure from civil society, also the ethically motivated norms set by the employees and managers within the company are increasingly playing a role. The growing interest in long-term strategic planning is supporting this development. For global buyers, the fact that many producers of agricultural commodities are poor is becoming a threat to their future supply. As a result the risk for supplier failure has become more prevalent over the last few years. This partly explains why the multinationals became more insistent on controlling product and process specifications further down the value chain. This insistence was reinforced by the increasingly competitive global markets, which required companies to constantly look for new supply regions and to invest in developing new value chains, as a means of risk-diversification. On the other hand, the growing attention paid to sustainability issues in commodity sector policies demands of multinational buyers and traders to play a more active role in improving supplying conditions.

How this works out for a particular chain in a particular country is the main question in this chapter. I will focus on the global cocoa value chain and the world's second largest cocoa producing country, Ghana. In addition to the volume traded, and the importance of cocoa for the Ghanaian economy, Ghana's particularities make it an interesting case study. Ghana is known for growing 'world's finest cocoa'[50] and is the only cocoa producing country in the region that has only partly liberalised its marketing and pricing system. So, while cocoa producers in Ghana are inserted in a global chain that is increasingly being governed by international buyers of cocoa, the Ghanaian government still plays a major role in the supply chain.

In this chapter, I will analyse how the main interests of global actors who currently govern the cocoa chain manifest locally, both through their involvement in local upgrading strategies in Ghana and through their building of more direct relations with cocoa suppliers and the formation of new public-private partnerships.

In this chapter I will use the Global Value Chain (GVC) analysis as a tool to 1) understand shifts in governance that occurred in the global cocoa value chain; 2) understand the main interests of the dominant drivers of the cocoa chain and the risk of supplier failure they face; 3) understand the role of institutions vis-à-vis these main drivers; and 4) analyse the responses of drivers to these risks of supplier failure in terms of (a) changing relationships between chain actors and (b) interventions that affect cocoa producers in Ghana and elsewhere.

4.2 Governance in the global cocoa chain

The global cocoa chain is increasingly driven by international buyers of cocoa. From the mid 1950s until the 1980s, cocoa chains were driven by associations of direct producers. Now all are under the leadership of international buyers, with the exception of Ghana. The contemporary 'global cocoa chain' is often typified as one characterised by bi-polar governance (Fold, 2002; Gibbon and Ponte, 2005; Kaplinsky, 2004; Losch, 2002):

One pole arises from the concentrated [concentration] amongst the grinders (processors/traders),[51] who increasingly have operations in both producing and consuming countries, and many links in the chain. The second pole is the large chocolate manufacturers; but their operations are much more limited along the chain, and their governance is much weaker than that of the grinders. In most cases it only extends to the relationship between the grinders and the chocolate manufacturers (Kaplinsky, 2004: 24-5).

There are different reasons for the increase in the global buyers' control over the chain. For example, global processors and manufacturers have become stronger actors in these chains due to takeovers and an increase in the scale of their operations (Gibbon and Ponte, 2005; Kaplinsky, 2004). In the cocoa sector, there are high levels of concentration among manufacturers and processors. Among chocolate manufacturers the seven largest companies constituted 40 per cent of the world market in 2006, with Mars as world number one, followed by Nestlé SA, Hershey Foods, Kraft Foods, Cadbury Schweppes and Ferrero SpA (Tropical Commodity Coalition, 2008: 6). Between 1970 and 1990, some 200 mergers and acquisitions took place, and as a result, by the mid-1990s, Archer Daniels Midland (ADM) became the world's largest cocoa processor with the take-over of Grace Cocoa and its purchase of the cocoa processing units of E.D. & F. Man in 1997. Recently ADM lost its leading position to Cargill. Barry Callebaut became the world's third largest processor in 1996, with the take-over of Callebaut by 'Cocoa Barry'. Together with Petra Foods and Blommer these three multinationals control over[52] per cent of grindings and liquid chocolate (see Table 4.1).

Table 4.1   The main processing companies and their volumes of production in 2006 and 2008

Company	Volume per 1000 tonnes in 2005/06	Volume per 1000 tonnes in 2007/08	Share in 2006
Cargill	480	520	15%
ADM	470	500	15%
Barry Callebaut	400	440	13%
Petra Foods	190	250	6%
Blommer	170	190	5%
Total	1710	1900	54%

Source: Tropical Commodity Coalition, 2008: 6.

While cocoa traditionally is processed in consuming countries, cocoa grindings are now increasingly shifting to cocoa producing countries. There are two reasons for this shift. Firstly, the emerging chocolate market in Asia and South America, and secondly, attractive investments proposals for outsourcing processing activities.[50] Nevertheless, the Netherlands is still the leading cocoa processing country and Amsterdam is the world's leading cocoa import and distribution point.[53] The next table illustrates the location of cocoa processing activities in season 2005/06.

Table 4.2   Location of cocoa grinding activities in season 2005/06

Region/country	Processing volume (in thousands tonnes) 2005/06
Europe	1462
Germany	302
Netherlands	470
Others	690
Africa	507
Côte d'Ivoire	360
Others	147
America	856
Brazil	223
United States	426
Others	207
Asia & Oceania	651
Indonesia	120
Malaysia	250
Others	281
World total	3476

Source: ICCO, 2007: 14.

These concentrated multinational corporations have a growing potential to limit competition and influence prices (Gilbert, 2000). Like other commodities, annual fluctuations in cocoa bean prices are caused by changes in the world markets' supply and demand for the product. Historically, cocoa bean prices have fluctuated in tandem with the availability of stocks of cocoa beans in relation to the annual world grindings (which measure the world demand for cocoa beans). When the 'stocks-to-grindings' ratio declines, the price of cocoa beans rises. Stabilisation of the cocoa stocks used to be regulated by the International Cocoa Agreement, but this system was abandoned in 1994 due to a shortage of funds to finance the buffer stocks. The private sector, however, has no shortage of funds and owns a giant stock of cocoa, equalling two third of total demand. The main owners of these stocks are cocoa processing companies and traders who hold cocoa in stock for the futures market (personal communication Cargill 2005).

But the increased governing role of global cocoa processors and manufacturers is not only related to increasing power, but can also be explained by increasing interdependency in the chain and the increased risks for supplier failure, which is visible at different levels. At the global level the risks for supplier failure increased due to changes in demand that favoured sustainable cocoa production methods.[54] Advocacy movements placed these 'process-quality' standards high on the agenda and increasingly confront multinationals with demands for corporate social responsibility at local and international levels. The increased involvement of international buyers in 'process upgrading', as opposed to 'product upgrading', can be also attributed to the new International Cocoa Agreement (UNCTAD, 2001b), which gave the private sector a large role in supporting a sustainable cocoa economy. As a result, international traders and chocolate manufacturers have become more dependent on the local suppliers who operate at the bottom of a chain. This also entails greater responsibility to provide producers with the information as well as the new technologies they need to comply with the new production and process standards. Because such standards usually do not (yet) apply to local domestic markets, and/or require substantial investments (Keesing and Lall, 1992), producers need financial and other support to improve their operations.

Although there is a trend among global buyers to increasingly become involved in sustainable cocoa production, this does not mean that they always appreciate the role of the advocacy movements, which often put more sensitive issues on the agenda, such as child labour. The following response by a representative of a German confectionary association illustrates this point:
Q: 'What do you see, in general, as the main threats to the cocoa sector?'
A: 'Unfounded and disqualifying accusations by consumer groups, NGOs and others concerning the supply of cocoa or other ingredients used in the production of chocolate products'.[55]

At the national level, marketing reforms in cocoa producing countries had quite an impact on the organisation of the cocoa chain. Prior to the Structural Adjustment Programmes the marketing boards (or stabilisation funds) governed the supply chain. The state determined 'who participates in the chain and to what standards they perform, and in activities designed to upgrade performance amongst chain members' (Kaplinsky, 2004: 22). According to Losch (2002: 225), 'the old national standards have now been replaced by grinders' reputation for compliance with the (demanding) specifications of chocolate manufacturers (concerning timing, volume and quality)'. While reforms are often evaluated positively (they abolished inefficient marketing boards and initially increased the producer price) (cf. Akiyama et al. 2001; Gilbert and Varangis, 2003), their negative impacts in terms of farmer income and conditions under which cocoa producers operate gives reasons for concern. For example, prior to the reforms the state was responsible for quality control procedures. After the reforms most countries privatised their quality control system. This, together with the entrance of many unprofessional buyers, adversely affected the quality of the supply. Also, as a result of reforms, tracing the cocoa back to the cocoa buyer became (even) more problematic.

There are also local and regional factors that are a threat for global buyers. For example, the concentration of cocoa production in West Africa is perceived as a threat, especially with the recent political crisis in Côte d'Ivoire. Heavy rains, or conversely water shortages, adversely affected the volume of cocoa production. Particularly damaging are the outbreaks of pests and diseases, such as Witches Broom, Black Pod and the Swollen Shoot Virus Disease. Other local risks for supplier failure have to do with the old age of both farmers and their tree stock. A serious problem for global buyers of cocoa is the low productivity levels, which make cocoa farming an unattractive business for current and future farmers. While the global demand for cocoa increases, the supply tends to fall down.

These developments have increased the interdependency among global buyers and local suppliers. Global buyers responded in different ways. I asked representatives to score different developments in the sector in terms of threats and opportunity for the sector. The next table (Table 4.3) gives an overview/summary of the risks, as perceived by global buyers. The table represents the developments that received the highest scores. I made a distinction between global, regional, national

Table 4.3   Threats, opportunities and strategies of global cocoa buyers

	Global	Regional	National	Local
Threats	- Changes in demand; - Low world-cocoa price and price-fluctuations; - Obesity; - Cocoa over-supply; - Legislation on cocoa butter substitute; - Harkins Engel protocol.	- The high level of concentration of cocoa production in West Africa; - Weather conditions	- Political instability (Côte d'Ivoire); - Weak credit facilities; - Lack of strong farmer organisation; - Subsidies on input (e.g. chemicals)	- Low farmer-income; - Pests and diseases; - Soil degradation; - Average old age of farmers; - Old trees; - Working-conditions on farms; - Child labour; - Increase use of pesticides, fungicides, fertilizer.
Opportunities	- (new) niche markets; - Expansion of consumer markets; - Scientific evidence of health benefits of cocoa; - Certification; - Public-private cooperation.	- New supply from Asian countries; - Increased grinding in cocoa growing countries.	- Increase in producer-price; - Full liberalisation of marketing and pricing systems; - Strong farmer organisations; - Farmer-Field Schools; - Public-private cooperation; - Traceability.	- Increase in producer-price; - Direct trade with farmer-groups; - Cooperation with farmer-groups; - Training of farmers; - Diversification of cocoa farm; - Increase level of technology used on farms; - Increase use of pesticides, fungicides and fertilizer; - Establishment of large-scale farms.
Strategies	- Looking for new markets (new consumers); Making new cocoa-products; - Investing in image building; - Distinguishing itself from other industries; - Investing in new technologies/innovation; - Creating niche markets; - Working together with other manufacturers/ processors.	- Looking for new suppliers (Asia); - Looking for new consumers (Asia).	- Strengthening relations with the government of the producing country; - Outsourcing grinding in origin countries.	- Strengthening relations with current suppliers; - Working directly with farmer groups; - Investing in new technologies for farmers.

Source: Industry survey, 2005.

and local ('on-farm') threats, and linked these threats to the global buyers' perception of opportunities and strategies (based on the industry survey, 2005).

Looking at Table 4.3 two main observations can be made. A first observation involves the direct interest that buyers have in the well-being of their suppliers and to make sure cocoa farming is a profitable business. It has become strategically important to make on-farm investments; however, a second observation is that at the same time the exact owner and country location of the farm seem to become less important. For example, in response to the quality problems processors are searching for technological innovations which make quality characteristics less important for the processing process. According to the world's largest processor ADM (1999 in Fold, 2002: 233), 'it has become technically possible to compensate for variations in bean quality without compromising customer demand for intermediate goods with specific properties'. These technological innovations enable cocoa processors to generate 'variety/quality' in the liquoring process (processing), instead of in the earlier growing stage (production). This means that processing quality no longer requires sourcing multiple varieties of beans from different regions (at least not for quality reasons). Another response to problems with quality performance has been investments in quality control system in newly cocoa producing countries. For example, Cargill and Mars started to invest in quality control systems in Vietnam, using the Ghanaian quality control system as a template (PSOM, 2004)[56]. Cargill and Mars supported and established plantations in Vietnam that can start producing 100,000 tonnes of high quality cocoa in 2012 (TCC, 2009: 4-5). For buyers this investment can help them to overcome quality problems and functions as a remedy to the anticipated shortfalls in West Africa. In Vietnam this creates immediate opportunities for farmers, but what will be the impact on the longer-term for producers in other regions?

The next section will discuss some other mitigating responses against changes in consumer demand and risks for supplier failure, as utilised by global buyers and their representative organisations.

4.3 Responses to change

The next table gives an overview of the different actors in the cocoa chain and their representative international organisations. The farmers are visibly absent, as they have no representative body at the international level.

Global buyers and their organisations did respond to the changes in consumer demand. For example, in the late nineties, the Association of the Chocolate, Biscuit and Confectionery Industries of the European Union formulated quality criteria for cocoa beans that would continue to enable chocolate manufacturers to produce chocolate of the quality needed to satisfy both prevailing consumer tastes and the evolving more stringent legislation on food hygiene and safety. Within the International Cocoa Organisation (ICCO), the response meant a shift in attention from 'product' to 'process attributes'. For this purpose ICCO was involved in a

Table 4.4   Actors in the cocoa chain and their representative international organisations

Processors and manufacturers	International Confectionery Association (ICA)	A worldwide forum where more than 2,000 companies in 23 countries are represented. The members include the CMA (Confectionery Manufacturers of Australia), the ABICAB (Brazilian Chocolate, Cocoa & Confectionery Manufacturers Association), CAOBISCO and CMA.
	Chocolate Manufacturers Association (CMA)	An organisation of the processors and manufacturers of chocolate in the USA, with the goal to carry out joint research (American Cocoa Research Institute; ACRI), to provide information about chocolate and to stimulate its consumption. The members represent 90% of the trade volume in the USA.
	Association of the Chocolate, Biscuit and Confectionery Industries of the EU (CAOBISCO)	A sector association at the European level for manufacturers of chocolate, cake, pastry and sweets. Together, its members process 50% of all the cocoa beans that are produced.
Traders	1. London International Financial Futures Exchange (LIFFE) 2. Options Exchange New York Coffee, Sugar and Cocoa Exchange (NY CSCE)	Commodity exchanges. The cocoa contracts are meant to eliminate the price risk (hedgers) and are bought and sold for purposes of speculation (speculators). Only 1.5-2% of the total number of contracts bought and sold on the LIFFE result in a physical delivery of cocoa.
	1. The Federation of Cocoa Commerce (FCC) 2. Cocoa Merchants Association of America (CMAA)	A European system for closing cocoa contracts. Its purpose is to monitor the cocoa trade via the harmonisation of cocoa contracts and the provision of arbitration services. In the USA, the CMAA is the agency to contact for the international trade.
	European Cocoa Association (ECA)	A relatively new trade organisation representing the European cocoa sector. Its members are engaged in the trade of cocoa beans, the storage and distribution of cocoa beans and their processing into paste, powder and butter, and the production of chocolate.
Producing countries	International Cocoa Organization (ICCO)	An intergovernmental organisation set up in 1973 to implement the international cocoa agreement. At present, 42 countries representing 80% of the worldwide production and 70% of the worldwide consumption are signatories to this agreement.
	The Cocoa Producers Alliance (COPAL)	An intergovernmental organisation that unites the cocoa producing countries. Its members are: Ghana, Nigeria, Brazil, Côte d'Ivoire, Cameroon, Dominican Republic, Gabon, Malaysia, Sao Tomé, Principe and Togo. COPAL focuses on sufficient supply at good prices, technical information, improvement of the mutual socio-economic relations and the promotion of cocoa consumption.

Source: TCC, 2008, modified by author.

project on 'Supply Chain Management for Total Quality Cocoa', moving away from physical quality and more towards securing 'total quality' (ICCO, 2007b). Also, the 2001 International Cocoa Agreement encouraged its Members to 'give due consideration to sustainable management of cocoa in order to provide fair economic returns to all stakeholders in the cocoa economy' (Article 39). The increasing demand for organic cocoa stimulated the Market Committee of ICCO to execute a study that evaluated the organic cocoa market. In 2008, the Common Fund for Commodities (CFC) financed a feasibility study on organic cocoa production in Cameroon and Togo (KIT et al. forthcoming). At international level, the Codex Alimentarius Guidelines (FAO/WHO,[57] 1999) defined the general principles and requirements pertinent for the production and labelling of organic products.

Global chocolate manufacturers and traders also responded to recent accusations of child labour abuse in the West African cocoa sector. A BBC documentary, broadcast in September 2000, claimed that '90 percent of cocoa plantations in cocoa-exporting African nations use forced labour, suggesting that a significant percentage of this forced labour is children'.[58] Although these allegations were highly exaggerated, it resulted in mass protests against chocolate makers by consumer organisations, NGOs and policymakers. In response to these accusations, U.S. members of Congress and global chocolate manufacturers announced a comprehensive plan to address child slavery on West African farms and in the cocoa-chocolate sector worldwide.[59] The chocolate industry responded as well, by organising themselves under the auspices of the World Cocoa Foundation (WCF). WCF provides funding to the International Labour Organisation (ILO) to study the labour situation on cocoa farms. Also it supported the creation of the Sustainable Tree Crop Programme (STCP), a public-private partnership between industry, producers, researchers, government agencies, public sector institutions and conservation groups[60] (Abbott et al., 2005: 9). The STCP focuses on issues of sustainable rural development in cocoa-producing areas in West Africa. Also, the industry participated in several public-private initiatives, aimed at improving the quality and accessibility of education, and implemented individual programmes. There are a number of individual training programmes planned for the coming years.[61]

Besides training programmes there is another trend taking place among buyers of cocoa. Cadbury and Mars have announced a long-term programme, where they commit themselves to sustainable sourcing of cocoa. Cadbury plans to achieve Fairtrade certification for Cadbury Dairy Milk by the end of summer 2009 for the whole of the British and Irish markets.[62] The announcement from Mars follows a similar line. By 2020, all of Mars chocolate candies will use sustainably sourced cocoa. Mars will use the Rainforest Alliance program to certify the bulk of its sustainable cocoa.[63] Sustainable cocoa sourcing is also the aim of a mainstream cocoa certification scheme 'Utz Certified Good Inside'. In 2009, major cocoa traders (such as Cargill) and Manufacturers (such as Mars) together with a Dutch NGO (Solidaridad) started with testing this Code of Conduct for cocoa. Very recently, in September 2009, two first cocoa cooperatives in Côte d'Ivoire obtained an Utz certification (press release, 10 September 2009).

Another set of responses is linked to the reforms in cocoa-producing countries. For example, in response to the fluctuating cocoa prices in cocoa-producing countries, ICCO and the Common Fund for Commodities (CFC) started to invest in price risk management instruments. The aim was to reduce the vulnerability among suppliers and to 'help small-scale cocoa farmers' co-operatives to improve their capacity to manage price risks and to offer them opportunities to access futures contracts and options available on the world's cocoa markets and commodity exchanges' (ICCO, 2007a). Also, as a response to negative outcomes of the reforms, in 1999, ICCO launched a large project in Cameroon, Côte d'Ivoire and Nigeria to improve the quality of cocoa, to facilitate trade financing, to provide market information and to address trade and price risks.

In response to local and regional risks of supplier failure, ICCO is actively funding research projects and organising conferences on preventing the global spread of cocoa pests. Recently under the umbrella of ICCO, Nestle and Ecom worked on a project 'optimizing farmer income via sustainable and yield-increasing cocoa husbandry techniques' (ICCO, 2007a). ICCO is also involved in developing cocoa varieties resistant to the witches' broom disease. This project started in 2000 and is already producing results, e.g. farmers in Brazil have already started reactivating their abandoned farms.

4.3.1 Interventions in the cocoa sector in Ghana

The increasing risk of supplier failure forced international buyers to actively search for new alliances with local suppliers, and to start offering assistance to their suppliers in optimising their operations. However, at the same time, global buyers continually are searching for new suppliers, especially new Asian countries (Kaplinsky, 2001; industry survey, 2005). In order to understand the risks and opportunities for producers, due to their insertion in a value chain increasingly driven by multinational buyers, I took a closer look at the involvement of global buyers in local upgrading strategies in Ghana and how these reflected the main interests of the buyers.

Involvement of global buyers in local upgrading strategies takes place at different levels. First of all, buyers are involved in setting production standards (traditionally fermented and dried cocoa receive a premium price) and in controlling the quality of exported cocoa against excess levels of pest residues (product upgrading).[64] There is a global tendency of declining demand for 'basic performance requirements' and functional capacities. Ghana is the only country where cocoa beans are still consistently separated by national origin for grinding purposes (Gibbon and Ponte, 2005: 135-6). These changes in demand, which are related to innovation processes, demonstrate that grinders are increasingly becoming the main gatekeepers of knowledge, thus setting production standards.

Players who operate outside of the chain also have considerable influence on the development of production standards. For example the shift in performance requirements for cocoa from product quality to the 'quality' of production processes demonstrates the power of advocacy movements. The growing number of consumers interested in the production process is partly the result of a growing concern among consumers for sustainable development issues. Within this context, also the reputation and the legitimacy of multinational operations started to play a role, because they are considered responsible for local production conditions to some extent (Abbott et al., 2005). In the cocoa sector, as a result of pressure by the media, critical consumers and NGOs, multinational buyers became involved in setting process standards (process upgrading), such as the (national) certification scheme on child labour (Harkins-Engel Protocol) and fair trade certification.[65] The new certification scheme of Utz Certified, which is being piloted in Côte d'Ivoire, is another example of the trend towards mainstreaming of sustainable cocoa sourcing. The recent programmes aiming at sustainable sourcing of cocoa by Cadbury and Mars will also address the direct engagement of buyers in cocoa-farming activities. According to Cadbury their programme will triple the sales of cocoa under fair trade terms for cocoa farmers in Ghana. Cadbury is investing £45 million over the next ten years to secure the sustainable socio-economic future of cocoa farming in Ghana, but also partly in India, Indonesia and the Caribbean. Mars builds mainly upon already existing and new certification schemes, so far its programmes have not included Ghana.[66]

Also the New International Cocoa Agreement is stimulating global buyers to become more involved in process upgrading. This agreement gave the private sector a large role in supporting a sustainable cocoa economy, by encouraging new and ongoing projects in the field: the creation of farmer cooperatives, research on more efficient farm practices and training farmers to sort and grade their own cocoa for export and eventually to market the cocoa themselves. According to the Executive Director of ICCO 'the resulting value added could boost prices paid to farmers by up to 20%'.[67] I have found no data on projects in Ghana that were connected to this initiative.

The buyers' involvement in local upgrading strategies is also reflected in the increased grinding operations in Ghana; the main processing companies have established (or are about to establish) processing factories in Ghana. Barry Callebaut is already involved in cocoa processing in Ghana for some years now. In 2004 their installed capacity was 75,000 tonnes. In 2006, Cargill started to build a large cocoa processing plant in Tema, on the coast. In June 2007, ADM announced its plan to build a cocoa processing plant more inland, in Kumasi, aiming at processing cocoa products from a single source origin. In Octobre 2009 that plant has opened. According to a representative of ADM, this investment illustrates Ghana's growing importance in the cocoa processing value chain.[68] The Ghanaian government developed attractive conditions to stimulate international companies to settle down in Ghana. Cocoa processing companies use small (high quality) beans from the low season, which they are allowed to buy at a 20 per cent discount. Currently in Ghana there are four operational processing factories. Outsourcing to Ghana is generally perceived as one of the main ways of adding value to Ghana's cocoa production. The impact of this intervention is analysed in Chapter 7.

The increasing presence of the foreign private sector is not limited to marketing activities and grinding; it is also apparent in the provision of services and the strengthening of farmer-based organisations, mainly through the public-private partnership STCP. In 2003 the STCP opened a country office in Ghana. Currently it is experimenting with the provision of farmer-based extension services by founding Farmer Field Schools (FFSs) focused on fostering environmentally friendly and socially responsible farming practices. The school activities deal with farmer-based extension services, training of trainers, diversification, environmentally friendly practices (IPM) and improving labour conditions.[69]

In other (fully liberalised) cocoa producing countries (such as Côte d'Ivoire, Cameroon and Nigeria) buyers are likewise involved in setting standards, increasing grinding operations, community development and research. The SCTP is also, as in Ghana, active in these West-African countries by establishing FFSs, which actually have a more diverse set of activities than the schools in Ghana. In addition to farmer-based extension services, training, the introduction of Integrated Pest Management (IPM) and labour conditions, the FFSs aim to improve farmers' marketing and entrepreneurial skills and systems of information and quality control. In addition, in these countries buyers are increasingly involved in providing services to producers, such as local transport and storage of beans, and establish more direct trade relations with cocoa producers. For example, in Côte d'Ivoire the world's largest processors already claim to buy more than 50 per cent directly from farmer cooperatives. Furthermore, through 'civil-private partnerships', such as the 'Upcocoa project'[70] in Cameroon, buyers are actively looking for ways to strengthen partnerships with cooperatives. This multi-stakeholder initiative which came in existence in 2006,[71] includes representatives of many stakeholders: ISCOM[72], ADM Cocoa B.V., Masterfoods B.V/Mars Inc., RIAS[73] (a consultancy organisation within the Rabobank) and IITA/STCP. It is a four-year project that focuses on capacity building of cocoa farmers and their organisations. It addresses problems in areas such as marketing, institutional and social capacities, quality management, productivity, and sustainable agriculture. The project will (initially) work with eight cooperatives, comprising of about 1600 farmers.[74]

Involvement of global buyers in local upgrading strategies takes place at different levels; table 4.5 lists the interventions of global buyers in Ghana.

4.3.2 Changing relations

The industry survey indicated that in order to guarantee future supply and demand for cocoa, the main opportunities include 'working directly with farmer groups', 'trading with cooperatives' and 'strengthening relations with suppliers' (as identified by international buyers). Processors look for direct ways of interaction (buying directly from farmer cooperatives), while manufacturers often engage with farmers in a more indirect way (through membership of WCF, ICCO, STCP etc.). International buyers argue that farmer cooperatives produce better quality cocoa than individual farmers and that it is more efficient to buy directly from cooperatives (instead from middlemen) (interviews with cocoa processors, 2005). Although this development is taking place in Ghana's neighbouring cocoa-growing countries (a shift from arm's length relations towards semi-hierarchical relations), in Ghana it is difficult to establish direct relations with farmers. In Ghana farmers not only lack organisation, but also direct marketing with farmer groups (or with a LBC) is prohibited by law. Consequently, buyers have limited their activities in Ghana to initiating (or supporting) small-scale programmes aimed primarily at promoting the use of more environmentally friendly practices, at addressing the problem of

Table 4.5   Interventions/activities of global buyers in Ghana

Type of upgrading	Activities/interventions affecting cocoa producers in Ghana	Actors/institutions involved
Capturing higher margins for unprocessed commodities	Setting production standards (traditionally fermented and dried cocoa receive a premium price)	Private sector organisations
	Controlling the quality of exported cocoa against excess levels of pest residues	EU legislation
	Innovation (compensate for variety in bean quality)	Private and public sector
	Research on product upgrading	Private sector, Research institutes, Public sector
	Setting process standards (social/environmental certification schemes)	Harkin-Engels Protocol (Public Private Partnership (PPP); World Cocoa Foundation)
	Supporting farmers in meeting process standards	Sustainable Tree Crop Programme (STCP) (PPP)
	Providing farmers with bonuses, credit, (subsidised) input, awards, etc., through set-up LBC	Licensed Buying Company (LBC)
	Investing in social capital with local buyers	LBC
	Introducing integrated pest management (IPM)	STCP
	Training farmers in agroforestry	STCP
	Improving labour conditions, through (individual) educational programmes and community development	Civil-private partnerships
	Research on process upgrading	Private sector, Research institutes
	Bulk transport	Private sector
	Provision of farmer-based extension services and training in Farmer Field Schools	STCP
Producing new forms of existing commodities	Training	STCP
Localising commodity processing, marketing and consumption	Increased grinding operations in Ghana	Private sector, National legislation
	Establishing local buying centres	Private sector

Source: Composed by author.

child labour and at community development. Physically, in Ghana international traders did move 'closer' to the farmers, but grinding in sourcing countries has not yet resulted in direct relations between buyers and cocoa farmers.

4.4 Reflections

The global cocoa chain has become increasingly buyer-driven. Global buyers are powerful entities in the chain, able to reshape institutions (e.g. by lobbying and through their strong ability to innovate and to function as the main gatekeepers of knowledge).

Despite the strong position of global buyers, they have become more vulnerable to the behaviour of their suppliers. International manufacturers, traders and processors respond to global threats by diversifying supplier networks across countries. Working together with other actors is also seen as a way of spreading risks and public-private partnerships have been launched as a result. The STCP is an example of such a public-private partnership, which serves public as well as private interests. Recently, foreign cocoa processors and manufacturers have jointly started to form alliances with cooperatives in Côte d'Ivoire and Cameroon.

The 'risks' of supplier failure were the main driving force behind the shifts in governance that made the cocoa chain increasingly buyer-driven. At the same time, they also formed the main barrier for effective governance of the chain. The increasing risk of (local) supplier failure opened up a common field of interest in promoting the profitability of cocoa farming. The interdependency among the actors in the chain makes the establishment of more direct relations between buyers and suppliers very desirable. However, this holds not true for every country. For example in Ghana the strong role of the state obstructs direct relations with cocoa farmers.

While in GVC theory a distinction is made between buyer and producer-driven chains, there seems to be a need to discuss other more hybrid types of 'overall chain governance'. The role of the state could also be more addressed in discussions on various forms of 'chain coordination'.

5
THE ROLE OF THE STATE IN A LIBERALISED COCOA SECTOR

5.1 Introduction

The international process of liberalisation reduced the direct involvement of the state in economic activities in developing countries. It was expected that in these countries 'market-oriented structural reforms (...) [would] boost growth by reducing distortions and encouraging greater private sector participation' (IMF, 1997: 85-8 quoted in Fernández Jilberto and Hogenboom, 2007: 2). However, increasingly it is recognised that neo-liberal restructuring does not automatically lead to development (Fernández Jilberto and Hogenboom, 2007: 10; Singh, 2002; Stiglitz, 2002), giving impetus to new discussions on the role of the state in sectors in transition. Considering the importance of agriculture (for generating income, for producing exports, for harvesting foodstuff, for generating employment opportunities and for poverty reduction), national agricultural policies are a central policy concern in these countries.[75] Instead of direct involvement, neo-liberal thinking sees the state as an enabler that creates more favourable conditions for the development of the agricultural sector (Gibbon, 2001: 353; Hamdok, 2003: 15-7; Joosten and Eaton, 2007: 1[76]).

Although it is important to reconsider the role of the public sector, it is also important to recognise that the capacity of the developing country's government to successfully manage interventions geared at pro-poor agricultural development is generally very limited, especially in the African setting. This low capacity is linked to trade liberalisation and structural adjustment, which reduced the mandate and the ability of the public sector to make specific interventions. Insufficient capacity also plagues the other actors involved in this market, such as producer organisations, private companies and non-governmental organisations. These actors also need to be strengthened in order to enable them to overcome pervasive market failures and to secure the desirable social outcomes in their countries (World Bank, 2007).

The cocoa sector in West Africa is an interesting case for examining the effects that liberalisation has on the role of the state and its capacity to manage interventions, as cocoa production is concentrated in this region. Also it can yield useful insight in the links between this influence and the role and capacity of other (new) public, private and civil actors. Neo-liberalisation in cocoa producing countries in West Africa started with the introduction of the structural adjustment programmes (SAPs) by the World Bank in the late 1980s. The reforms stipulated a reduction of state involvement in the provision of marketing channels and services for cocoa, in order to open these markets to competition (Akiyama et al., 2001).

All major West African cocoa-producing countries implemented some reforms. Cameroon and Nigeria initiated drastic reforms, while Côte d'Ivoire and Ghana chose a more gradual approach to liberalisation. In these countries, where the export earning of cocoa contributed significantly to the foreign currency earning and the Gross Domestic Product (GDP), governments are reluctant to relinquish control and seek to continue their role as active intervener and 'chain actor'. Ghana did not liberalise as much as the other countries and 'systematically has tried to protect its effective system of parastatal-based governance' (Kaplinsky, 2004: 25). Compared to other cocoa-producing countries in the region, Ghana's position seems quite exceptional. However, there are a number of other sectors, countries and regions, where the government successfully retained a strong steering role in economic development. For example, the rapid economic growth of the Asian Tigers (South Korea, Taiwan, Hong Kong and Singapore) in the 1980s is (partly) explained by the guiding role of the state (Harris, 1986: 30-69; Lall, 2005). Also, in the United States and the member-countries of the European Union 'agricultural protectionism' (through tariffs, nontariff barriers, and subsidies) is still widely used to protect their farmers and to guarantee state revenues (Gibbon and Ponte, 2005).[77]

The idea that the state is more than an enabler and instead plays the role of governor of trade relations and competitiveness is not new; nevertheless, it is not reflected in the value chain and cluster literature. The political economy perspective did embrace a focus on the governing role of the state, and called this 'state governance' (Chapter 2). Griffiths and Zammuto (2005) proposed an integrative framework, combining political economy with strategic management literature. This approach can be utilised to analyse public and private interactions in the cocoa sector over time (Ton et al., 2008).

In this chapter, I will take the liberalisation of the cocoa sector as a key-turning point and analyse the changing role and capacity for intervention by the government and the other actors involved in the sector. I am especially interested in understanding the changing conditions under which cocoa producers operate and ultimately to discern the upgrading opportunities that evolve from this change. In order to evaluate the changes in Ghana, I will make some institutional comparisons with Côte d'Ivoire, Nigeria and Cameroon and some comparisons with the conditions in Ghana prior to the reforms.

5.2 A sector in transition: the experience of West-Africa

Before the market reforms in West Africa, cocoa was produced and marketed under state controlled systems, with significant variations between countries. Anglophone countries produced under the marketing board system, while Francophone countries used stabilisation funds. The low cocoa prices of the mid-1980s were the incentive for liberalisation; it was hoped that reforms would increase producer prices, by improving the efficiency of the cocoa related activities and by reducing the costs of inefficient marketing and pricing systems.

The next table presents the World Bank 'view' regarding cocoa-marketing and cocoa-pricing systems at that time (2001). It highlights the differences between a free market system, the stabilisation fund and the marketing board system (Akiyama et al., 2001).

Table 5.1   The World Bank 'view' on differences in cocoa marketing and pricing systems in 2001

Characteristic	Free market	Stabilisation fund	Marketing Board
Legal ownership of crop	Traders, exporters	Traders, exporters	Marketing Board
Physical handling of crop	Traders, exporters	Licensed private agents	Marketing Board
Domestic price setting	Market forces	Stabilisation fund	Marketing Board and government institutions
Price stabilization	None	Yes	Yes, but not explicit
Taxation	Absent or very low	Mainly explicit	Implicit
Marketing costs and margins	Low	Medium to high	High
Producer prices	High	Medium to low	Low

Source: adapted from Akiyama et al., 2001: 41.

In addition to the criteria used by the World Bank to evaluate the reforms, this study includes factors that frame the enabling environment for producers (for example quality control, extension services, credit supply and formal farmer organisation). These are important factors that support the farmers' response to the challenges of meeting current and future demand for their produce.

In this section I will briefly discuss the experience of Côte d'Ivoire, Cameroon and Nigeria. Consequently, I will analyse the experience of Ghana more in-depth and make some comparisons.

5.2.1 Côte d'Ivoire

The world's largest producer, Côte d'Ivoire, provides almost 40 per cent of the world's supply (ICCO[78], 2006). In Côte d'Ivoire the process of liberalisation was gradual. Prior to the reforms the Caisse de Stabilisation et de Soutien des Prix de Produits Agricoles (Caistab) coordinated the supply chain. Private exporters were allowed to operate under this state-controlled governance system. In season 1995/96, the role of Caistab in marketing cocoa was reduced. In Côte d'Ivoire initially the producer-price remained stable and rather low (usually between 45 and 55 per cent of the world price) (Beuningen, 2005: 69). The reforms did have an impact on the number of buyers and cooperatives; the cooperatives' share in volume of purchased cocoa grew from 22 to 29 per cent during the 1997/98 cocoa season. Trading through cooperatives made it possible to sell directly to exporters or large traders (Amezah, 2004).

In 1999 the role of Caistab was further restricted. As part of the reforms the system of export marketing changed, eliminating the minimum producer-price and privatising quality control. Also the *barème* (a schedule of costs, prices and margins that regulated the entire marketing chain) disappeared along with the elimination of public forward sales. After this second round of reforms, which coincided with a dramatic fall in the world price of cocoa, the producer-price fell to such an abysmal level that farmers refused to sell their cocoa beans (and even threatened to burn their produce) (LMC International and the University of Ghana, 2000: A2.6). In addition, the quality of cocoa suffered in this period, with a downward trend from 1997 to 2000 (Losch, 2002: 222).

World cocoa prices started to recover in 2000 (ICCO, 2007a: 22). But, even in combination with the general recovery of the country's economy, cocoa farmers in Côte d'Ivoire remained poorly paid (Akiyama et al., 2001). This spurred massive smuggling of cocoa across the border to Ghana. Another consequence was that some cocoa farmers were forced to look for alternative incomes. As a result of the economic difficulties labour conditions worsened and the incidents of abuse of child labour increased (STCP and IITA,[79] 2002).

In order to cope with the changes and corresponding problems, the government introduced stimulating packages to facilitate farmers in organising themselves. In 1997 the government reinforced a cooperative law, which encouraged cooperatives to operate as business entities and thus improve their credit worthiness with commercial banks (Republic of Côte d'Ivoire, 1998: 25). Hundreds of *Groupements de Vocation Coopérative* were created throughout the country; however, most were not operational because they often did not have enough members and lacked equipment and funds.[80] The government also implemented a series of other measures to encourage farmer cooperation and offered to support organisations that represent farmers' interests. These included the renovation and construction of new warehouses, which provided the potential for the development of a warehouse warranty scheme. The long term goal was for cooperatives to become the owners of the warehouses.

In 2000 a third round of reforms took place, as Côte d'Ivoire tried to bring coordination of the production chain back into the hands of government, by establishing two new structures: the Autorité de Régulation du Café et du Cacao and the Bourse du Café et Cacao. In addition, in 2001 four new institutions were set up to regulate the cocoa trade and to provide support to cocoa farmers. According to Ton et al. (2008) the newly established institutions overlapped and the system as a whole was not transparent. The country did not strengthen the autonomy or the economic capacity of the farmers, who are independent of the intermediaries and thus more vulnerable to changes in the market (ibid). Another consequence of this new round of institutional reforms was that farmers continued to pay high taxes and pay for export levies, that were transferred from exporters onto farmers. In 2008, taxes on cocoa accounted for nearly 40 per cent of the export price (ibid). The revenues collected from cocoa were not reinvested in the sector. According to a representative of the World Bank, these revenues were used to finance the armed conflict (quoted in Global Witness, 2007: 24; see also Ton et al., 2008).

The concentration of cocoa production in Côte d'Ivoire and the problems that occur in the sector are perceived as risks by global buyers. Global processors have responded by looking for ways to diffuse their risks, for example through direct trading with farmer groups that they directly support. In 2005, global processing companies claimed that in Côte d'Ivoire they purchased around half of the beans directly from cooperatives (industry survey 2005). Global buyers also got involved in local buying practices and started taking large chunks of the local buyer's market share (Kaplinsky, 2004: 24). Besides providing additional services, buyers also made sure that they paid their farmers promptly. Global manufacturers and processors, often through public-private partnerships, also started to introduce environmentally friendly practices and invested in strengthening farmer groups, providing educational programmes and improving information systems.

5.2.2 Cameroon

Cameroon is recognised as one of the main cocoa producers in Africa. The production of cocoa gradually increased during the past years; the capacity of 185,000 tonnes per annum (in 2008) positions Cameroon as the fourth largest cocoa producer in Africa (National Cocoa and Coffee Board [NCCB]).[81] In Cameroon around 420.000 hectares are used for growing cocoa. In 2008 cocoa accounted for 14 per cent of the country's total export income, while processed cocoa products (such as paste and butter) accounted for around 15 per cent of cocoa export earnings (KIT et al., forthcoming)

In Cameroon prior to reforms the Office National de Commercialisation de Produits de Base (ONCP) governed the supply chain from a monopoly position. The reforms were introduced rapidly and were characterised by different phases. In the first phase the system of internal marketing was liberalised and the ONCP was disbanded and replaced by the Office National du Café et du Cacao (ONCC). Furthermore, the price stabilisation mechanism was reviewed and cocoa buyers were no longer required to obtain licenses. Initially the government wanted to continue with some kind of price stabilisation, but this system was abandoned in the second phase of the reforms. In this period, the responsibility for quality control and organisation of the marketing chain shifted to the private sector and price formation was left to private forces (Akiyama et al., 2001; LMC International and University of Ghana, 2000; Jong and Harts-Broekhuis, 1999: 96-8).

The increase in producer price was the main immediate impact of these reforms. However, this positive effect was mitigated by the loss of quality and unreliable delivery. In 1996 and 1997, large quantities of cocoa were classified as below export quality; this affected adversely the good reputation of Cameroonian cocoa and its price. This loss in quality was a direct result of the removal of restrictions that limited cocoa trading, which precipitated the entry of a large number of unprofessional new (foreign) traders and middlemen. Also the ONCC took serious heat for the drop in quality (Jong and Harts-Broekhuis, 1999: 98). The absence of local funds capable of strengthening the domestic market resulted in the total takeover of the domestic market by international traders, 'the part of nationals in export went from 80 per cent before the liberalisation to less than 20 per cent today' (COPAL[82], 1998). The number of licensed exporters increased rapidly (from around 60 to over 300) and the licensing criteria was rarely respected (LMC International and University of Ghana, 2000: A1.2). In response, the industry association issued a so-called *carte professionelle*. This voluntary card indicated to farmers and international buyers that the card holder was a reliable company. However, as this was a voluntary measure it offered no concrete guarantees. In 1997 the situation stabilised; the number of active exporters had declined to around fifty, with the ten largest companies accounting for over 70 per cent of total export. Four of the five largest exporters were foreign owned.

The last phase of the reforms, involved some public reinvestments in information systems and the state restructured its extension services, initiating the provision of unified extension. Credit facilities generally diminished, and while well-developed arrangements for the export sector existed, there were only scarce credit facilities that catered to farmers. According to representatives of producing countries, who joined forces within COPAL, this was problematic because it made it hard to pay for the higher costs of production (COPAL, 1998).

Global buyers became involved in the provision of extension services, in setting-up information systems and in strengthening farmer organisations (mainly through public-private partnerships). The Sustainable Tree Crop Programme (STCP) was involved in developing a production information system in order to provide information essential for developing business plans at the farm level, to build capacity of farmer organisations and to develop targeted extension approaches to address specific market demands, such as Farmer Field Schools (FFSs) and integrated pest management (IPM). The STCP aimed also at setting-up an information system to track production practices and product attributes for marketing and/or environmental purposes.[83] In 2006 the 'Upcocoa project' came into existence. This multi-stakeholder initiative focuses on capacity building of cocoa farmers.

5.2.3 Nigeria

In Nigeria prior to the reforms (until mid 1980s), the Nigerian Cocoa Board (NCB) coordinated the production chain. Nigeria liberalised over-night: dismantling the NCB, deregulating the internal and external marketing system and abolishing its price-controls. In Nigeria buyers of cocoa no longer require a license and quality control was abandoned.

Reforms in Nigeria had different short-term and long-term impacts. The price increase did boost cocoa production and exports (Akiyama et al., 2001; LMC International and University of Ghana, 2000); however this produced only a very limited rise in income due to the sharp drop in the exchange rate of the local currency (naira) in 2001. Price fluctuations, linked to sales on the spot market (instead of forward sales) and to increased production costs, made farmer income less secure. Prior to privatisation, inputs were generally free or heavily subsidised, but privatisation of input distribution caused enormous production cost increases in Nigeria (Haque, 2004; Walker, 2000: 163-4). This situation was exacerbated due to several adverse factors: the old tree stock, the frequent incidence of pests and diseases, labour shortages, the farmers' average high age and the lack of access to formal credit[84] (Ogunleye and Oladeji, 2007). This hindered investments in increasing productivity of farm operations and resulted in the poor maintenance of farms.

The subsidies on extension services were also abolished with liberalisation and consequently the quality of services declined. As illustrated above, in Nigeria the private sector took over part of these responsibilities. The opening of the market led to the entrance of a large number of inexperienced buyers; this, together with the abandoning of the quality control system, greatly contributed to the quality decline. Nigeria lost its quality premium on the world market and, as a direct consequence, faced a decline in demand (COPAL, 1998; Haque, 2004; LMC International and University of Ghana, 2000). After some time, external marketing consolidated and the quality of cocoa recovered to some extent. However, prices remained instable. In an attempt to bring the coordination of the cocoa supply chain back into the hands of the state, in 2000 the government set-up the Cocoa Development Committee (CDC). The CDC is chaired by the Ministry of Agriculture, with the Deputy Governors of the thirteen cocoa producing states as members. The CDC aims at increasing cocoa production, partly through the provision of grants to new seedlings and through the sale of chemicals, fertilisers and other inputs at a 50 per cent discount[85] (Ogunleye and Oladeji, 2007: 15).

5.2.4 *Responses to reforms*

All three countries took numerous actions to recover from the overall negative experiences from liberalisation, such as working on the consolidation of buyers, redefining the role of the government and strengthening farmer groups. Especially in Côte d'Ivoire, due to its strong dependence on cocoa exports for its foreign exchange, the government aggressively tried to reclaim the coordination of the supply chain.

In the current process of liberalisation, most cocoa-producing countries are trying to strengthen their farmer cooperative structure, weakened under state-owned marketing boards and stabilisation funds. In general it is acknowledged that organised farmers negotiate better prices and services, ultimately also producing better quality cocoa beans (COPAL, 1998; industry survey 2005; Ogunleye and Oladeji, 2007; STCP, 2005). In Cameroon the number of farmers organised in formal groups is small but growing. In Côte d'Ivoire most farmer groups are malfunctioning or still not yet operational. It is generally recognised that farmer organisation should not be imposed upon farmers, as was the case in the past; rather the state should support existing movements rooted in the rural community.

Farmers on the other hand responded to reforms by shifting cultivation and by migration. For example in Nigeria, due to the worsening economic conditions in rural areas, the youth migrated to cities; thus, the farming population is on average very old. The ageing of farmers and their farms is perceived as a serious threat by global cocoa buyers (industry survey 2005). In Côte d'Ivoire the poor production conditions for cocoa farmers resulted in a shift to cultivating other crops. Also, in Côte d'Ivoire the political conflict forced the migrants who worked on cocoa farms to abandon these farms.

5.3 A sector in transition: partial liberalisation in Ghana

The Ghanaian government has always been actively involved in the development of its cocoa sector. During colonial times, public involvement was initially combined with private efforts aimed at stimulating cocoa production and improving quality. From the late 1940s onward, a system of 'state governance' was put in place, which was further consolidated during the early years of independence.

With the introduction of structural adjustment programs (SAPs) in the late 1980s, the control over governance processes in other cocoa producing countries shifted from the state to multinational buyers of cocoa. In Ghana, however, the state continued to play a major role. The introduction of gradual reforms did lead to other (private) actors entering the sector and taking over some of the state's previous responsibilities. Also, new public-private partnerships were developed in this period. These developments fell in line with the general shifts in the global cocoa chain governance, which became increasingly driven by international processors and at the expense of chocolate manufacturers (Kaplinsky, 2004).

In order to describe the developments in the cocoa sector, I will start by discussing the period between 1920 and 1980, when a shift moved the sector away from a fairly liberal economy towards a state-controlled economy. Subsequently, I will cover the period from 1980 until 2008.

5.3.1 From a fairly liberal policy to a more state-controlled economy: 1920-1980

Cocoa has dominated the political economy of Ghana (formerly Gold Coast) since 1920. Between 1923 and 1932, cocoa accounted for an average of around 77 per cent of Ghana's total exports (Department of Cooperatives, 1990: 11). In the cocoa season 1920-21, Ghana became the world's largest producer, reaching its peak with a total production of around 560,000 tons. It kept this leading position for more than 55 years, until in 1977-78 when Côte d'Ivoire took over. In the early years, cocoa production was concentrated in the Eastern Region (Ministry of Finance, 1999: 6). In the mid 1940s, the centre of production shifted to the Ashanti and Brong Ahafo regions, where cocoa was produced on virgin forest. Forty years later, it shifted to the location where it remains concentrated to this day – the Western region. The migration and settlement by small-scale farmers was fundamental for the expansion of cocoa in Ghana (Cocobod, 2000a; Hill, 1963).

During the colonial period, the cocoa sector initially operated under a free market system. European companies controlled both local buying and exporting, with Cadbury as the leading British cocoa manufacturer. The (colonial) Department of Agriculture steered the public interventions, which encouraged the establishment of cooperative enterprise in Ghana's cocoa industry. Organising farmers into cooperatives was seen as a way to ensure the production of good quality cocoa for export. Moreover, by means of combined sales, cooperative members could also demand higher prices. Around the same time, cocoa societies became involved in the provision of credit, farmer inputs (for example chemicals) and consumer goods. Many European traders were hostile to the promotion of cooperatives with government backing, because they were worried that this would eventually eliminate them from the cocoa trade (Department of Cooperatives, 1990: 9-16).

The strength of the cooperative movement was clearly demonstrated with the first cocoa 'hold-up' in 1930-31. The action achieved its goal – both the supply and the price steadily increased (Milborn, 1970: 59). In 1937, a second cocoa hold-up was organised in response to a drop in the global price of cocoa and the concomitant introduction of the 'cocoa buying agreement', which introduced the idea of paying a uniform price to all farmers. Farmers and brokers (both of whom had not been consulted by the foreign buyers) initially rejected this agreement. Both the internal and the external marketing of cocoa stagnated, until March 1938 when the parties signed a 'truce' and trading continued as usual (Department of Cooperatives, 1990: 42-4).

In response to the hold-up, a royal assignment called for a 'commission' to report on the condition of the cocoa economy in West Africa. The commission without a doubt recommended to establish public support for cocoa marketing operations (Graue, 1950: 259). Consequently, the West African Cocoa Control Board was founded in 1940, replaced in 1942 by the wider West African Produce Control Board (WAPCB). In 1947, the Cocoa Marketing Board (Cocobod) was installed with the British Ministry of Food as the only seller of Ghanaian cocoa. The existing (mainly expatriate) cocoa buying companies were appointed as purchasing agents of the WAPCB (Anin, 2003: 15).

Cadbury supported the argument that West African cocoa economies should be rehabilitated and took the position of 'general acceptance of the marketing boards' (Beckman, 1976: 43). However, no one had foreseen that Ghana would gain its independence so soon and that the marketing board would be affected by strong nationalisation pressures so soon after its installation (ibid: 44). In 1953, the United Ghana Farmers Cooperative Council was founded as 'a general farmers' organisation with political objectives'. It became the so-called 'farmers' wing' of the Convention People's Party (CPP), led by President Nkrumah (Beckman, 1976: 11). As the head of the cooperative movement, the Farmers' Council was given sole responsibility for cooperative development, thus making it into an economic and political force to be reckoned with. In 1961, four years after the country gained its independence, the Farmers' Council took over the cocoa trade and became the monopoly buyer of Ghana's cocoa, effectively ending the direct involvement of foreign buyers in cocoa marketing. The same year saw the founding of the Cocoa Marketing Company (CMC) Ghana as a commission agent of Cocobod. CMC Ghana became responsible for several key aspects of the cocoa trade: registering buyers, appointing local (Ghanaian) buying agents to facilitate foreign buyers' operations in Accra, marketing of cocoa, recording of cocoa sales and shipping of cocoa (Amoah, 1998: 78-104).

The full-scale nationalisation process was a significant turning point. The cocoa sector was transformed from being governed by fairly liberal economic policies into a completely state-controlled system, now completely independent from colonial ties. In 1963, two cocoa processing factories were built; both were owned by Cocobod and operated by one of its subsidiaries, the Cocoa Processing Company limited (CPC) (Ministry of Finance, 1999: 71-2). In 1982, Cocobod took over a third factory, the formerly British processing company West African Mills (WAM) at Takoradi.[86]

When in 1966 the Nkrumah Government was overthrown by a military coup, the Farmers' Council was dissolved and banned (Beckman, 1976: 11-7). During the following decade (of political instability and mismanagement) cocoa prices fell and cocoa production was halved. Competition among local buyers was re-introduced, but this did not last long due to problems with delayed payments. A single buying system was reintroduced and from 1977 onwards the Produce Buying Company (PBC) (another Cocobod subsidiary) controlled internal marketing. PBC hired purchasing clerks (PCs) who bought cocoa at the community level from farm-owners and/or their caretakers. In this period, different subsidiaries of Cocobod provided support and services to the farmers. The Cocoa Services Division (CSD) had the monopoly on the procurement and distribution of inputs. In addition to its role as input provider, the CSD was also responsible for cocoa extension services: advising farmers in cocoa and coffee production, the production and distribution of planting materials, and the control of pests and diseases (Amezah, 2004: 1). The Quality Control Division (QCD) was responsible for executing a strict control on the quality of cocoa. The Cocoa Research Institute Ghana (CRIG) was the national centre of excellence for the study and cultivation of cocoa (Figure 5.1).

Figure 5.1   Organisation of the cocoa sector in Ghana in the 1980s

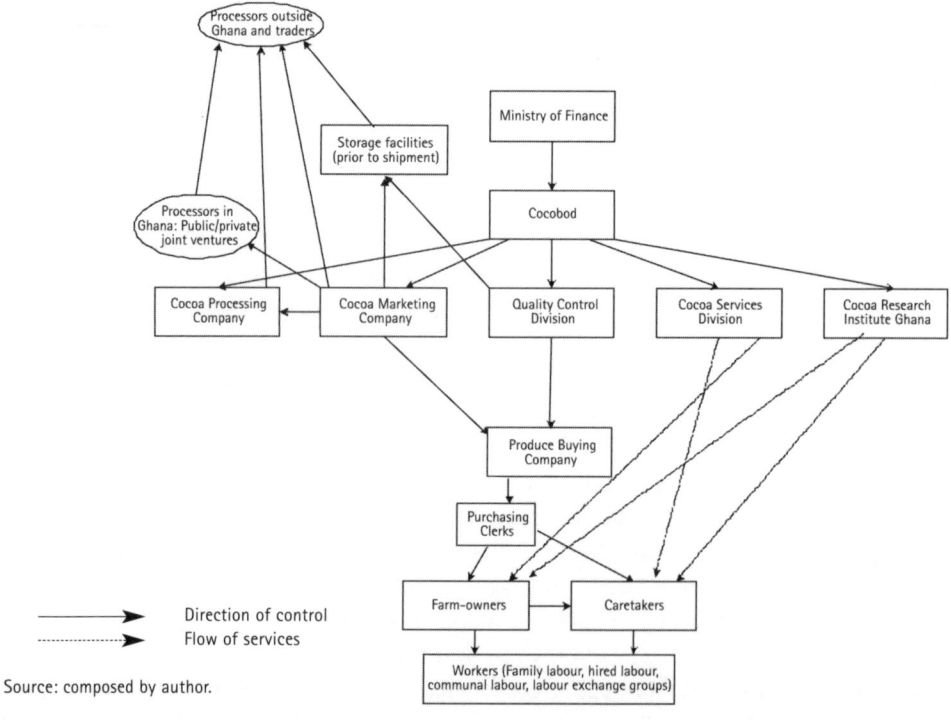

Source: composed by author.

By the early 1980s, Ghana's economy was in an advanced state of collapse. In 1981, Lieutenant Rawlings came to power and recognised the complexity of the economic crisis. When the situation worsened due to a severe drought in 1983, his government accepted the intervention of international donor organisations. The Economic Recovery Program (ERP) introduced reforms in the cocoa sector through its Cocoa Rehabilitation Project (CRP) and the Agricultural Sector Adjustment Programme (ASAP) (Fold, 2002; Ministry of Finance, 1999).

The next section will describe the introduction of gradual reforms and discuss the consolidation of the strong position of the state, both domestically and internationally. It will also describe the shift towards chain integration, resulting ultimately in a system of 'joint governance'.

5.3.2 The introduction of gradual reforms: 1980-2008

In order to avoid the generally negative experiences of cocoa producing countries that liberalised over-night (such as Nigeria) the Ghanaian government opted for the gradual introduction of reforms in the cocoa sector. To date, Ghana has implemented the following reforms: the liberalisation of internal marketing, privatisation of input distribution, reform of extension services, reorganisation of processing activities and a drastic reduction in Cocobod's staff-level.[87] Cocobod continues to control external marketing. The QCD remains responsible for the final quality checks of cocoa beans and Ghana continues to deliver consistent supplies of relatively good quality cocoa, cashing in premiums on the world market. The CRIG is still involved in cocoa research (see picture 5.1). Through a system of forward sales, Cocobod managed to secure pre-financing for local purchasing of cocoa the cocoa and preserved the price stabilisation intact (Ministry of Finance, 1999).

Picture 5.1 The Cocoa Research Institute Ghana

Internal marketing

The liberalisation of internal marketing started in 1992 with the introduction of private Licensed Buying Companies (LBCs) as competitors to the state-owned monopoly (picture 5.2). During 1996-97, sixteen LBCs obtained the permission to buy cocoa alongside the PBC and an additional four received provisional licenses (Ministry of Finance, 1999: 45). The number of buyers fluctuated as did their active involvement in buying cocoa. In 2005, the number of licensed buyers stood at twenty-five, of which nine are responsible for selling more than 90 per cent of all Ghanaian cocoa (Table 5.2) on the international market. The majority of LBCs are Ghanaian, only two are foreign-owned (Olam and Armajaro).

Picture 5.2   A district buying store in Asankrangwa (Western region)

One LBC is farmer-owned (Kuapa Kokoo Ltd owned by the Kuapa Kokoo Farmer Union) and was set up with the support of international NGOs. One new (still small) farmer-owned LBC, Ghana Sompa Kokoo, was recently set up by the Kuapa Kokoo Farmer Union. PBC is still the major buying company (buying almost 33 per cent of the cocoa in 2005-06), although its buyer's share is declining.[88] Recently, the PBC was partly privatised and its shares are now traded on the stock market, with Cocobod as the company's majority shareholder.

The majority of the LBCs are Ghanaian. LBCs are formally organised in a buyers' association but in practice each LBC operates independently. An exception is a group of three LBCs (Cocoa Merchants, Transroyal and Fedco), which are all owned by the same shareholder, the transport company Global Haulage.

Table 5.2   Regional cocoa purchases by LBCs (in tonnes) for season 2005-06

LBC	Total
Produce Buying Company	205,602
Olam Ghana Ltd	85,576
Akuafo Adamfo Marketing Co	74,824
Adwumapa Buyers Ltd	55,369
Kuapa Kooko Ltd	42,676
Federated Commodities Ltd	41,804
Armajaro Ghana Ltd	34,833
Transroyal Ghana Ltd	30,132
Cocoa Merchants Ltd	15,063
Dio Jean Company Ltd	9,099
Diaby Company Ltd	8,741
CocoaExco Ltd	7,826
Sika Aba Buyers Ltd	5,509
Sompa Kokoo Ltd	3,834
Royal Commodities Ltd	3,666
West African Exchange Co Ltd	3,577
Chartwell Ventures Ltd	786
Fereday Company Ltd	173
Total	629,090

Source: Vigneri (2007).[89]

It is important to realise that despite the introduction of competition in Ghana, LBCs do not compete on prices – all buyers pay the floor price. Some LBCs have introduced small bonuses for their buyers. LBCs receive a yearly fixed 'buyers-margin' set by the government. In 2002-03 this margin was set at 9 per cent of the FoB price and was somewhat reduced for the subsequent 2004-05 period. LBCs use this margin to pay their purchasing clerks (PCs) on a commission basis, thus encouraging them to buy as much cocoa as possible from the farmers in their communities. According to farmers, LBCs almost never reject the cocoa they offer. This is probably also due to the existing social relationships between farmers and buyers (FS 2005). During the group discussions, one farmer succinctly illustrated this shift from stringent towards less strict quality control procedures:

> In the previous years there was only one cocoa buying company. It had very strict rules and regulations regarding the quality of the cocoa. When we sent our cocoa they tested it and made sure the beans were dry enough. If not, they refused to buy it and advised you to go back and dry it well before you bring it. Now we have so many buying companies, there is a lot of competition these days. Companies are competing with each other to buy the cocoa and don't really care if the cocoa is dry enough. That also explains the decline in the quality of cocoa.

In the absence of competition through prices, LBCs developed other ways to ensure that farmers sell (only) to them. The two strategies frequently go hand-in-hand:

- **Building trust** – Investing in local purchasers of cocoa and making sure that the PC is capable, trustworthy and motivated to serve the farmers' needs. LBCs often select their PCs with the help of community representatives.
- **Building social capital** – Investing directly in maintaining durable social relationships with suppliers (for example, by attending funerals) and providing them with prompt payments, bonuses, gifts, rewards, inputs, credit and training (based on interviews with LBCs in 2003 and 2005).

In 2003, farmers that participated in the farmer survey (FS 2003) selected LBCs mainly on the basis of their prompt payment. Social relationships with the PC and the provision of credit were ranked second and third (see Figure 5.2). In 2005, for this same group (n = 103) the social connection was the main reason to sell, followed by prompt payment and trust (FS 2005). It is likely that this change in selection criteria is linked to the fact that in the mean time prompt payment became common practice among local buyers.

Figure 5.2   The main reasons for farmers to select an LBS in season 2002-03 (n=173)

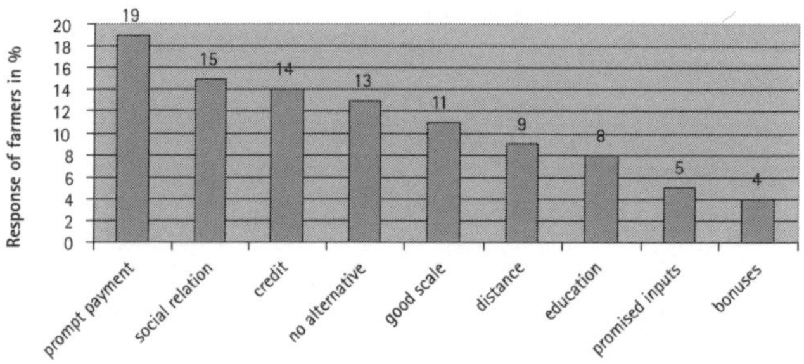

Source: FS 2003.

Even though the introduction of competition in internal marketing did not result in price differentiation, the farmers appreciated the liberalisation of internal marketing and claim that 'it saved them from a lot of hardship' because payments are now made on time. Significantly, if LBCs are unable to pay promptly, farmers always have the option of selling their cocoa to another buyer. Yet, some research showed that farmers did not fully benefit from the liberalisation of internal marketing and that more can be done (FS 2005). Firstly, despite the promises, only a small number of farmers received any services or bonuses from LBCs. In 2005 almost 88 per cent of the interviewed farmers indicated that they recived no support from the LBCs (FS 2005). Also, because PCs' earnings are small they are induced to cheat farmers with fraudulent scales used for weighing cocoa (personal observation;[90] farmer profiles 2005). Secondly, in my sample (FS 2003), farmers living in the Ashanti and Brong Ahafo regions have rather limited choices between LBCs, in contrast to farmers living in the Western region and Central region. While the former state-owned buying company is still obliged to operate in every cocoa growing district, private buying companies can choose where they open their buying depots (picture 5.3). They flock to communities with high cocoa production and easily accessible roads. Consequently, the PBC has become a 'buyer of last resort' (Ministry of Finance, 1999). In the communities visited in the Ashanti Region, the PBC was often the only local buying agent. This was also the case for some communities in the Brong Ahafo Region, although most communities had two or three local buyers. In the Western and Central Regions, farmers reported a minimum of four local buyers (FS 2003).[91]

Picture 5.3 A buying station in the Western region

Surprisingly, the introduction of competition did not result in farmers negotiating with local buyers to sell their produce collectively or negotiating for extra services as a group (for example through contract farming). This could be explained by the farmers' preference to sell to someone from their social network, or by their unfamiliarity with the available negotiating options in the new marketing system. Moreover, cocoa farmers lack farmer organisations that could provide additional negotiating power; a major legacy of the former state-marketing system. This prevents them from taking full advantage of the (potential) benefits of liberalisation of internal marketing. Also, the outcomes of the implemented reforms have not been optimal for LBCs. Due to the slow and gradual pace of reforms, LBCs are locked into a system that offers few incentives for high performance and little financial scope for establishing strong relationships with farmers.

External marketing
Following the liberalisation of the internal marketing of cocoa, the government decided to allow qualified LBCs to export part of their cocoa purchases, implemented from October 2000. Officially LBCs are allowed to export 30 per cent of their domestic purchases, as long as they meet the conditions set by the Ghanaian Ministry of Finance. A company wishing to get involved in the external marketing of cocoa must:

- be a LBC and must have participated in the internal marketing of cocoa for a minimum of two cocoa crop years;
- have purchased a minimum of 10,000 tons of cocoa per year over the immediately preceding two consecutive crop years;
- have personnel who possess the relevant technical know-how and experience in international commodity marketing or can demonstrate access to such adequate human resources; and
- demonstrate access to adequate financial resources (Adapted from Cocobod, 2000b: 2).

The reasoning behind the gradual pace of the reforms was to provide a transition period in order for LBCs to become familiar with and acquire the necessary skills for effective external marketing (Ministry of Finance, 1999). The transition period was scheduled to finish in 2003, with a final decision on whether or not to proceed with full liberalisation of the external market. But the process is stuck. No formal decision on full liberalisation has been taken and there is a sense that the current status quo of partial liberalisation is the desirable 'end-stage' of the reforms. Government officials argue that the current system works well and that LBCs are unwilling or not ready to enter into direct exporting. A delegate of the CMC shared his view:

> A lot of them [LBCs], they are about 20, 25 now. ... they hand it over to CMC and we do the export. ... there was a policy that they should export 30 per cent of their purchases, and that offer has been there for the past two, three years. And nobody has come forward to take that opportunity.[92]

Some of the larger LBCs contest this view, arguing that Cocobod obstructs their involvement in external marketing. Despite the fact that a number of larger LBCs meet the requirements for exporting licences and some of the smaller LBCs also indicated that they could meet requirements by joining forces, they claim that Cocobod obstructs their involvement in external marketing. As one LBC representative shared, 'we are stuck, even though we qualify and [...] they gave us a provisional license about four years ago... nothing has happened.' ... 'because Cocobod is not letting go, nobody is pushing it'.

The stagnation in the transition process is also illustrated by the reduction in the number of international buyers who directly contacted LBCs to arrange trades of cocoa. This number was very high in 2003 but two years later it plummeted as international buyers no longer approached LBCs and seemed to have completely lost interest.[93]

LBCs do not openly complain about the practices of Cocobod. After all, it is a hierarchical relationship where buyers depend on Cocobod to issue them operating licenses. In addition, many LBCs (especially smaller) benefit from the current system; they can take advantage of the marketing expertise and pre-financing of local purchasing.

The partially liberalised system in Ghana is also beneficial for the global buyers of cocoa. Thanks to the reliability of the marketing system, Ghana enjoys a high reputation for honouring contract and providing good quality. This resulted in an alliance between global buyers, the Ghanaian government and Cocobod, all three parties seek to maintain the present system.

For the Ghanaian government and Cocobod there are strategic reasons why the cocoa sector is not fully liberalised. Cocobod still has a huge staff of some 5,500 employees (IMF, 2007). Already for almost a century, the Ghanaian economy is largely based on cocoa exports and there is no real substitute for cocoa in terms of generating domestic tax revenue. In other words, the government has a stake in retaining control over cocoa exports. It is also in the interest of Cocobod to remain the sole exporter of cocoa, for which it receives a significant extra margin (difference between the CIF price of exported cocoa and the Free On Board (FOB) price). Table 5.3 shows the composition of the 'net' (FOB) price during the 2002-2003 season. That cocoa season the margin equalled around 35 per cent, while in cocoa season 2001-2002 it climbed as high as 45 per cent (Abbott et al. 2005).

Table 5.3   The composition of the Net FOB price in 2002-03

Component	Mainstream Cocoa USD/tonne 1 = 8700 cedis	Distribution in % Net FoB
Producer price	976	68,11
Buyers' Margin	128	8,93
Domestic transport costs	32,2	2,26
Storage and shipping	18,4	1,27
Disinfectation costs	9,66	0,67
Crop finance costs	33,3	2,3
Government Tax	236	16,44
Net FOB price	1433,56	99,98 %
Extra margin	± 770	
Export Value	± 2200	

Source: Cocobod, 2003 (white cells) and personal communication industry (dark cells), 2003.

The producer price is adjusted to the level of the current market price through a yearly review of prices and margins. A bonus is paid for cocoa supplied by farmers through the LBCs. The producer bonus is calculated by using the policy defined percentage of the FOB price (in 2007 it totalled 70 per cent). The calculation of the bonus and the distribution of the bonus to the cocoa farmers is an innovative institutional arrangement that influences price stability and fairness within the cocoa chain. Since producer prices are announced prior to the actual purchase or export of cocoa, a situation may arise where ex-post payments could exceed 70 per cent of FOB price. In such a situation, the over-payment to the farmers should be borne by stakeholders other than the farmers themselves. However, in a situation where the exchange rate and the world market price exceed their projected values, the arising surplus should be shared between the government and the farmer as 'bonus' or 'compensation' in order to bring the producer price to the programmed FOB percentage (Ministry of Finance, 1999: 85).

Despite this compensation mechanism, the difference between the export value of cocoa and net FOB price raised questions about the allocation of this 'extra margin' that Cocobod receives. Personal communication with Cocobod revealed that its officials do not know the allocation mechanism.

Cocoa exports are subject to taxes and repatriation of export revenue (which is converted into local currency). The tax rate on exports of cocoa beans is determined annually by the Minister of Finance and Economic Planning (set at 11.1 per cent of the FOB price for the crop year 2007-08). The effective tax rate varied significantly over the past several years. Another part of this margin is reinvested in the cocoa economy. This lack of transparency on the re-allocation of these funds into the sector is problematic and undermines the credibility of a partial liberalised system.

My fieldwork indicated three different types of reinvestments. *First*, Cocobod provides farmers with additional income through the payment of government bonuses and rewards to successful farmers. These measures are intended to stimulate them to continue increasing the production volumes of premium quality cocoa. Also, Cocobod provides scholarships and houses to some farmers and their children. *Second*, Cocobod stimulates upgrading process by providing 'free pesticides' and fertilizer on credit. Although this type of reinvestment was initially intended to stimulate reinvestment by farmers, they primarily perceived it as a reduction in production costs, i.e. they need to buy fewer pesticides. Cocobod also invests indirectly in process upgrading by investing in research and infrastructure. *Third*, Cocobod invests strategically in functional upgrading, by attracting foreign cocoa processors to establish processing factories in Ghana. Stimulating processing activities within Ghana contributes to maintaining a continuous demand for Ghanaian cocoa while providing an added value to cocoa production. Currently, there are four operational processing factories, with two additional ones to follow soon.

The impact of some of these investments on the level of producer will be discussed in Chapter 7.

Institutional reforms

The reforms in Ghana included a number of institutional changes that affected the enabling environment for producers of cocoa. In this section I will discuss the impact of reforms on the quality control system, extension services, input distribution and application, credit facilities and farmer organisation.

Quality

The reported quality losses that the fully liberalised countries suffered after the introduction of reforms were a main reason why the Ghanaian government opted for the introduction of gradual reform. The Ghanaian government wished to maintain its reputation as producer of good quality cocoa beans, in order to maintain a strong negotiating position for securing its premium on the world market. Therefore, final quality control remained firmly in the hands of Cocobod. And yet, despite the gradual introduction, reforms resulted in (temporary) problems with quality, also in Ghana. As already illustrated earlier in this section, it is widely believed that the fault lies with local buyers responsible for the first quality check. Because they did not encourage farmers to continue with their traditional good farming practices, high percentages of inferior beans entered the market. This adversely affected Ghana's high reputation (Asenso-Okeyere, 1997). The following statement by a CMC representative illustrates this point: '[...] But it is all the fault of the LBCs [...] they tell the farmer "bring it and we will pay you", whether it is dry or not, and they take it from them. And they are also in a rush, they want to come and collect their money'.[94]

It is plausible to conclude that this lack of strict quality control reduced the farmer's incentives to invest in the pre-selection of beans and traditional farming practices, both labour-intensive processes. Consequently, there is a risk that quality performance may continue falling in a negative spiral. Cocobod responded in 2005 by declaring all bags of cocoa with more than 25 per cent 'purple beans' as substandard and paid local buyers only half the producer price. By doing so, Cocobod ignored the other possible explanations for the decline in quality, such as the smuggling of beans from Côte d'Ivoire and the problems faced by the QCD (a subsidiary of Cocobod) in controlling the higher volumes of cocoa during this period. These problems, combined with logistical issues and recent shortages of cocoa bags, increased the pressure on LBCs (ICCOa, 2007). In a response, LBCs made more active use of their buyers' association to counterbalance Cocobod's policies (personal communication industry, 2007). Despite these (temporary) problems and thanks to worse quality losses in neighbouring countries, Ghana continues to receive a premium for its cocoa beans.

Extension services
The export of premium quality cocoa does not only require a decent quality control system and economic incentives, but also institutional support. According to farmer responses, they obtain know-how on producing good quality cocoa mainly from their families,[95] but also through radio broadcasts and newspapers (FS 2005). Public extension services (which provide farmers with advice on good farming practices, quality issues, etc.) traditionally play an important role in providing knowledge and technologies to Ghanaian farmers; 15 per cent of the farmers that participated in the 2005 survey mentioned extension services as one of their sources of knowledge. The Cocoa Service Division (CSD) was responsible for providing cocoa extension services up to 1999, when the extension services of the CSD merged with those of the Ministry of Food and Agriculture (MoFA). CSDs non-extension functions were further reorganised in two units: the Cocoa Swollen Shoot Virus Disease (CSSVD) Control Unit and a unit in charge of seed production/distribution (Ministry of Finance, 1999: 23-4).

The reduced staff and costs for Cocobod provided the possibility to offer more cost-effective agricultural extension services to farmers. Since most cocoa farmers also cultivate other crops (and some of them keep livestock) it was thought best to consider them as 'general farmers' and therefore to provide services under a unified extension services system (Ministry of Finance, 1999). In practice, the new unified system appeared problematic and was heavily criticised, mainly for its lack of adequate personnel and expertise (interview MoFA extension directorate, 2003; 2005). In response, the government initiated a review of the system.[96]

Due to the problems with extension services, other actors have entered the scene. Private input providers started to accompany the selling of their chemicals and spraying equipment with advice on good farm practices. There are also multi-stakeholder initiatives that focus on training and advice, for example the STCP.

Input distribution and application
The reforms ended CSD's monopoly on the procurement and distribution of inputs. In 1995, the Ghana Cocoa, Coffee and Sheanut Farmers Association (GCCSFA) took over this responsibility. The objective of privatising the input supply was to increase competition. It was hoped that this increased competition would increase in the timely availability of the right quantity of inputs and reduce the cost of inputs.

Although the availability of inputs did increase, its adoption rates remained very low. In 2004 only between 3.5 and 7 per cent of cocoa farmers 'adopted pest and disease control technologies developed by CRIG' (Ayenor et al., 2004: 262). Two probable explanations for these low adoption rates are the lack of funds and the relatively high (world-market) prices of new technologies. One glaring problem is that a significant number of farmers do not make any or only a miniscule profit (interview MoFA, 2005; Mehra and Weise, 2007). Compounded with a general lack of savings and credit facilities, it poses a major barrier that prevents investment in adequate pest management. Lately the adoption rates have sharply increased. CODAPEC,[97] a subsidiary of Cocobod, provides mass spraying to the majority of the farmers. There seems to be a trend that farmers themselves also increasingly apply technologies to combat pests and diseases (Chapter 6 and 7 elaborate in more detail on the use of technologies).

Credit
Prior to the establishment of the Gold Coast Cooperative Bank in 1946, 'agricultural credit' was almost entirely provided by relatives, friends and money-lenders. Starting in the mid 1940s, efforts were made to channel rural credit to farmers, but these failed due to problems with loan recovery and mismanagement. In the 1980s, some innovative attempts were made to improve lending practices for the rural sector. Lending experiences from the informal sector were used, such as door-to-door services (mobile banking) and lending out small amounts of money without charging interest. However, these innovative practices only reached a small number of farmers (Palmer, 2004: 15). It is argued that as a side-effect of the reforms in the late 1980s, access to formal credit became (even) more restricted.[98] Up until today, Ghanaian cocoa farmers cannot easily gain access to formal credit; the main barriers indicated by farmers are a lack of savings, high interest rates,[99] the collateral requirement and a lack of trust. This lack of trust is twofold: banks do not trust farmers to pay back their loans and farmers indicated that they do not trust banks. Farmers prefer to save their money at home and borrow money from friends or relatives. Another option for farmers is to borrow money from local buyers of their cocoa. Alternatives for farmers include borrowing money from private money-lenders, who charge excessive interest rates (annually between 50 and 100 per cent) or borrowing money through credit unions. Recently new experiments were

introduced that provided 'inputs on credit' to groups of farmers. In this situation farmers obtain chemicals to spray their farm from input providers, and the costs involved are paid back to the input provider after the cocoa is harvested and sold. The group is collectively responsible for the payback. Private initiatives, (see the example with Wienco outlined in Chapter 7) in particular turn out to be more successful than the ones of the public sector, as 'public money' tend to be associated with 'free money' (interview CODAPEC, 2005).

Farmer organisation
In discussing institutional reforms, it is important to look also at some of Ghana's 'missing reforms'. For example, liberalisation did not provide the incentives for institutional reforms that could have empowered farmers. According to Tiffen (no date):

> the concepts underpinning the liberalisation process ignored the institutional framework in Ghana and the severely disadvantaged position of farmers which made them vulnerable and therefore likely "losers" in the process. From the institutional perspective small-scale farmers appear to have been invisible to the designers and implementers of Structural Adjustment Programmes in Ghana.

Tiffen rightly posed the question: Why 'in the vacuum created by the abolition of the state marketing boards, [...] weren't new forms of institutions, for example farmers co-operatives, considered, given the context of a rural-based activity like commodity crop production?'

In Ghana, there is one farmer association (the GCCSFA) and one large farmer union (Kuapa Kokoo Farmer Union). The one cocoa cooperative registered by the Department of Cooperatives, the Ghana Marketing Cooperative Association (GCMA), is not operational since 1984 due to financial and managerial problems (Department of Co-operatives, 1990; GCMA, 2005).

Membership in GCCSFA happens automatically with registration; in 2003 around 360,000 farmers registered themselves as cocoa producers (interview GCCSFA, 2003). In 1995, the GCCSFA became responsible for the procurement and distribution of agro-chemicals and spraying machines. Through its network of seventy-five shops, GCCSFA was able to sell chemicals relatively cheap as the association 'is not interested in making profit' (interview GCCSFA, 2003). It also tried to assist farmers by distributing inputs on credit, but this proved rather unsuccessful (Ministry of Finance, 1999: 34-5). Furthermore, GCCSFA assisted farmers by giving scholarships to their children. Nevertheless, the main objective of GCCSFA is the setting of cocoa prices. GCCSFA is one of the members of the Producer Price Review Committee (PPRC), which has the sole responsibility for fixing cocoa producer prices and the other rates and fees related to the purchasing and marketing of cocoa. Other members of the committee include the representatives of licensed cocoa buyers, cocoa transporters, the Ministry of Finance, the Bank of Ghana, ISSER of the University of Ghana and Cocobod officials. The Minister of Finance is the Chairman of the PPRC.

The establishment of the Kuapa Kokoo Farmer Union (KKFU) was an initiative of Nana Frimpong Abebrese, an influential cocoa farmer who was also a farmers' representative on the Cocobod Board of Directors. He 'saw the potential for farmers to benefit through the liberalisation reforms' (Tiffen, no date). Together with the British NGO Twin Trading and supported by the Dutch NGO SNV, he took the initiative to set up a LBC, Kuapa Kokoo Ltd.[100] Ten years later in 2004, the union had approximately 50,000 members, from 6 different cocoa growing regions in Ghana (Kuapa Kokoo Annual Report, 2004: 38). In principle, membership of KKFU is open to farmers who sell all their cocoa beans to Kuapa Kokoo Ltd.; caretakers need written permission from the owners of the farm (for more details see Chapter 7).

Recently some new attempts were made at organising farmers, for example through Farmer Field Schools (FFSs). FFSs were initially set-up in the Central Region around the Kakum park area. An international NGO – Conservation International Ghana (CI Ghana) – took the lead, in partnership with the research department of Cocobod, MoFA and KKFU. In 2003, FFSs were also established through public-private partnerships in the Ashanti Region, under the STCP. These schools provide extension services on a 'learning by doing' basis. They focus on environmentally friendly farm practices, labour conditions and empowerment. While governmental services have the tendency to be top-down, these alternative extension services, provided by NGOs and public-private partnerships, generally are more farmer-driven.

Although many farmers in Ghana are not formally organised, informally they join forces on a regular basis, mainly by means of labour exchange groups (the so-called *nnoboa*).[101] Farmers who participate in these groups share labour and knowledge. The farmers who opted not to join forces usually indicate that there are no real incentives to work together and that a lack of mutual trust prevents them from trying (Figure 5.3). In the 2003-04 season, around 65 per cent of the farmers participated in labour exchange groups (FS 2005).

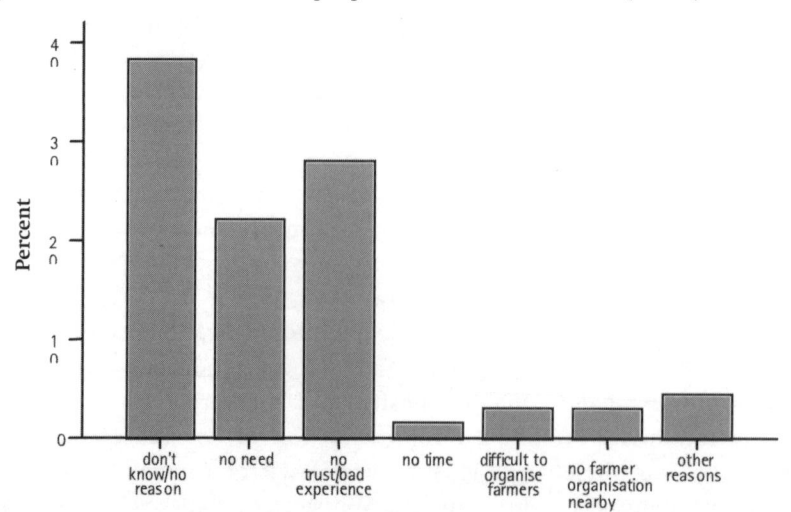

Figure 5.3   Reasons for not working together in the 2003-04 season (n = 68)

Source: FS 2005.

Shifts in governance in Ghana

Reforms in the cocoa sector had a strong impact on the governing of the cocoa sector and on the conditions under which cocoa producers worked. Looking at shifts in governance illustrates that already before World War II Ghana started with coordinated action to stabilise prices. This was done in direct negotiation between import companies (buyers), exporters and organised producers. This resulted in corporate governance systems which governed transactions in the cocoa sector (the 'cocoa buying agreement'). However, after the war transactions and price setting were increasingly regulated by the state, amid decreasing chain integration. The recent periods of gradual reform during the 1990s and the current re-affirmation of the state control in the cocoa chain, illustrate the high influence of the state in governing transactions. Influenced by the overall increasingly trader-driven cocoa chain Ghana gradually moves to a joint governance system. In Ghana there is an alliance between international buyers and the government (see also Fold, 2001); they both share an interest in maintaining the current system, as it guarantees a consistent supply of premium quality beans. In the joint governance system the state plays an active role and global traders and manufacturers play a more passive role. The shifts in governance are illustrated in figure 5.4, making use of the integrative framework of Griffiths and Zammuto (2005).

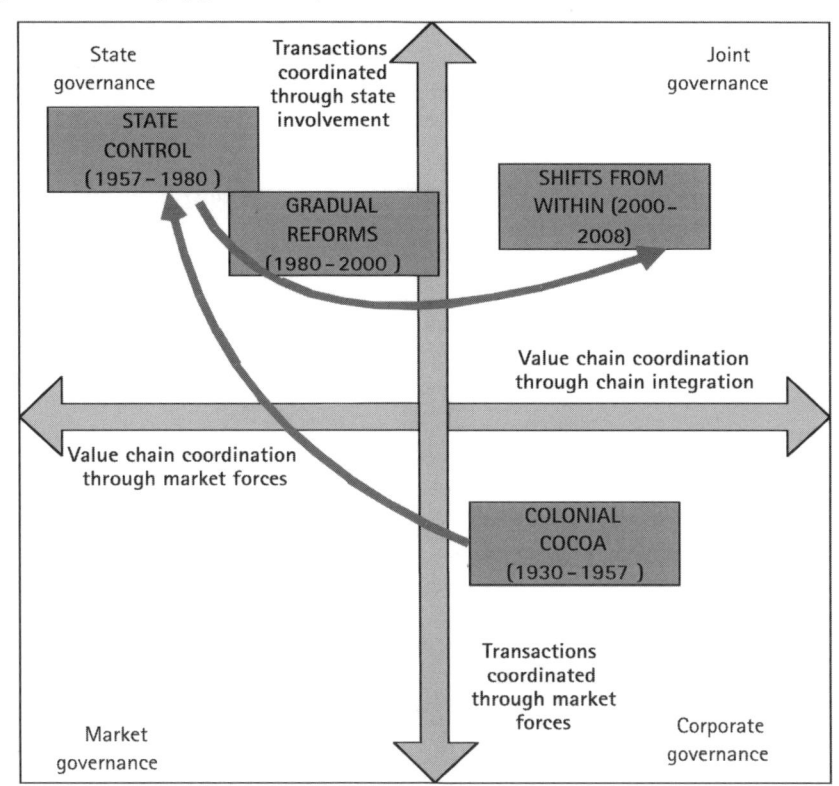

Figure 5.4   Changing governance systems in the Ghanaian cocoa sector

Source: Ton et al., 2008.

5.4 Discussing the impact of reforms: comparisons with fully liberalised countries in West Africa

In the last section I described the changing conditions under which cocoa producers operate, using a comparison over time. In order to validate the impact of the reforms, I will also make some comparisons with the current conditions in fully liberalised cocoa-producing countries in West Africa. I will compare the following indicators: (1) price-developments, margins and taxes; (2) volume of production; (3) farmer income; (4) quality and services; and (5) farmer organisation.

5.4.1 Comparison 1: price-developments, margins and taxes

Earlier studies that evaluated the impacts of liberalisation mainly focused on producer price (cf. Akiyama et al., 2001; Gilbert and Varangis, 2003). They concluded that the liberalisation of state-marketing systems increased the farmers' share of the FOB price, thus positively assessing the reforms in these countries. In Cameroon and Nigeria, the producer price initially increased due to reduced marketing costs and taxes, thus producing significantly higher prices than in Côte d'Ivoire and Ghana (Gilbert and Varangis, 2003; Haque, 2004). However, the undermining of national and international price stabilisation mechanisms negated the impact of the initial price-increase on farmers' incomes. Higher price fluctuations and the generally low world cocoa price made farmer income less secure.

Also, in Ghana the objective of liberalising internal marketing was to improve the operational and financial performance of its marketing system, in order to secure higher producer prices. The specific aims were to gradually reduce the export tax and marketing margins (Table 5.4).

Table 5.4 Objectives of gradual reforms: producer-prices, margins and taxes

Crop year	1998-99	1999-00	2000-01	2001-02	2002-03	2003-04	2004-05
Producer price	56.0	60.0	62.0	64.0	66.0	68.0	70.0
Marketing/ Cocobod operations	18.2	16.5	16.2	15.9	15.6	15.3	15.0
Government tax	25.8	23.5	21.8	20.1	18.4	16.7	15.0
Total	100.0	100.0	100.0	100.0	100.0	100.0	100.0

Source: Ministry of Finance, 1999: 90.

In terms of increasing producer prices and reducing margins and taxes, the reforms in Ghana were successful. Already for some years, farmers receive around 70 per cent of the net FOB, equalling around 995 USD/tonne in 2006-07. The state officially announced that it will increase the share of the Net FOB price to over 72 per cent for the 2007-08 season.[102] Despite this increase in producer price Cocobod's share of the export value remains high.

The next figure, adapted from Abott et al. (2005), illustrates the margins in the cocoa supply chains for the major cocoa producing countries in West Africa during the 2001-02 season. Compared to Côte d'Ivoire, Ghana pays farmers a higher price and has relatively low export taxes. However, in Ghana the margin for exporting (controlled by CMC, a subsidiary of Cocobod) is very high. In contrast to Cameroon and Nigeria, which have a much lower dependence on cocoa exports, Ghana's margins for producers, purchasers and traders are relatively low, while margins collected by Cocobod and the Ministry of Finance are relatively high (Figure 5.5).

Figure 5.5   Margins in the cocoa supply chain 2001-02 main crop

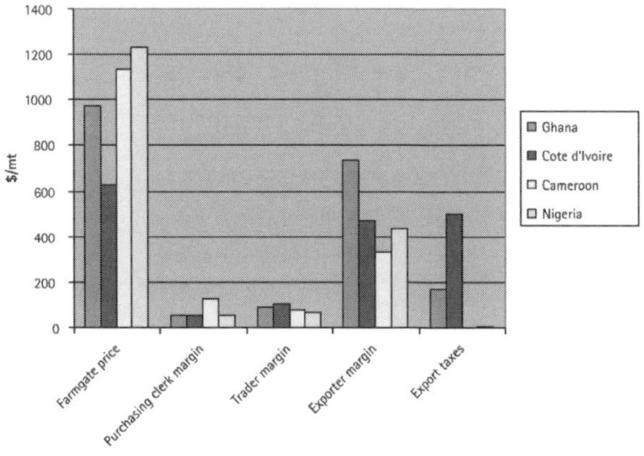

Source: adapted from Abbott et al, 2005: 44.

5.4.2 Comparison 2: volume of production

In the beginning of the 1980s cocoa production in West Africa suffered. Besides drought and fires also political factors played a role, especially in Ghana (Ruf, 2007a: 2).[103] Since then cocoa production has been gradually recovering, partially in response to gradual increases in producer prices. In Côte d'Ivoire there was a 'massive boom' in cocoa production from the 1970s until the late 1980s (ibid: 3). In Cameroon cocoa production stagnated. The reforms in Cameroon put an end to input subsidies and farmers reduced their investments in pest management.[104] In Nigeria, production initially stagnated but picked up from 1999 onwards, with increased production volume. This increase is attributed to the incentives provided by the Cocoa Development Committee (CDC).[105]

Also in Ghana the higher producer price did, as intended, contribute to the recovery of cocoa production. For Ghana as a whole, it is estimated that between season 1990-91 and season 1997-98 the harvested areas increased with 73 per cent (from 707,000 hectares to 1,220,000 hectares). But, the increase in cocoa production was 'almost entirely due to the traditional method of expanding output by means of additional land' (Teal and Vigneri, 2004: 8-12; Ruf, 2007a). Later, in 2003 there was a boom in cocoa production, attributed to the cocoa trees that were planted in the Western Region in the mid to late 1980s. Also farmers, stimulated by the government and private extension providers, started to intensify their methods of cultivation, by using more labour, new seed varieties, with mass spraying of their farms and with increased application of fertilizer (FS 2005; Ruf, 2007a; Ruf, 2007b; World Bank, 2007b). Smuggling from Côte d'Ivoire (and the end of smuggling from Ghana into neighbouring countries) also contributed to the increase of the processed volume of cocoa beans (GAIN report, 2005; Ruf, 2007b; informal discussions with industry, 2005).

The targets of the government in terms of production, as formulated in the Cocoa Strategy (published in 1999), were set at 500.000 tonnes for 2004-05 and 700.000 tonnes for 2009-10 (Ministry of Finance, 1999). Although these targets were initially regarded as ambitious, already in 2003 Ghanaian cocoa farmers produced almost 500,000 tonnes of cocoa, which is the second highest harvest ever in Ghana. In 2003-04 total cocoa output increased considerably (including smuggling up to 736,000 tonnes). In 2004-05 cocoa production decreased to 600,000 tonnes and in 2005-06 it increased again to 705,000 tonnes (based on first estimations of ICCO, 2006 and Ed&F Man, 2004 in Ruf, 2007b).

5.4.3 Comparison 3: farmer income

The increase in producer prices did not automatically result in an increase in remunerative farmer income, which also depends on the costs of cocoa production, productivity and on diversification of income. Moreover, inflation and the terms of currency exchange also adversely affected farmer income. During the group discussions, one Ghanaian farmer clearly highlighted the deterioration in terms of trade:

> [...] when you compare the price of cocoa to the cost of production I can say the price increase is insignificant. During the time of our fathers a bag of cocoa could buy twenty bags of cement, these days you cannot buy even ten bags from that. My father's proceeds from four bags of cocoa could buy six aluminium roofing sheets; nowadays you cannot get four aluminium roofing sheets from ten bags.

The reforms abolished the input subsidies, which resulted in enormous price-increases, especially in Nigeria and Ghana. Haque (2004) suggests that in Ghana production costs have tripled between 1989 and 1999. It is argued that the increases in production-costs resulted in 'self-exploitation' among farmers. Farmers worked longer hours, mobilised family members, and participated more in labour exchange groups (Blowfield, 2003).

In recent years both in Nigeria and Ghana the (indirect) provision of subsidies on inputs came back on the agenda. Interestingly, the push for this initiative does not only come from the government but also from the private sector. The main aim is to increase farmers' productivity, which is especially low among Ghanaian farmers. In 1999, cocoa yields (300 kg/ha) in Ghana were about 13 per cent below the African average and 30 per cent below the yields in Côte d'Ivoire (Ministry of Finance, 1999: 8-9). Between 2001 and 2005 the average yields in Ghana increased somewhat higher (377kg/ha), but are still considered very low (Ayenor, 2006).

5.4.4 Comparison 4: quality and services

Prior to the reforms the state was responsible for quality control procedures, while after most of the quality control system were privatised. Also, after marketing reforms reduced the restrictions on cocoa trading a large number of inexperienced buyers entered the field. Nigeria, Cameroon and Côte d'Ivoire all faced quality losses. For Nigeria and Cameroon this meant losing their premium and facing a decline in demand. Ghana was the only country where a public system of quality control remained in place. Nevertheless, Ghana had some problems with quality, mainly blamed on the LBCs. Cocobod's responded by sanctioning LBCs, an action that constrained their financial performance and seriously affected the livelihoods of many farmers.[106] No longer able to sell (or store) their cocoa, they (temporarily) lost their main source of income (farmer profiles, 2005).

In addition to reducing quality, reforms also changed the organisation of extension services. Since the introduction of the reforms there is a general problem in all the countries with under-funded cocoa extension services and poor institutional support for extension (Amezah, 2004). Weak (unified) extension services render it more difficult to bring new knowledge and technology to the farmers. This adversely affects their ability to upgrade the product and the production process, and has a negative impact on their ability to anticipate market changes.

Also in Ghana extension services worsened, as also reported by more than half of the farmers interviewed (FS 2003). In response, NGOs (together with the private and public sectors) got involved in the provision of farmer-based extension. These services were appreciated by the farmers. But, farmer-based extension, such as FFSs incur high costs and the services reached only a limited number of farmers (interview CI Ghana and Wienco, 2005) (Chapter 6 and 7). During the pilot of CI Ghana between 120 and 150 farmers were trained as Trainer of Trainers (ToTs). In 2005, the STCP trained around 15,000 ToTs in FFSs in West Africa. Inside Ghana the training focused on environmental friendly agricultural practices and the challenges of child labour. Outside Ghana, farmers also received training on marketing techniques and information systems.

In Ghana poor public extension services produced fragmentation. This, together with the privatisation of input distribution, brought different suppliers of inputs on the (black) market, resulting in conflicting advice and inefficient application of chemicals. Farmers indicated that they did not receive proper information on how to apply the chemicals. The majority of the farmers did use toxic chemicals with adverse side-effects on their health and the environment. The limited use of protective clothing and the inadequate application of chemicals further exacerbated the health problems.

Farmers also complained about the lack of credit facilities, another side-effect of the reforms. In all countries formal credit facilities are weak and reach only a limited number of farmers. In Côte d'Ivoire the consolidation of foreign buyers made it more difficult for local financiers to disperse their risks. After liberalisation, Cameroon had well-developed arrangements for the export sector, but according to the members of COPAL, fewer credit facilities were available for farmers and local buyers (COPAL, 1998). Also in Nigeria there is a complete lack (or irregular provision) of adequate credit granting services for investments in farm productivity. In Ghana formal credit facilities are marginal, of the twenty-five farmers (n = 200) with access to credit in 2003-04, 10 farmer received a loan from a LBC, 4 farmers from a friend or colleague, and only 2 farmers from a bank (FS 2005).

5.4.5 *Comparison 5: farmer organisation*
In Côte d'Ivoire, Cameroon and Nigeria the reforms gave an incentive for farmers to set up cooperatives. The literature shows that in West Africa these initiatives still seldom come exclusively from the farmers themselves. Many weak cooperatives require institutional support. Nigeria approached foreign donors for financial assistance in order to strengthen 'producer groups' (COPAL, 1998). In Côte d'Ivoire the government and the global buyers supported cooperatives as their (new) trading partners (industry survey 2005). In addition to this emerging direct (trading) relationship between buyers and suppliers, global buyers (indirectly) also started to support the farmers' acquisition of marketing expertise and improved information systems through public-private partnerships.

In Ghana there are no incentives for farmers to set-up a formal farmer organisation, partially due to the indifferent attitude of the government towards the development of cooperatives (Department of Cooperatives, 1990: 130-2). The development of cooperatives requires a supportive environment and facilitators who can help farmers to start a farmer organisation. It also requires an environment where 'organisation' provides farmers with tangible benefits, for example higher incomes or better access to services.

There are currently only two farmer organisations formally registered and operational in Ghana: the GCCSFA and KKFU (Section 5.3.2). Farm-leaders within GCCSFA are chosen democratically, at district, regional and national level. At district level, chief farmers are regarded as important leaders of the cocoa community. Farmers know that these chief farmers report to district farmers, who in return report to the regional chief farmer (farmer profiles 2005), but farmers do not consider themselves to be a member of GCCSFA or even claim that they have never heard of this farmer association. As it already has a structure in place, the GCCSFA certainly has the potential to adequately represent farmers' interests. However, such a role would require greater transparency about its exact functioning and also more direct involvement of farmers in decision-making processes, both within GCCSFA and Cocobod. Also, GCCSFA leaders should be held accountable, especially the national chief.

The Kuapa Kokoo Farmer Union is organised according to cooperative principles. Membership provides better access to community development, services, training (farmer field schools) and credit schemes. This farmer union is often presented as a successful farmer cooperative (cf. Mayoux, no date; Tiffen, no date; Vuure, 2006) but it had serious problems with its management team, such as embezzling by ex-leaders who disappeared with money belonging to the union. It also proved difficult to exercise effective democratic control in such as large cooperation. The impact of membership in the KKFU on the producer-level is discussed in Chapter 7.

In Table 5.5 I give an overview of the characteristics of the different cocoa producing countries. This evaluation of reforms presents another picture than the evaluation of the World Bank (Table 5.1), where the free market was presented as optimal. The next table shows that a 'free' market (Côte d'Ivoire, Cameroon and Nigeria) has more serious drawbacks then a partially liberalised system.

Table 5.5   The impact of reforms in the major cocoa-producing countries

Characteristics	Prior to reforms	Ghana	Côte d'Ivoire	Cameroon	Nigeria
Price developments	Price stabilisation	Price stabilisation Producer-price gradually increasing	Price fluctuations Price decline, later small increase	Price fluctuations Initial increase, later decline	Price fluctuations Initial increase, later decline
Taxes	High	Decrease	Still high	Absent or very low	Absent or very low
Exporter margins	High	High, but (partly) reinvested in sector	High, little reinvestment in cocoa sector	Considerable	Considerable
Farmer income	Medium/low	Medium/low	Medium/low	Medium/low	Medium/low
Volume of production	Decline, only in Côte d'Ivoire boom	Increase (boom in 2003)	Stagnation	Stagnation	Stagnation, later small increase
Quality	Good	Good, comparing to neighbouring countries	Mediocre (quality losses)	Loss in quality (loss of premium)	Loss in quality (loss of premium)
Public extension services	Good/ medium	Medium/fragmentation of services	Bad	Bad	Bad
Credit	Some countries weak, some medium	Weak and insufficient	Weak and insufficient	Weak and insufficient	Weak and insufficient
Farmer organisation	No formal farmer organization	Lack of formal farmer organisation (one farmer union; one association)	Large number of farmer groups, but many of them not operational and weak	Small number of farmer organisations	Attempts to strengthen producer groups

Source: created by author.

5.5   Public and private interventions in Ghana

As a result of reforms in the fully-liberalised countries of West Africa, national institutional mechanisms lost several functions. Global buyers and organised suppliers, together with NGOs and overseas governments, tried to fill the vacuum by creating new institutions and new ways of diffusing their risks. In Côte d'Ivoire and Nigeria the state has been trying to reclaim its coordination over the supply chain. While it did manage to protect its income through high taxes and levies, it is not investing any part of its cocoa revenues back into the cocoa communities.

The partially liberalised system in Ghana did not reduce the role of the government as much as in the fully liberalised countries like Cameroon, Nigeria, and to a lesser extent, Côte d'Ivoire. The Ghanaian state still plays a major role in the cocoa sector. Table 5.6 below lists most of the important state interventions. Only a limited number of NGOs and private actors are actively intervening in the sector. In some cases this is due to low need for interventions as the cocoa sector in Ghana is well organized, but in other cases it is clearly because Cocobod's strong coordinating role constrained the performance of some actors or obstructed interventions.

Table 5.6   State interventions in the cocoa sector 2005

Type of upgrading	Activities/interventions
Capturing higher margins for unprocessed commodities	Forward sales
	Strict system of quality control (QCD)
	Sanctioning LBCs selling inferior cocoa
	Extension services through broadcastings (radio/newspapers)
	Increase producer-price, bonuses, rewards etc.
	Rehabilitation of old-cocoa farms
	Mass-spraying programme (CODAPEC)
	Fertilizer on credit (Public Private Partnership)
	Public extension services (MoFA)
	Research (CRIG)
	Farmer-based extension services (CRIG is partner in FFSs)
Producing new forms of existing commodities	Research (CRIG), e.g. on new varieties/pests and diseases
	Development and marketing of cocoa by-products
Localising commodity processing, marketing and consumption	Attracting foreign processing companies (20% discount on beans)
	Development and marketing of cocoa waste products

Source: Composed by author.

The partially liberalised system in Ghana does not allow direct (trading) relationships between global buyers and local suppliers. Nevertheless, international buying companies intervene at different levels, through strengthening relationships and/or direct investments. First, major processing companies (cf. Barry Callebaut, Cargill) have outsourced part of their processing activities to Ghana. Second, processing companies and chocolate manufacturers invest time in maintaining a good relationship with Cocobod and support it in planning measures against the (future) risks of supplier failure. Third, global buyers intervene on the level of internal marketing. Currently there are two foreign buyers who established an LBC and thus became internal players in the sector and created a direct links between farmers and their companies. Fourth, global buyers support cocoa farmers directly through participation in public-private partnerships, sometimes together with other buyers and NGOs, sometimes with public players, or a combination of different actors (cf. STCP). Fifth, buyers develop individual programmes for farmer support (cf. Cadbury invests in war wells).[107] A new trend is that global manufacturers (such as Cadbury and Mars) start to develop programs for sustainable sourcing of their cocoa. A part of this cocoa is to be sourced from Ghana. How this affects cocoa farmers is not yet known.

Local private actors also became involved in Ghana's cocoa sector. First, LBCs started to issue credits, bonuses and services to farmers, as a way of competing with other local buyers. However, LBCs are constrained by Cocobod, because they receive a fairly low buyers' margin and carry the risks of quality declines. Also, LBCs are unable to expand their marketing activities from internal to external marketing. This environment reduces both the incentives and the opportunities to establish strong alliances with other buyers and farmer groups. Local buyers also lack incentives that would stimulate them to contribute more to enhancing quality performance. In the partially liberalised system LBCs have little negotiating power. Also, other private actors, such as input providers, got involved in public-private partnerships, by providing inputs on credit. One private input provider, Wienco, took the initiative to provide farmers with extension services, thus helping farmers to apply inputs more adequately while at the same time increasing the sales of their inputs. This was a costly exercise for Wienco (interview Wienco, 2005). While Cocobod does appreciate that input providers have assumed this role, it still does not officially support these types of private interventions. The Agricultural Development Bank was involved in different attempts to supply farmers with inputs (fertiliser) on credit.

NGOs are not actively involved in the cocoa sector in Ghana. This is due to the fact that cocoa farmers are considered to be better off than other farmers, since they have Cocobod to protect them from fluctuations in world market prices. There are some exceptions: the international NGO Conservation International Ghana initiated the FFSs and also introduced farmers to nature conservation practices. In addition CI Ghana linked up different service providers and established a partnership with the public and private sector. But even with this partnership CI Ghana feels constrained in its performance. For example, an attempt to introduce organic cocoa in Ghana, which would have provided farmers with a new marketing channel for niche markets, was obstructed by Cocobod on the grounds that the organic pesticides lacked official CRIG approval. This decision was the result of 'legitimate concerns and unfortunate misperceptions' (CI Ghana, no date: 4). According to a representative of Cocobod, the main concern was the economic viability of organic cocoa: 'the question is if the premium paid for organic cocoa covers the costs of crop loss'.[108] In 2006 an attempt of Agro Eco/Louis Bolk Institute was more successful. In partnership with Cocobod the first organic cocoa project was realized. Another NGO, mentioned by farmers, is Action Aid. Together with Cocobod they provide farmers with more resistant seedlings and inputs (farmer profiles, 2005).

Interventions by organised farmers are exceptional, because, as argued earlier, the current system does not provide farmers with incentives to organise themselves. Due to this lack of organisation, the farmers have not been able to fully benefit from the reforms. Instead they are increasingly confronted with the risks of the partially liberalised system. Farmers, individually, have adopted many different strategies to cope with the changes and improve their performance. These strategies will be discussed in more detail in Chapter 7.

Figure 5.6   Existing interventions in the partial liberalised cocoa sector in Ghana

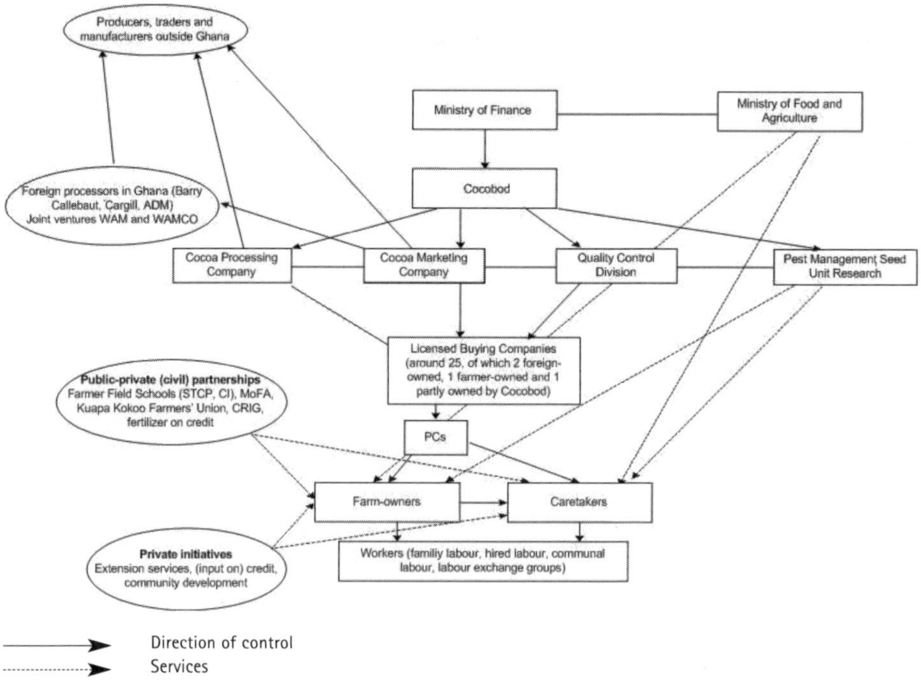

Source: figure by author.

5.6 Conclusions

Reforms in cocoa producing countries had a severe impact on the conditions under which cocoa producers operate. This is especially acute in countries that liberalised over-night, such as Nigeria, but also in countries that liberalised more gradually, such as Côte d'Ivoire. In most countries the quality of cocoa declined, high numbers of unprofessional private buyers dominated cocoa marketing, production costs increased and prices started to fluctuate. For farmers the risks involved in cocoa production increased. The vacuum left by the government was not filled with actors who had the capacity to take over governmental tasks, because the private buyers were unprofessional and farmer organisations were lacking.

In countries with a high stake in cocoa, such as Côte d'Ivoire and Ghana, as is expected the state did not want to leave its strategic position in the sector. This confirms my hypothesis that perceiving the public sector as 'enabler' does not reflect the reality on the ground in the economies of all developing countries. Even within a fully liberalised setting, governments can continue to represent the interests of certain economic sectors and groups within society (not to mention the strong personal stake of the involved officials). Governments in developing countries should not be considered neutral players in their economy. In Côte d'Ivoire, despite the fully liberalised sector, taxes on cocoa export equalled nearly 40per cent of the export price. For Ghana, it was quite shocking to see that the government is making a difference between the gross and net FOB price, capturing a margin of 45 per cent in cocoa season 2001-2002 and 35 per cent in season 2002-2003. Although the Ghanaian government reinvested part of this margin back into the cocoa economy (unlike Côte d'Ivoire where there is no reinvestment of taxes), it is not clear exactly how this money was allocated and how much of it disappeared in the pockets of Cocobod officials.

Ghana is a particularly interesting case because, in response to the negative experiences with full liberalisation in the region, the Ghanaian government opted to introduce reforms gradually and partially. It is an example of a government retaining a key steering role, together with the private sector. This hybrid form of governance ('joint governance') proved as quite favourable for cocoa producers and for the other actors involved in the sector. Due to the reliable marketing system, Ghana enjoys a high reputation for honouring its contract and offering relatively still high quality produce. Other benefits of the partially liberalised systems include the intact price stabilisation, the gradual price increases, tax decreases and increased production volume. Also the services provided to farmers are generally better than in fully liberalised countries. Farmers appreciated foremost the prompt payment of LBCs for their cocoa.

The Ghanaian government still controls external marketing and regulates internal marketing, pricing systems, processing activities, research, quality control and the provision of services. In short, it retained its role as the coordinator of the cocoa supply chain. Nevertheless, with the reforms the government abandoned some of its former duties, which were taken over by other public, private and civil actors. Extension services were merged with the services of MoFA, the input distribution system was privatised and internal marketing was liberalised. The opportunities and incentives for the actors to assume their new roles were sometimes still limited by the state, in some cases resulting in serious drawbacks. Also, the state had difficulty in successfully managing the cocoa sector. For example, the quality control system was pressured by the increased volumes of cocoa. Farmers were made particularly vulnerable through the fragmentation of extension services and the lack of bargaining power by local buyers and farmers. As a result farmers were unable to fully benefit from the reforms. The reforms were also not optimal for licensed buying companies. The reforms made it possible for them to enter domestic marketing but they were not allowed to play a role in external marketing of cocoa. Another weakness of the partially liberalised system is that the export margins received by the state are still high; Cameroon and Nigeria have lower government margins. Although part of this money is reinvested in the cocoa sector, reinvestments do not go without problems and furthermore are not transparent.

Despite the noted tensions and weaknesses of the Ghanaian system, this experience shows that partial liberalisation may indeed be a viable alternative model to full liberalisation. The country's recent achievements in terms of increased producer prices and volume of production, coupled with the negative experiences in fully liberalised countries, seems to have dissuaded the global buyers and international donor organisations against pushing for more liberalisation in Ghana. The Ghanaian government and Cocobod will not object to this trend, because they have strategic (economic) reasons to resist full liberalisation. However, it is not unthinkable that changes in preferences of international buyers or pressures from the World Bank and international donors may eventually bring another wave of liberalisation. The changing governance structures in Ghana (figure 5.4) already indicated that governance systems do tend to change over time. It is important that Ghana shows more openness to global developments and to the changes in demand in global cocoa markets. It is also important the Ghanaian government invests more in the capacity of other actors, especially the farmers and the private sector, empowering them to contribute more to building a strong cocoa sector.

6
WHO ARE THE COCOA FARMERS?

6.1 Introduction

Insertion in a global value chain or clustering does not automatically lead to upgrading. Upgrading can be enabled or hindered by powerful players in a chain, but also by governments and social structures. Moreover, the gains from upgrading are often unequally distributed (Giuliani et al., 2005; Tiffen, no date). In the value chain literature, this recognition led to a debate on social inclusion and exclusion of groups of the producers and other actors who operate at the beginning of a chain (Altenburg, 2006; Gibbon and Ponte, 2005; Kaplinsky, 2001; Knorringa and Pegler, 2006). It is reflected in the cluster literature in the increased emphasis it places on 'inclusive economic growth' targeting the poor (Cortright, 2006; Nadvi and Barrientos, 2004: v).

The general notion that 'value chains become more exclusive as small-scale producers fail to meet rising scale and standard requirements' Altenburg (2006: 508) is also observed for producers inserted in the global cocoa chain (Kaplinsky, 2001; Gibbon and Ponte, 2005). Small-scale cocoa producers have more difficulty with increasing their production levels than larger farmers, who are constrained due to high production costs (Teal and Vigneri, 2004).[109] In the long-term it expected that a smaller group of more innovative and, as a result, bigger cocoa farmers will produce cocoa. Marginal, non-competitive producers will have to look for alternative means of generating income.[110]

The scenario of a 'structural transformation' of the cocoa sector is rather abstract for cocoa farmers in Ghana, where 80 per cent are smallholders (Mehra and Weise, 2007). In Ghana, the state-owned marketing Board (Cocobod) still keeps the monopoly on the external marketing of cocoa, which allows suppliers to benefit from forward sales of large quantities of quality cocoa. Because Cocobod functions as a 'lead firm' (or large farmer) for Ghanaian smallholders, there are generally few problems with selling the right quantities of cocoa. Ghana's position is also special regarding standard requirements, as it produces cocoa of a relatively high quality. While for other countries process requirements have become increasingly more important, in Ghana this is not the case (yet). Therefore, at least in the short run, there are no acute risks for Ghanaian cocoa producers to become excluded from the global cocoa chain for being not able to reach scale and product-quality requirements. However, it is important to consider the changing conditions under which cocoa producers are included in the chain and the different outcomes for producers. For the case of Ghana it is relevant to analyse the heterogeneity among farmers and to assess how this relates to the changing conditions. Such insight in the characteristics and social structures that explain different outcomes for producers helps policy-makers and NGOs develop more effective interventions that can reach the most vulnerable farmer groups. Moreover, it also helps the private sector to think more strategically for realising some key goals, such as securing delivery of large quantities of good quality beans from Ghana and the strategic targeting of specific groups for assistance in developing sustainable sourcing of cocoa.

An understanding of the differences among cocoa producers and the social structures in which they are embedded is also important to identify inclusive upgrading strategies. I assume that in the long-run, the risks of being excluded depend at least partly on the effectiveness of current interventions taking place in the cocoa sector in Ghana. It will also depend on capacity of the Ghanaian state to continue managing the supply chain successfully and on the flexibility of the Ghanaian system to respond to changes in the environment.

6.2 Different outcomes for different types of producers

In Chapter 5, I discussed the shifts in governance structures in the Ghanaian cocoa sector, emphasising the role of the state. Two comparisons (1) Ghana before and after the reforms, and (2) Ghana and other cocoa producing countries in the region, provided insight in the effect of the reforms on the sector and the position of cocoa farmers in Ghana. However, there is great internal heterogeneity among cocoa farmers, which leads me to the following question: What are significant differences among cocoa producers that influence the impact of shifts in governance and the success of upgrading strategies?

In this chapter, I will use earlier studies that sought to shed light on why partial liberalisation in Ghana produces different outcomes for cocoa producers (Vigneri, 2007; Teal et al., 2006)[111] as a starting point. They will help to understand the relationship between national governance structures, heterogeneity and upgrading in the Ghanaian cocoa sector. These studies focus on the impacts of reforms on the scale of production and how the location of the production region explains possible differences in outcomes. Some authors have searched for explanatory factors other than 'geography'. For example, Takane (2002) looked specifically at the role of institutions, focusing on differences in types of contracts between farm owners and caretakers and how this explained the variety in farmer responses to the incentives for increasing volumes of cocoa production. Takane (2002: 391) demonstrated for cocoa production in Ghana that price is not the only incentive for the production of higher volumes of cocoa beans, highlighting the long-term and inheritable use-right to land. This point was also raised by the farmers, who shared that one of the main reasons for getting involved in cocoa farming was the potential for acquiring land and inheritance (a kind of social security) (farmer profiles 2005).

Previous studies showed that use-right of land and the location of the farm are two characteristics that explain different outcomes for producers. In considering differences among farmers and their correlation, the cluster literature provides valuable additional input. Contrary to the GVC approach, which does not explicitly analyse local power relations, the cluster approach emphasises the role of social structures in determining different upgrading outcomes. This literature highlights the difficulty in separating economic systems from the social system in which they are embedded (Cortright, 2006: 25; Granovetter, 1985). Another valuable contribution of the cluster literature is its emphasis on 'joint action'. Joint action as a potential upgrading strategy is relevant for small cocoa producers and can help to achieve economies of scale, higher quality production and increased bargaining power.

In order to analyse differences among farmers I made sure that I gathered data on the location of the farm (following Vigneri, 2007 and Teal et al, 2006), ownership of a farm and the type of contract under which caretakers work (following Takane, 2002), patterns of domestic migration (following Hill, 1963),[112] production level and size of a farm (following Altenburg, 2006; Mehra and Weise, 2007) and the extent to which farmers worked together with other farmers (following the findings of different authors from the cluster literature). In order to capture some information on the social structures in which the farmers are embedded, I also included questions in my farmer survey on gender, the position of farmers in the community or chain, age and level of education. As already pointed out in Chapter 3, I clustered the 'characteristics' around different themes: (1) Location, migration and farm-ownership; (2) Cocoa production, size of farm and productivity; (3) Position in the community or chain, age and education; and (4) Gender and the level of cooperation.

6.2.1 Location, domestic migration and landownership

The migration and settlement by small-scale farmers was fundamental for the expansion of cocoa in Ghana. In the early cocoa growing years, cocoa production was concentrated in the Eastern region. In the mid 1940s, the centre of production shifted to the Ashanti and Brong Ahafo regions, where cocoa was produced on virgin forest land. In the mid 1980s, it shifted again to the Western region, where currently there is intensive cocoa growing on forest lands by migrant farmers from the Ashanti, Brong Ahafo and Eastern regions (Cocobod, 2000a: 3). Currently cocoa production is concentrated in the Western region. In 2005/06 the Western region accounted for 57% of total cocoa production (Gockowski, 2007).

I conducted field work in the Western region, Central region, Brong Ahafo region and Ashanti region.[113] Of the 210 farmers who participated in the 2005 survey, more than 55 per cent lived and worked in the Western region, with another 20 per cent in Brong Ahafo, 15 per cent in the Central region and almost 10 per cent in the Ashanti region. Of all the respondents, around 37 per cent were (domestic) migrants. The field data indicated that the interviewed migrants coming from the Ashanti region mainly settled down in Brong Ahafo, while the Western region attracted farmers from a wider variety of regions. A large part of the interviewed migrants currently living in the Central region were also born in this area. For the respondents who settled in the Ashanti region their number is too low to draw any conclusions (Figure 6.1).

Figure 6.1   Patterns of domestic migration

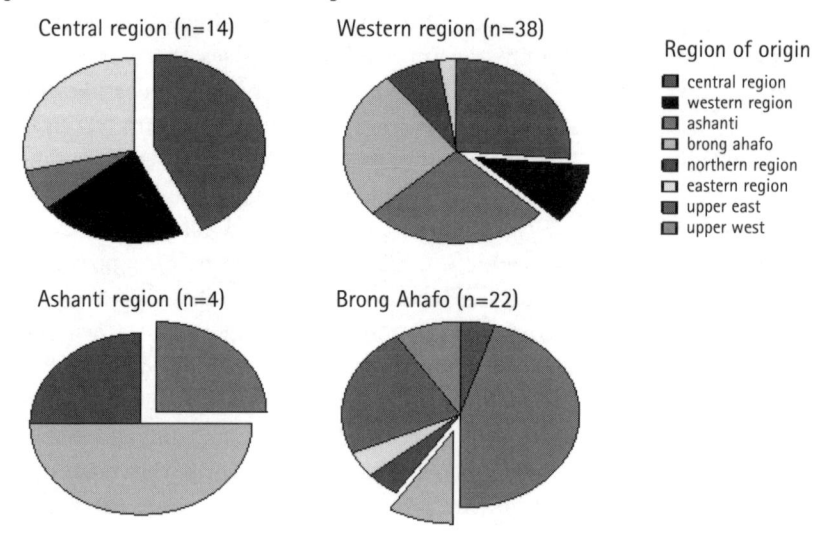

Central region (n=14)
Western region (n=38)
Ashanti region (n=4)
Brong Ahafo (n=22)

Region of origin
- central region
- western region
- ashanti
- brong ahafo
- northern region
- eastern region
- upper east
- upper west

Cramer's V 0,483*** (FS 2005).

It is generally assumed that a severe outbreak of the Cocoa Swollen Shoot Virus disease (CSSVD) was the main driver of the first wave of (domestic) migration (Cocobod, 2000a). However, Hill (1963) opposes this view and argued that CSSVD, which was 'solved' by cutting down the trees, actually had the opposite effect; it limited migration instead of stimulating it. The main barrier to migration was the lack of financial means to purchase land elsewhere. She demonstrated that Ghanaian migrant cocoa-farmers are 'remarkably responsive to economic incentives, [and] remarkably dedicated (within the framework of cocoa-farming) to the pursuit of economic ends' (Hill, 1963: 3). In the mid-late 1980s and the 1990s another massive migration to the south of the Western region took place. This settlement of new farmers contributed to a production growth (Ruf, 2007b). Of the migrants I interviewed, approximately 35 per cent owned the farm[114] they worked on, and around 62 per cent worked under contract.

Cocoa harvesting land in Ghana is mainly family-owned and cocoa growing regions have a matrilineal system of inheritance. This land tenure system encourages land fragmentation; land has to be divided among all the family members (Adusei, 1993: 9). The small size of land plots constrains farmers in making cocoa production a profitable business. It is even more limiting when they do not own any land, but have to work under contract on land owned by other farmers. There are two types of contracts between farm-owners and caretakers, Nhwesoo and Yemayenkye. Under a Nhewsoo contract caretakers manage already established cocoa farms and in return receive a share of the profit (usually one third, which is known as the Abusa system).[115] In Yemayenke (do and let share) caretakers are responsible for all the farm tasks and in return receive a half share of the harvest (this type of contract is also known as Abunu), but the caretaker has to wait until the trees start bearing fruit before receiving a share of the harvest (Takane, 2002: 382-3).

A big difference with the Nhewsoo contract is that in Yemayenkye a tenant establishes cocoa farms with his/her own labor and expenses (Takane, 2000: 383).

My own fieldwork provided some additional information on the division of responsibilities and costs between caretakers and farm-owners. Generally, caretakers that were interviewed were responsible for the main activities on the farm, including training and hiring labour, applying input, buying equipment and farm management. They claimed that owners make a considerable contribution only to buying input. An outcome of these interviews was that there is a significant relationship[116] between the type of contract the caretaker has with the farm-owner and the extent to which responsibility for paying inputs is shared. While for caretakers working under an Nhwesoo contract the owner purchased the bulk of the inputs, most caretakers working under a Yameyenke contract had to provide the inputs largely themselves and shared the responsibility only in one quarter of the cases. In terms of responsibilities and costs, the Yemayenkye system puts a higher burden on the caretaker. But, Takane's study shows that 'a Yemayenkye tenant acquires a stable and inheritable use-right of land, and may even have an opportunity to become a landholding farmer' (2002: 390). Rights to land depend very much on the type of arrangements[117] between the farm-owner and the caretaker and on the long-term investments of the caretaker in the respective farm. Also for farm-owners, investment in 'trees' is a way of claiming one's land right (different authors in Takane, 2002: 391; IFPRI, 2002). Under an Nhwesoo contract, becoming a land-owner is not a likely prospect. Another constraint of this shareholding system is the absence of the owner, which often leads to the farm being neglected (Adusei, 1993).

Of the caretakers interviewed, almost 30 per cent worked under an Nhwesoo contract, 52 per cent under a Yemayenkye contract and almost 19 per cent work under both types of contracts on different farms. Working for different farmers is perceived by the interviewed caretakers as a good way to manage risk and to increase their income (FS 2005). The type of contract associates significantly with the location of the farm.[118] In the Western region and Brong Ahafo more than 70 per cent of the respondents working under contract signed a Yemayenkye contract. In the Ashanti region this was less than half of the small number of contract-farmers. In the Central region no caretakers were interviewed.

Location, migration and land-ownership are inter-linked and relate significantly with some of the other characteristics. For example, location associates significantly with the number of bags produced on a farm (Figure 6.2). In Brong Ahafo most respondents produced only a very small number of bags; for these farmers cocoa production does not seem a very profitable business. In the Western region the opportunities for cocoa farmers seem to be better; in this region a sizeable number of respondents produced more than 30 bags per year.

Yield also varied with land-ownership (see Table 6.1).[119] The number of farmers being 'caretaker and owner' is too low to draw any conclusions.

The next Table (6.2) gives an overview of respondents according to: ownership, the region where they currently live and work, and whether they are a migrant.

Figure 6.2   Cocoa yields for cocoa season 2003/04 in different locations

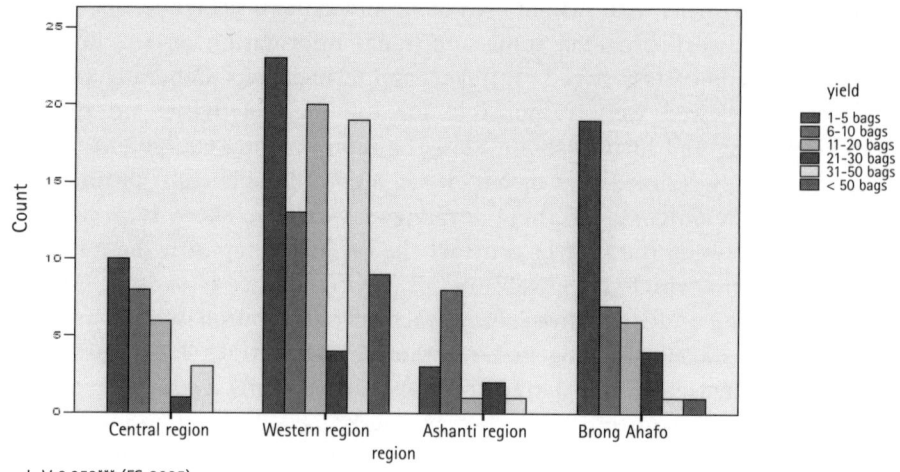

Cramer's V 0,259*** (FS 2005).

Table 6.1   Cross-tabulation between position farm and yield; horizontal percentages

		Yield						n
		1-5 bags	6-10 bags	11-20 bags	21-30 bags	31-50 bags	> 50 bags	
Position Farm	caretaker	48,3%	22,4%	12,1%	6,9%	1,7%	8,6%	75
	caretaker and owner	33,3%	33,3%	,0%	33,3%	,0%	,0%	3
	owner	23,4%	19,6%	25,2%	5,6%	21,5%	4,7%	130
Total		32,1%	20,8%	20,2%	6,5%	14,3%	6,0%	208 (100,0%)

Gamma 0,375*** (FS 2005).

Table 6.2   Ownership, region and migration

Heterogeneity	Categories	Number of respondents
Ownership	Farm-owners	130
	Farm-owners and caretaker of other farm	3
	Yameyenke	39
	Nhwesoo	22
	Both Yemayenkye and Nhwesoo	14
	Total	208
Region	Central region	31
	Western region	116
	Ashanti region	18
	Brong Ahafo	44
	Total	209
Migrant-status	No migrant	105
	Migrant	78
	Total	183

Source: FS 2005.

6.2.2 Cocoa production, size of farm and productivity

The majority of Ghanaian cocoa farmers are smallholders. According to official data of the World Cocoa Foundation in Ghana more than 80 per cent of the farmers produce on farms smaller than 12 acres (around 5 ha) (WCF, 2007).[120] As Polly Hill (1963) already indicated in her early work on migration in Ghana, it is rather difficult to estimate the size of a cocoa farm (see also Ruf, 2007b). On two occasions (2003 and 2005) I asked farmers about the size of their plot. Without exception farmers owned and/or worked on more than one plot of land. Often farmers had a separate plot for the production of food crops. Despite some difficulties with estimates, the data gathered on farm size corresponds reasonably well with the national data (my data was a bit lower). For all respondents interviewed in 2005 the median number of acres was 9,[121] with some regional differences. In the Central region the median number of acres was 9,[122] in the Western region 10 acres,[123] the Ashanti region had a relatively large median of 11 acres,[124] while in Brong Ahafo the size of a farm was generally smaller, namely 6 acres.[125] Not surprisingly, the size of the farm correlated positively with the yield;[126] thus also yield varied per region (see also Section 6.2.1). The relationship between yield and region was also demonstrated in Teal et al (2006), in a report they submitted to Ghana's Cocobod. In this report, they compared the 2002 and 2004 harvests, showing that the yield increase varied widely across regions. The Western region had the highest increase in yield due to a relatively large expansion in the area of cocoa cultivated land.

In Ghana average productivity levels are low (Ayenor, 2006; Ministry of Finance, 1999: 8-9; Teal et al., 2006). Between 2001 and 2005 the average yields in Ghana were 377 kg/ha, while for the same time period the national average yield in Côte d'Ivoire was 744 kg/ha (Ayenor, 2006). Teal and Vigneri (2004: 14) provided data on productivity levels in Ghana and their high variations per region. Table 6.3 shows this data for cocoa seasons 1990/91, 1997/98 and 2003/04. The data for 1990/91 and 1997/98 is based on Teal and Vigneri (2004), while my own fieldwork provided the data for estimating productivity in bags per acre for cocoa season 2003/04.[127]

Table 6.3   Productivity (in bags per acre) (mean)[128]

Productivity bags/acre	Western region	Ashanti	Brong Ahafo	Central region
1990/91	2,30	1,41	3,27	1,49
1997/98	3,19	1,70	1,86	1,50
2003/04	2,27	1,41	1	1,76

Source: Teal and Vigneri, 2004: 14 and FS 2005.

Reforms in Ghana did lead to increases in national yields. This is partly because of increased productivity levels (thanks to an increase in the use of fertilizer, the planting of new tree varieties and the use of more labour). The increase in yields is also the result of increases in the land area used for cocoa cultivation, especially in the Western region (Vigneri, 2007).[129] In addition, the muggling cocoa from Côte d'Ivoire contributed to higher volumes of Ghanaian cocoa 'production' (Ruf, 2007b).

Of the farmers I interviewed in 2005, around 27 per cent had some fallow land available that could be used to expand cocoa production. In this survey, more than 80 per cent of the farmers who had fallow land were farm-owners; the remaining 20 per cent were caretakers. The next table illustrates the number of respondents according to different yields.

Table 6.4   Volume of production in 2003/04

Heterogeneity	Categories	Number of respondents
Yield	1-5 bags	55
	6-10 bags	36
	11-20 bags	34
	21-30 bags	11
	31-50 bags	24
	> 50 bags	10
	Total	17

Source: FS 2005.

6.2.3 *Position in the community or chain, age and education*

Ghana is a hierarchical society, characterised by strong inequalities. In an attempt to unravel some of them, I asked farmers whether they fulfilled a special position in their community. A list of almost thirty different positions was produced. Although it is rather difficult to make a valid categorisation of these different positions, I took on this challenge because social relationships can play a major role in facilitating or constraining farmers in the process of upgrading. In an attempt to analyse the farmers' position, I made a distinction between 'no special position/status in (cocoa) community', 'moderate-strong position/status' and 'very strong position/status'. The first category covers the farmers who either have no position in the community or chain or who do not benefit from an enhanced status due to their position. Examples include membership in funeral committees, fire watchdog committees and town committees. Approximately 53 per cent of the respondents indicated that they belong to this first category. The second category, farmers with a moderate-strong position, includes Purchasing Clerks (hired by Licensed Buying Companies to buy cocoa at the community level) and Society members (treasures, secretaries of cocoa buying societies) who play a significant role in the cocoa supply chain. It also includes members of the 'mass spraying gangs', hired by the government to spray cocoa farms with pesticides (almost 6 per cent held such a position). The farmers with a moderate-strong position deal directly with farmers in marketing and the provision of services. The last category, 'very strong position/status' accounted for around 36 per cent of my respondents. Farmers who obtain such a position are chief farmers, sub chiefs, village leaders, opinion leaders, women leaders, spiritual leaders, queen mothers, trainer of trainers, elders and linguists. Farmers from this last category perform the role of leader in their community or for other farmers.[130]

Picture 6.1 left: Regional Chief Farmer of Western region (Enchi)
right: spiritual leader and large farmer (Asangkrangwa)

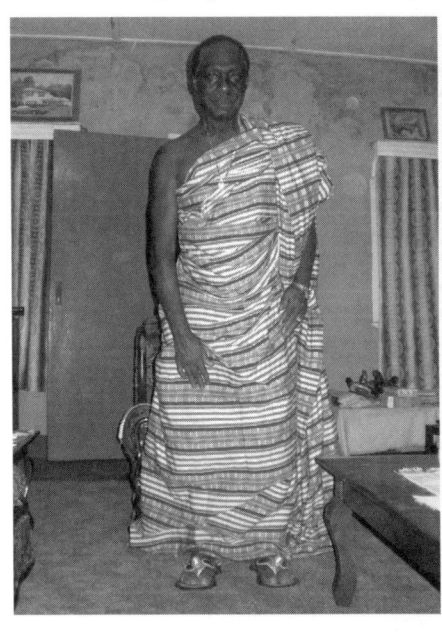

Not surprisingly, the farmer's position in the community correlates with ownership of a farm (see Table 6.5). The farmer's position in the community was also important in terms of yield.[131] For example, the main criteria for selecting chief farmers included: 'high yield', together with size of farm and 'good leadership' (FS 2005).

Table 6.5 Cross-tabulation between position farm and position in community; horizontal percentages

		Position in community			n
		no (significant) position/status	moderate-strong position/status	very strong position/status	
Position on farm	caretaker	72,2%	5,6%	22,2%	75
	caretaker and owner	,0%	,0%	100,0%	3
	owner	46,7%	7,4%	45,9%	130
Total		55,3%	6,6%	38,1%	(178)
					100,0%

Gamma 0,436*** (FS 2005).

The status of farmers is also related to their age (Table 6.6). One of the stronger positions that a farmer can hold in their community is that of elder. The concept of elder (opanyin) and their position in Ghana is the topic of an ongoing anthropological research (cf. Van der Geest, S.).[132] The next picture shows the portrait of an elder.

Ghana has a relatively young population, about 41 per cent of the population is younger than 15, and the dependency ratio is around 87 per cent (Census, 2000: 112).

Picture 6.2   An elder selling cocoa to the Yukwa society (Central Region)

Table 6.6   Cross-tabulation between age and position in community; horizontal percentages

		Position in community			n
		no (significant) position/status	moderate-strong position/status	very strong position/status	
Age	20-39	68,2%	6,8%	25,0%	45
	40-54	61,2%	3,0%	35,8%	69
	55-70	44,4%	6,7%	48,9%	48
	71-90	40,0%	,0%	60,0%	17
Total		56,7%	4,7%	38,6%	179 (100,0%)

Gamma 0,315*** (FS 2005).

However, with a mean of almost 50 years of age, cocoa farmers that were interviewed in 2005 were generally old (Figure 6.3). This average of 50 years corresponds to other estimates made by Vigneri (2007), who interviewed almost 500 farmers in the same cocoa growing regions.[133] The ageing of cocoa farmers is seen as a major threat for future cocoa supply.[134]

Age also matters in terms of education (Figure 6.4), with older farmers having on average lower levels of education. The level of education also reveals other information indicative of the farmers' ability to upgrade. For example, illiteracy can

Figure 6.3   Average age of respondents

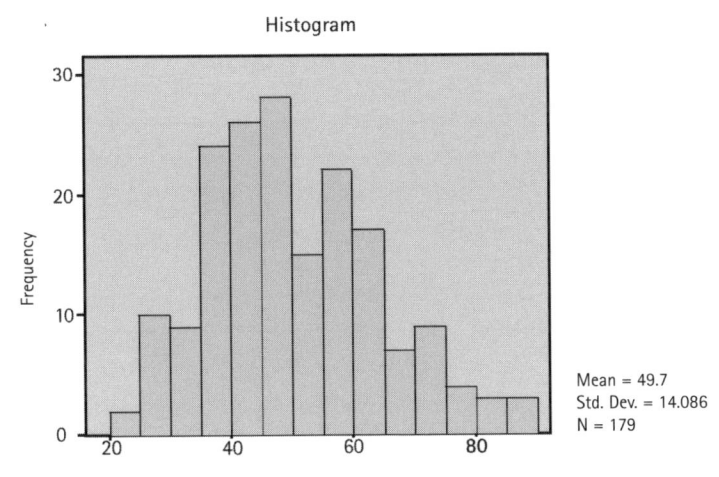

Histogram

Mean = 49.7
Std. Dev. = 14.086
N = 179

Source: FS 2005.

Figure 6.4   Age and education

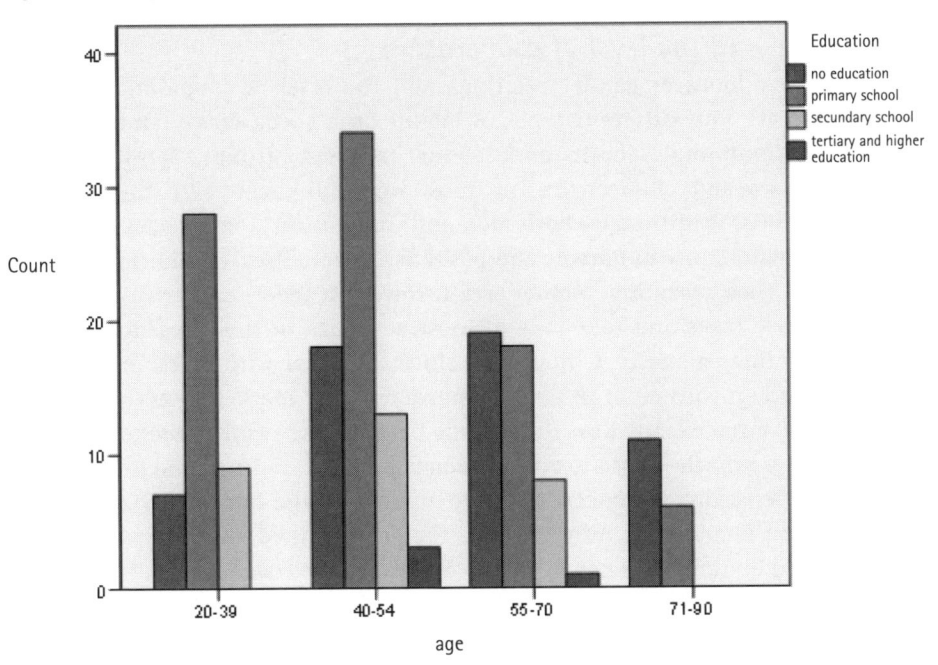

Gamma -0,321*** (FS 2005).

be a handicap as it restricts access to information on for example farm practices, adequate use of inputs and quality issues.

Just like older respondents, for women higher levels of education are also low (see Section 6.2.4).[135] The next table (6.7) illustrates the number of respondents per category.

115

Table 6.7   Status, age and education

Heterogeneity	Categories	Number of respondents
Position in the community	No special position/status	111
	Moderate-strong position	13
	Very strong position	76
	Total	200
Age	20-39	45
	40-54	69
	55-70	48
	71-90	17
	Total	179
Level of education	no education	57
	primary school	90
	secondary school	30
	tertiary and higher education	4
	Total	181

Source: FS 2005.

6.2.4 *Gender and the level of cooperation*

The last category looks at gender relations and the level of cooperation among farmers. There are four different types of labour on a cocoa farm: family labour, hired labour, communal labour, and labour exchange groups. Family labour, especially spouses and children, is the most important source of farm labour (Takane, 2002: 381). Traditionally both men and women work on a cocoa farm. The women assist men mainly in harvest and post-harvest activities. In addition to their involvement in cocoa farming, women are involved in food crop production and have reproductive tasks and take care of most activities in the household. Young girls often help their mothers, while boys help their father with cocoa production. Women have little involvement in farm management and marketing activities. It is difficult to determine exactly how the income between men and women is divided. Male respondents said they tend to give women 'a part (for trading)' and to 'take care of her needs'. Generally, income is used for 'upkeep of the family' (FS 2005). In a group discussion (2005) with only women, they complained that men confiscate their money and do not share any of the income from the cocoa farm.

   Traditionally in many parts of the world women have not been able to own land. Different studies, which examined gender in cocoa growing communities, stressed that without land rights women are unable to take complete responsibility for their own and their children's well-being. Although in Ghana there have been some changes in these traditional practices (Box 6.1), restrictions still exist, for example: illiteracy, lack of education and lack of support to help women exercise their rights (IFPRI, 2002). Land-ownership is often a precondition for membership in official farmer organisations and for participation in training activities. Land is also necessary to apply for credit at a bank (as collateral). As a result women are less targeted by interventions, are less involved in decision-making processes, are less informed about market developments and effective ways of farm management and have even less opportunity to invest in their farms than men.

> BOX 6.1 LAND-RIGHTS FOR WOMEN IN GHANA
>
> In Ghana in 1985, passage of the Interstate Succession Law changed the inheritance system and provided greater security for widows and children. This law makes land available to women in two ways. First, since the passage of the Succession Law in 1985, women have been able to inherit a third of their husband's land by law (1/3 land for widow; 1/3 for children and 1/3 for extended family). However, many are not aware of this law as many Ghanaian women are illiterate. The average age of women involved in cocoa farming is old. A second way for women to obtain land rights is through 'gifting'. Husbands give their wives land in return for their help in cultivating it. This land cannot be taken from the women, which increases their financial security and provides them with something that they can leave to their children. Additionally, if they help their husbands with the planting of cocoa trees, women are also entitled to a portion of the land should the marriage end in divorce (IFPRI, 2002).

In the farmer survey (2005) around 15 per cent of the respondents was female, of which the majority were farm owners. There are two main reasons for this over representation of farm owners among the interviewed women (around 80 per cent).[136] First, women that assist their husbands on their farm are not regarded as cocoa farmers (but rather as the *spouse* of a cocoa farmer). A result was that most of these women were not participating in farmer group meetings (where part of the selection of the respondents took place). Second, it turned out to be very difficult to locate female caretakers. Nevertheless, some observations on differences between male and female cocoa producers can be made. For example, our survey showed female farmers produce significantly less cocoa than male farmers.[137] The small size of female operated farms, 70 per cent of our female respondents produce cocoa on a farm smaller than 9 acres, partially explains this low yield. Our survey also showed that compared to male farmers women seem to cooperate relatively little with other farmers (Table 6.8). The physical differences in strength between men and women are one possible explanation: 'I am a woman and I cannot do this heavy job. I don't have the strength to do such jobs. I would rather hire labour to work on my farm.'[138]

Table 6.8 Cross-tabulation between gender respondent and work together with other farmers; horizontal percentages

		Work together with other farmers		n
		no	yes	
Gender respondent[139]	female	65,6%	34,4%	32
	male	28,8%	71,2%	178
Total		34,4%	65,6%	210
				(100,0%)

Gamma 0,650*** (FS 2005).

In discussions on upgrading through horizontal networks and relations, joint action and collective efficiency are central concepts. Joint action can take different forms. In discussions on joint action among producers of cocoa, farmer cooperatives are emphasised, which neglects the great diversity in farmers' organizations and in the ways they represent farmers, which vary per sector, country and even locality (Wennink et al, 2007: 27).

In the previous chapter, I elaborated the different organisation forms in Ghana, making a distinction between a farmer association, the Ghanaian Coffee Cocoa Sheanut Farmer Association (GCCSFA); a large farmer union, Kuapa Kokoo Farmer Union (KKFU), which is at the same time a LBC; and a cocoa cooperative, the Ghana Marketing Cooperative Association (GCMA[140]). Fieldwork indicated that there are a few smaller cooperative societies that focus on processing cocoa into soap (the Asikuma Cooperative Farmer Society) and on the production of other crops (such as plantain and maize). When applying for credit and extension services, some farmers work together in informal groups. Participation of farmers in Farmer Field Schools (FFSs) is also considered by the farmers as working together (farmer profiles 2005). The most common form of cooperation among farmers is the labour exchange groups, also known as *nnoboa*, where farmers help each other with harvest and post-harvest practices (knowledge exchanges are a positive side-effect). The main benefit from working together in these groups is lower costs and time saving. In terms of 'ad hoc' organisation, there were also some positive experiences. Generally, with consent of the chief farmer, it is relatively easy to bring together farmers for discussions, trainings and interviews.

In 2005, around one fifth of the respondents claimed to be a member of a Farmer Based Organisation (including membership of GCCSFA, KKFU, farmer societies, and farmer groups). These different organisations did not involve *nnoboa*; the majority of the respondents in the farmer survey (65 per cent) work together in these labour exchange groups. According to the survey, the region, gender and the type of contract are all important variables that influenced whether farmers worked together informally. For example, the farmers from the Central region work together much less than farmers from Ashanti, Western region and Brong Ahafo. As I showed earlier also the interviewed women were less involved in informal types of organisation, which is especially the case in Brong Ahafo and in the Western region (Section 6.3 will elaborate on the different types of farmer organisation and the accrued benefits for farmers).

Table 6.9 Gender and working together

Heterogeneity	Categories	Number of respondents
Gender	Male	178
	Female	32
	Total	210
Working together	yes	137
	no	72
	Total	209

Source: FS 2005.

6.3 When do differences matter?

So far, I have concentrated on explaining the differences between farmers and how they relate to each other.[141] In the next chapter this information will be used to interpret the outcomes of the interventions in the cocoa sector in Ghana. In this last section of this chapter, I will now take the discussion one step further and see how these farmer characteristics and social relations relate to their enabling environment, and how they affect the extent to which individual farmers have access to services, technologies, credit and membership in farmer organisations.

The selection of farmer characteristics and the discussion of the different clusters of characteristics already showed that land is not equally accessible to all farmers. Besides access to land, the economic feasibility of cocoa farming strongly depends on access to information, technologies and training. Traditionally extension services also play an important role in providing knowledge and technologies to farmers. But the quality and the reach of these services is under pressure (interview MoFA, 2003; 2005; FS 2003). In 2005, around 67 per cent of the respondents claimed to have received outside assistance for the production of premium quality cocoa, mainly from MoFA extension services. However, these services are not equally accessible to all farmers. First of all, the region of production is very important; almost three quarters of the farmers living in the more remote Brong Ahafo region did not receive any quality assistance. Second, the position of the farmer (farm-owner versus caretaker) influences whether or not assistance is provided. Farm-owners received more assistance than the caretakers.[142] Among caretakers also the type of contract mattered, surprisingly 70 per cent of the caretakers working under an Nhwesoo contract[143] received quality assistance, against only 32 per cent of the farmers working under a Yemayenkye contract.[144] A likely explanation is that of absentee farm-owners. Another possible explanation is that for most interviewed caretakers working under a Nhwesoo contract the production of cocoa provided a substantial part of their income. In contrast, almost half of the caretakers working under a Yemayenkye contract earned only a small part of their income with the production of cocoa. The survey also indicated that exercising a strong position in the community and/or chain determined the level of outside support. Around 70 per cent of the respondents who had a strong position in the community/chain received quality assistance, against 45 per cent of the farmers who had a weaker position. Furthermore, access to quality assistance is highly linked to yield. After correcting this relationship for region, the findings show that it is strongly significant for the Western region,[145] while in other regions there was no significant correlation.

Not everybody shared the view that weak extension services are the bottleneck for good quality performance. A MoFA administered survey concluded that the reduced contact between formal cocoa extension officers and cocoa farmers did not affect the farmers' awareness of the recommended practices, rather 'the main problem is...how do we [MoFA] help farmers to adopt these technologies?' (interview MoFA, 2005).

At farmer level a distinction is made between three types of 'technology':
1   *Low Technology* – only labour input: weeding, harvesting, pod breaking, fermenting, drying and marketing; technology: cutlasses;

Table 6.10 Access to quality assistance

		Quality assistance		n
		no assistance	assistance	
region	Central region	33,3%	66,7%	31
	Western region	36,6%	63,4%	116
	Ashanti region	23,5%	76,5%	18
	Brong Ahafo	74,4%	25,6%	44
	Cramer's V 0,330***			209
position on farm	Caretaker	57,1%	42,9%	75
	Caretaker and owner	66,7%	33,3%	3
	Farm owner	33,6%	66,4%	130
	Gamma 0,442***			208
kind of contract	Nhwesoo	30,0%	70,0%	22
	Yemayenkye	67,6%	32,4%	39
	working under different contracts (both Yemayenkye and Nhewsoo)	64,3%	35,7%	14
	Gamma -0,440***			75
position in community	No significant position/status	52,9%	47,1%	111
	Moderate-strong position/status	23,1%	76,9%	13
	Very strong position/status	33,8%	66,2%	76
	Gamma 0,354***			210
education	no education	51,9%	48,1%	57
	primary school	45,8%	54,2%	90
	secondary school	24,1%	75,9%	30
	tertiary and higher education	33,3%	66,7%	4
	Gamma 0,291**			181
yield	1-5 bags	51,0%	49,0%	55
	6-10 bags	51,5%	48,5%	36
	11-20 bags	27,3%	72,7%	34
	21-30 bags	20,0%	80,0%	11
	31-50 bags	26,1%	73,9%	24
	> 50 bags	11,1%	88,9%	10
	Gamma 0,289***			170

Source: FS 2005.

2   *Improved Technology* – labour input: generally higher for the basic activities, plus additional activities such as shade and mistletoe control and pruning; technology: mist blower and pruner;
3   *High Technology* – labour input: generally higher plus additional activities relating to black pod control, fertiliser application and capsid control; technology: mist blower, knapsack sprayer and pruner (based on CREM, 2002: 12).

Picture 6.3   Knapsack sprayer for sale in an inputshop

The farmer survey showed that most farmers have access to all three types of technologies. Almost 100 per cent of the respondents use cutlasses, which is a basic tool. Mist blower and pruner were accessible to the overall majority of farmers (80 per cent for mist blower and knapsack sprayer and nearly 90 per cent for the pruner). However, these positive figures are somewhat misleading. A more in-depth analysis showed that respondents' access to these technologies significantly depends on region, gender and position in the community. For example, in the Central region only one third of the respondents used a mist blower. In this same region around half of the respondents used a knapsack sprayer. Gender also is a determining factor. Relatively fewer women than men use the more sophisticated technologies, such as the knapsack sprayer, mist blower and pruner (although these are still accessible to the majority of female respondents). The next tables illustrate significant variables for access to more advanced technologies, the mist blower and the knapsack sprayer.

In addition to technology, there are other important elements for improving yields and farmers' incomes, such as access to pesticides, fungicides, herbicides and fertilizer. In 2005, almost 90 per cent of the respondents claimed to have used some kind of input in the 2003/04 season; however the use of fertilizer was particularly low.

Most farmers replied negatively to the question whether they had a higher income than the previous year. Obviously access to inputs does not automatically result in higher incomes, as it also involves costs, especially in the case of inadequate use. Especially women reported that their economic situation worsened.

Table 6.11 Access to a mist blower

		Use mist blower		n
		no	yes	
Region of respondent	Central region	67,7%	32,3%	31
	Western region	13,4%	86,6%	116
	Ashanti	5,6%	94,4%	18
	Brong Ahafo	9,3%	90,7%	44
	Cramer's V 0,507***			209
Position on farm	Caretaker	11,1%	88,9%	75
	Caretaker and owner	,0%	100,0%	3
	Owner	26,8%	73,2%	130
	Gamma -0,496***			208
Position in the community	No (significant) position/status	22,2%	77,8%	111
	Moderate-strong position/status	38,5%	61,5%	13
	Very strong position/status	10,8%	89,2%	76
	Gamma 0,291**			210
Gender	Female	45,2%	54,8%	32
	Male	16,1%	83,9%	178
	Gamma 0,622***			210

Source: FS 2005.

Table 6.12 Access to a knapsack sprayer

		Use knapsack		n
		no	yes	
Region of respondent	Central region	44,8%	55,2%	31
	Western region	20,5%	79,5%	116
	Ashanti	5,6%	94,4%	18
	Brong Ahafo	2,3%	97,7%	44
	Cramer's V 0,336***			209
Position in community	No (significant) position/status	27,8%	72,2%	111
	Moderate-strong position/status	23,1%	76,9%	13
	Very strong position/status	5,6%	94,4%	76
	Gamma 0,626***			210
Gender	Female	33,3%	66,7%	32
	Male	16,8%	83,2%	178
	Gamma 0,426**			210
Yield	1-5 bags	24,5%	75,5%	55
	6-10 bags	29,0%	71,0%	36
	11-20 bags	14,7%	85,3%	34
	21-30 bags	,0%	100,0%	11
	31-50 bags	,0%	100,0%	24
	> 50 bags	,0%	100,0%	10
	Gamma 0,523***			170

Source: FS 2005.

The next table provides an overview of the response according to the different relevant characteristics.

Table 6.13   Perceptions on the improvement of farmers' income

		Higher income than last year				n
		I don't agree	I agree a little	I agree	I fully agree	
Region respondent	Central region	63,3%	0,0%	26,7%	10,0%	31
	Western region	41,6%	2,7%	30,1%	25,7%	116
	Ashanti	25,0%	18,8%	12,5%	43,8%	18
	Brong Ahafo	39,5%	4,7%	20,9%	34,9%	44
		Cramer's V 0,195***				209
Gender	Female	69,0%	6,9%	17,2%	6,9%	32
	Male	39,1%	3,4%	27,6%	29,9%	178
		Gamma 0,550***				210
Education	No education	50,0%	5,4%	32,1%	12,5%	57
	Primary school	43,5%	3,5%	20,0%	32,9%	90
	Secondary school	40,0%	0,0%	26,7%	33,3%	30
	Tertiary and higher education	50,0%	0,0%	50,0%	0,0%	4
		Gamma 0,166*				181
Yield	1-5 bags	53,7%	1,9%	20,4%	24,1%	55
	6-10 bags	48,6%	8,6%	17,1%	25,7%	36
	11-20 bags	38,2%	2,9%	35,3%	23,5%	34
	21-30 bags	18,2%	0,0%	36,4%	45,5%	11
	31-50 bags	37,5%	0,0%	33,3%	29,2%	24
	> 50 bags	40,0%	0,0%	0,0%	60,0%	10
		Gamma 0,199**				170

Source: FS 2005.

Access to credit is problematic for farmers. In 2005, only 25 out of 208 respondents had access to credit. All farmers faced difficulty in obtaining credit.

Besides land, extension services, technologies and credit, another important enabler for upgrading are farmer organisations. The role of farmer organisations has changed over time; processes of liberalisation went hand-in-hand with the reform of cooperatives previously controlled by the state. Just like the private players higher up in the chain, they became responsible for their own management and often privatised (Wennink et al, 2007). In Ghana, however, the situation is different. In Chapter 5 I illustrated the collapse of the cooperative movement in Ghana. The introduced reforms did not provide the incentives for institutional reforms that could have empowered farmers (Tiffen, no date).

Policy-makers abroad, as well as inside Ghana, argued that some of the problems faced by farmers could be overcome through better farmer organisation. This is however not easy to realise. In one of the group discussions with farmers, it was mentioned that in the early years farmers successfully refused to sell their cocoa when prices were considered too low. Nowadays this is no longer possible as the need for immediate cash to cover daily expenses is high and storage facilities are lacking:

> [...] we the cocoa farmers don't have a choice when we are not satisfied with the price the government is giving us; we have to accept it whether we like it or not because we need money for our children's school fees, etc. Unlike fishermen, who have cold stores, we the cocoa farmers have no storage facilities. That is why the government decides how much it pays us. (Group Discussions, 2005)

Farmers also reported other reasons, such as no need and a lack of trust. Generally farmers and other actors involved in the cocoa sector seem to agree that farmers should organise themselves and should strengthen their (often) weak organisational structures. Also there seems to be agreement on the importance to avoid imposed forms of cooperation. However, there is no agreement on how to facilitate bottom-up organisation and who should be involved in this process. Looking at other cocoa-producing countries, one sees that external donors and other (private) agents play a role in facilitating (strengthening) formal farmer organisations. In Ghana there is nothing that can compare to such activities in neighbouring countries, e.g. the multi-stakeholder 'Upcocoa' project in Cameroon.

Recently there are some new attempts at organising farmers, through Farmer Field Schools (FFSs), credit groups and soap making cooperatives. While governmental services have a top-down tendency, these alternative extension services, provided by NGOs and public-private partnerships, are generally more farmer-driven. Farmers appreciate these services, but due to the high costs involved (mainly due to expensive consultant fees) they were mostly small scale and reached only a limited number of farmers, mainly in the Central and Ashanti region. It is challenging to effectively scale-up these approaches, which are adapted to coordinate and leverage the resources of many actors, while improving their cost-effectiveness.

In order to understand how different types of organisation enable farmers to upgrade, I looked at differences in: the goal of organisation, possible constraints for the different types of organisation, the type of upgrading and the level of farmer involvement in decision-making processes (Table 6.14). An important observation is that different organisations can help farmers achieve different types of goals. Also different types of organisation are not equally accessible to all farmers.

6.4 Reflections

In this chapter I analysed a number of differences between cocoa farmers that participated in the farmer survey (FS 2005), which produced two main observations. First, there are significant differences between the respondents. For example, the farmers in Brong Ahafo, a more remote region with lower population density, had less favourable opportunities for cocoa production. In this region 50 per cent of the respondents produced only 1-5 bags of cocoa and of which almost 65 per cent

Table 6.14 Types of farmer organisation: linking upgrading to empowerment

Informal ←——→ Formal

Type of cooperation →	Ad hoc organisation	Informal group (credit/soap making)	Nnoboa (labour exchange groups)	Farmer Field Schools	Cooperative	Association
Goal	Dissemination of information/extension	Obtain credit or generate extra income	Share labour, costs and knowledge	Integrated Pest Management	Several goals: functional upgrading, price increase, empowerment of farmers (especially women)	Registration Input distribution
Constraints	No constraints	Lack of trust (between bank and farmers as well as among farmers)	Majority of farmers work together in nnoboa. Although significantly less women and caretakers	Only small number of farmers participated in pilot. FFS is expensive. Caretakers have difficult access to these programmes.	There is currently only one (semi) cooperative. Organisational costs are high and starting a cooperative demands external investments and guidance. Caretakers have difficult access to cooperatives.	Caretakers are excluded
Type of upgrading	Capturing higher margins for unprocessed commodities	Capturing higher margins for unprocessed commodities Localising commodity processing	Capturing higher margins for unprocessed commodities	Producing new forms of existing commodities	Capturing higher margins for unprocessed commodities Localising commodity processing, marketing	Capturing higher margins for unprocessed commodities (better record keeping and access to input)
Level of involvement of farmers in decision-making processes	n.a No decision making-processes	High Initiative to organise comes from farmers, they make the decisions	High Initiative to organise comes from farmers, they make the decisions	Medium Training provided at school is partly farmer-based. Through experience farmers learn to improve their farm practices. Only a small group of already innovative farmers participates.	Medium Formally farmers are the owners of the cooperative and involved in local decision-making processes. However, there is an information gap between the central organisation and its members. It is not certain to what extent the farmers' wishes are being honoured by the management of the organisation and the extent to which members benefit.	Low Formal representation of farmers is organised on district, regional and national level. But it is not certain how the farmers' views are integrated in national policies. There are some complaints against this shortcoming.

Source: Composed by author.

Picture 6.4 A field visit to farmers that have been trained as trainers in the FFS (2003)

produced cocoa on plots smaller than 10 acres. There are few extension officers that travel to this region where farms are spread out far apart. Also the number of buying agents in the villages is low. It seems that farmers producing in the Western region are better of; the concentration of cocoa production in the Western region has attracted buyers and service providers. Besides region, the analysis makes it clear that the context played a major role producing different outcomes for different groups of farmers. Both 'exclusion' and 'inclusion under unfavourable conditions' are not natural processes, but an outcome of social structures, land and shareholding systems and interventions. For example, the matrilineal system of inheritance stimulates land fragmentation. This makes it difficult for farmers to make cocoa farming a profitable business. This system hit families of small-scale farmers and caretakers the hardest

A second observation is that differences between farmers are inter-related. For example, the previous sections showed that land is not equally accessible to all farmers. This is a problem, not only because land is needed for the production of cocoa, but also because land is often requested as collateral for obtaining credit at a financial institution. Without land the participation in farmer groups is also restricted. For example, the GCCSFA issues registrations only to farm-owners. Also the members of the farmer union, Kuapa Kokoo, are primarily owners. Caretakers have to obtain permission from their land-owner to participate in the farmer union. This has consequences on yet another level because members of the farmer union in turn participate in the farmer field schools. Consequently the majority of farmers who receive training are farm owners. Caretakers, many women, migrants and younger farmers have more chance of being are left out. This is naturally problematic, but even more problematic as without training and options to raise their productivity, cocoa production can become less attractive to young farmers, which threatens the future of this important economic sector for Ghana.

In discussions on inclusive upgrading and the development of interventions, it is important to be aware of the differences between farmers and how they interrelate. Different farmers require different interventions and can respond differently to interventions. Because upgrading is a selective process it is important to understand the different impacts that interventions have on those farmers who are included and who are left out.

7
THE RISKS OF INCLUSION

7.1 Introduction

It is clear that linking competitiveness with development demands a broader and more inclusive view on upgrading. In the value chain literature there are some different notions on what this concept entails and which elements deserve to be emphasised. For example, Gibbon and Ponte (2005) highlighted that the upgrading possibilities for most producers of primary export commodities are only marginal or sometimes even completely absent. Knorringa and Pegler (2006) worried about the lack of consideration for labourers in the upgrading debate. Others (for example KIT et al., 2006; Long, 2001) emphasised the importance of involving farmers in processes of chain management, which would contribute to empowerment of producers. In recent discussions 'inclusive upgrading' has also been linked to sustainable partnerships between public, private and civil actors (Vermeulen et al., 2008). Table 7.1 illustrates some the different notions on inclusive upgrading.

Table 7.1   Notions on more inclusive upgrading

Some notions on 'more inclusive' upgrading	Authors
Reaching more producers, including more vulnerable groups (such as workers and women)	e.g. Barrientos et al., 2001; 2003; KIT et al., 2006; Knorringa and Pegler, 2006;
More equal distribution of added value in the chain and including producer/farmer view on upgrading (which can be perceived as 'sub-optimal' or marginal by actors higher up in the chain)	e.g. Long, 2001
Involving social and environmental elements; i.e. sustainable production	e.g. Abbot et al., 2004; Bolwig et al, 2008; Daviron and Ponte, 2005
Involving 'institutional upgrading' and empowerment of producers	e.g. Daviron and Ponte, 2005; KIT et al., 2006; Wennink et al, 2007.
Involving 'diversification' (both as type of risk management and strategy to increase remunerative income)	e.g. Gibbon, 2001
Contribute to the ability to create and control value	e.g. Daviron and Ponte, 2005
Sustainable partnerships	e.g. Vermeulen et al., 2008.

Source: composed by author.

In my study I integrate most of these different notions, albeit at different levels. In analysing inclusive upgrading strategies, I unravel strategies by looking at sub-strategies and interventions. The main analysis takes place on the level of interventions, where I make a difference between their: scope, impact, farmers' perspective, constraints and trade-offs. In terms of scope, I analyse the interventions by looking at the number of smallholders that they reach. But in order to analyse levels of 'inclusiveness' it is not enough to look at the number of farmers; it is also necessary to look at who exactly are included. In the previous chapter I already showed that upgrading opportunities are not equal for all farmers; for example, among my respondents, caretakers and farmers without any status were more vulnerable and generally had more difficulty obtaining access to land and services.

In terms of impact, I do not only look at competitiveness and adding value, the items that conventional approaches present as the main goals of upgrading. The reasons for this are that on the farm-level in Ghana 'competitiveness' is not a goal as such (see Chapter 5) and 'adding value' does not necessarily compensate for the costs of upgrading.[146] Therefore, when assessing the impacts I make a distinction between 'competitiveness' (and adding value) and 'remunerative farmer income'. But, because inclusion issues are closely related to levels of empowerment, I include 'empowerment' as a third type of impact in evaluating interventions. Empowerment is about vulnerable actors taking increased control over their lives and destiny. People need to exercise their 'voice' (Bebbington and Thompson, 2004). Empowerment can also result in 'self-exclusion'; some groups of farmers for example may deliberately choose to remain outside a chain or intervention (Wennink et al., 2007). In terms of impact I will make a distinction between:

1 Competitiveness or adding value (for example strategies that support farmers in meeting [new] standards, increase the farmers' margin, add value to the bean etc.);
2 Remunerative farmer income (for example strategies that support farmer productivity, efficiency in terms of cost-benefit ratios, volume of production, diversification of income, improved risk management); and
3 Empowerment (for example strategies that increase the farmers' involvement in decision-making processes, provide trainings, that increase the farmers' negotiating power, collective action, improve labour conditions, etc.).

In addition to scope and impact, I also analyse interventions on the farmer-perspective, the possible constraints and trade-offs. To explain why some strategies are (expected to be) more successful than others, I included the farmers' view on the intervention (or on the problem the intervention seeks to address). In addition, I looked at possible constraints (for example institutional constraints) that made it difficult for farmers to benefit from specific actions. In terms of trade-offs, I refer to unexpected economic, social and/or environmental tensions that the intervention generates.

In my (mainly qualitative) analysis I will use two different matrixes. To illustrate the impact of an intervention on the farmers' position in a chain *(individual level)* I employ the empowerment matrix, developed by KIT et al., (2006: 20-1). I developed the scenario matrix to reflect on the cocoa sector in its totality *(collective level)*.

In the first matrix 'empowerment' refers to intervention strategies that enable farmers to strengthen their capacity to manage chains and to be involved in various chain activities. In addition to moving up in the chain, empowerment requires the farmers to obtain economic power by participating in chain management. This matrix, which should be regarded as a 'tool for strategic thinking about chain development', has two dimensions: who does what in the chain (vertical integration), and who determines how things are done in the chain (horizontal integration). There are four distinct positions within the matrix: chain actor, chain activity integrator, chain partner and chain co-owner (Figure 7.1).

Figure 7.1   Chain empowerment matrix

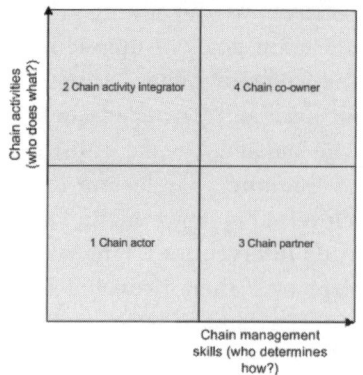

[Matrix with y-axis "Chain activities (who does what?)" and x-axis "Chain management skills (who determines how?)". Quadrants: 1 Chain actor (bottom-left), 2 Chain activity integrator (top-left), 3 Chain partner (bottom-right), 4 Chain co-owner (top-right).]

Source: KIT et al., 2006.

In this matrix there are four empowerment strategies:
1 *Upgrading as a chain actor* – the farmers become crop specialists with a clear market orientation;
2 *Adding value through vertical integration* – the farmers move into joint processing and marketing in order to add value;
3 *Developing chain partnerships* – the farmers build long-term alliances with buyers, centred on shared interests and mutual growth;
4 *Developing ownership over the chain* – the farmers try to build direct linkages with consumers.

Chain empowerment can be 'measured' by comparing the situation after an intervention with the situation beforehand, as visualised in the next matrix (Figure 7.2).

Figure 7.2   Understanding empowerment

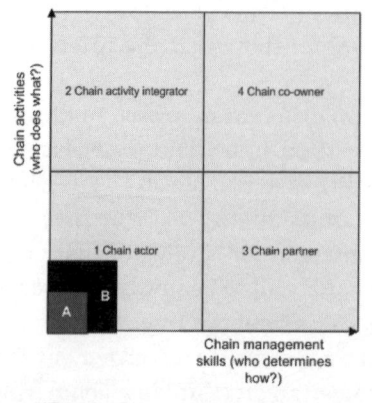

$(B - A)$ = *increase in economic rent* = *Chain empowerment*

[Matrix with same quadrants as Figure 7.1, showing positions A and B within the "1 Chain actor" quadrant.]

Source: KIT et al., 2006.

Although the matrix suggests that the ideal position for farmers is that of co-owner, that is not necessarily true. The best position (and the most effective intervention) depends on the specific context and may change over time (KIT et al., 2006: 23-4).

I use a second matrix to capture changes over time. Taking into account possible future scenarios will shed light on the current (and future) position of Ghanaian cocoa farmers in the global cocoa chain and the kind of interventions that can generate long-term benefits for the sector. This 'scenario matrix' is built around two dimensions: changes in demand, moving from 'product' to 'process' requirements, and the level of liberalisation (Figure 7.3).

Figure 7.3   Scenario matrix

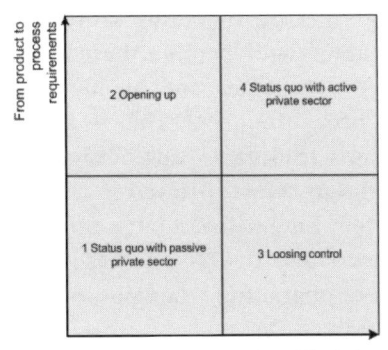

2 Opening up	4 Status quo with active private sector
1 Status quo with passive private sector	3 Loosing control

y-axis: From product to process requirements
x-axis: Level of liberalisation

Source: composed by author.

These are four different scenarios, which require different types of upgrading strategies, sub-strategies and interventions:

1   *Status quo with passive role of private sector* – This scenario reflects the current situation in Ghana, a status quo where international buyers and the Ghanaian government have a common interest in maintaining or perhaps only slightly changing the system. The sector is partly liberalised and the focus in on product quality and volume.
2   *Opening up* – This scenario reflects a shift away from primary demand for high quality cocoa beans ('product requirements') towards an increasing demand for example sustainable cocoa production ('process requirements'). This would require more transparency and enhanced levels of public-private partnerships.
3   *Status quo with active role of private sector* – This scenario is a continuation of focusing on product quality. Instead of the Ghanaian government, the private sector coordinates supply. Because the Quality Control System of Cocobod is quite successful in ensuring certain quality standards, Cocobod could continue to play a supportive role. Another option is that in this scenario QCD is privatised.
4   *Loosing control* – This scenario is the most radical, reflecting changes in demand and an increased level of liberalisation. The government is no longer in control. Marketing channels are privatised and in order to remain competitive on the world market producers have to focus on process quality instead of (only) product quality.

This matrix does not 'measure' shifts; it provides an enhanced understanding of the vulnerability of the current system by looking at changes in context. This contributes to the identification of more inclusive upgrading strategies that are (also) effective on a longer term. In the next section I will start to unravel the upgrading strategies in the cocoa sector.

7.2 Upgrading strategies, sub-strategies and interventions in the cocoa sector

In Chapter 2, I described the different views on upgrading and the way that these perceptions complement each other (by linking types of upgrading more to the process of upgrading and its outcomes). In this chapter I will focus on these linkages by looking at the impact of different upgrading strategies and by identifying structures of rewards for different groups of cocoa producers. This is a rather complex exercise; there are multiple interventions leading to upgrading, which interact with each other and are executed by different actors involved in the cocoa chain. In order to make an understandable overview, I identified a large number of interventions that affect Ghanaian cocoa producers and structured these around sub-strategies. These are in turn linked to the three upgrading strategies identified by Gibbon (2001: 352-4):

- Strategy 1: Capturing higher margins for unprocessed commodities;
- Strategy 2: Producing new forms of existing commodities; and
- Strategy 3: Localising commodity processing and marketing.

Sub-strategies for capturing higher margins for unprocessed cocoa are contributing to producing better quality cocoa, increasing productivity and the production of higher volumes of cocoa, and producing under more remunerative contracts. Sub-strategies for producing new forms of existing commodities are divided into producing for specialty/niche markets, development of non-traditional uses of cocoa and diversification into non-traditional products, and other (non-farm) income-generating activities. Sub-strategies for localising commodity processing and marketing are processing cocoa waste, processing cocoa beans and the marketing of cocoa beans. As Figure 7.4 shows, different types of actors are involved in different upgrading (sub-) strategies.

It will not be possible to discuss each (sub-) strategy or to analyse each intervention in detail. Table 7.2 bellow gives a clear overview of the interventions I identified between 2003 and 2005. In particular it was not possible to analyse all these interventions thoroughly because some of them were still in a pilot phase while for others I was not able to interview the target group. These interventions are discussed in a more descriptive way.

Appendix 7.1 provides a more extensive overview of the different interventions in the cocoa sector in Ghana and discusses their mechanism, target group, (expected) impact, identified constraints and trade-offs. Based on this exercise, I made a

Figure 7.4 Overview of upgrading strategies, sub-strategies and the involved actors

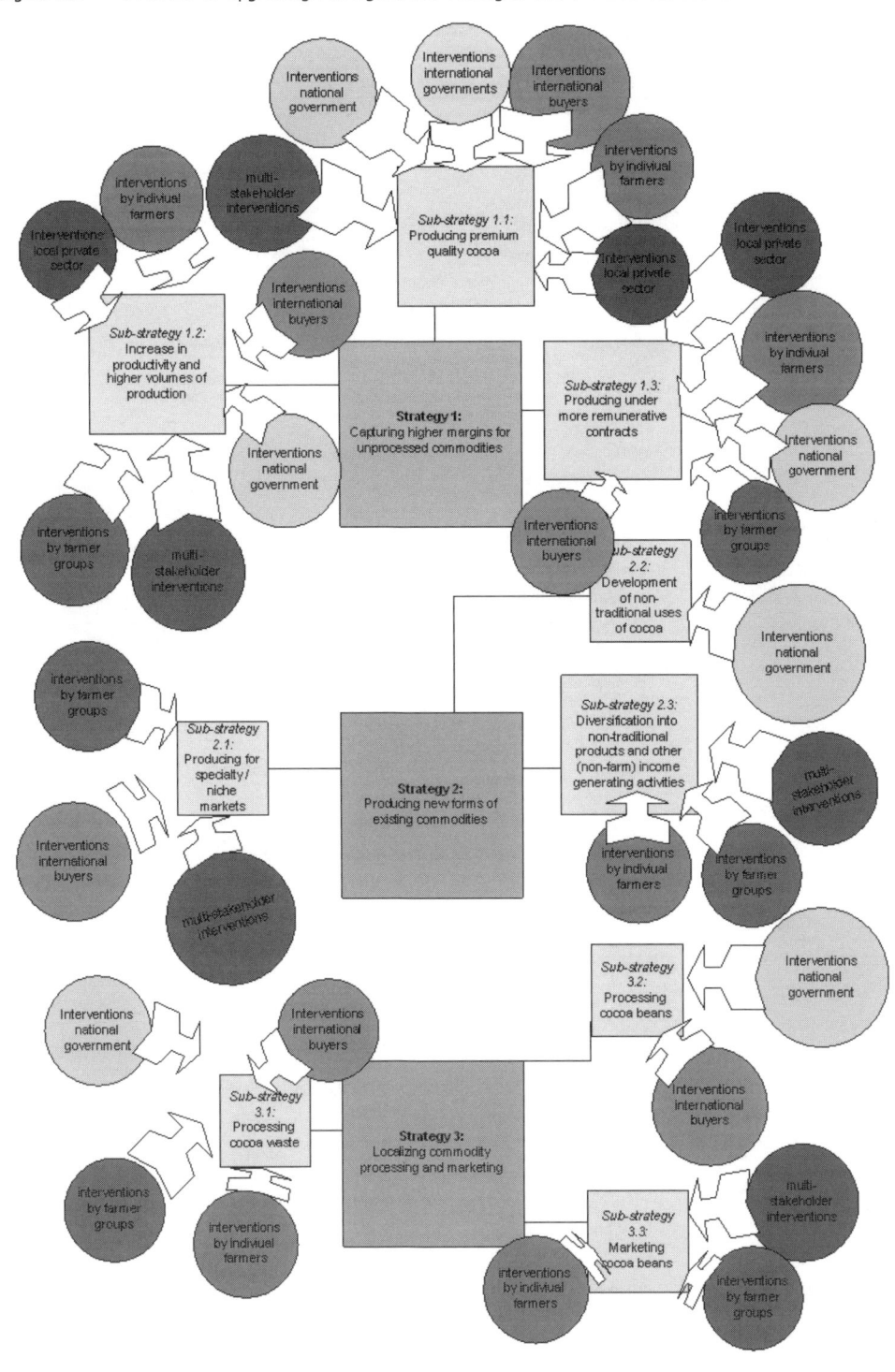

Source: composed by author.

Table 7.2   Identified interventions affecting cocoa producers in Ghana

Strategies	Initiator of Intervention (between 2003 and 2005)	Activity
STRATEGY 1: CAPTURING HIGHER MARGINS FOR UNPROCESSED COMMODITIES		
Sub-strategy 1.1 Capturing higher margins for unprocessed cocoa by producing better quality cocoa	International institutions	
	Food and Agricultural Organisation (FAO)/ European Union (EU)	Setting standards
	International buyers	Paying premium, rejecting beans
	Ghanaian government	
	QCD	Quality control
	MoFA	Extension services
	CRIG	Research/listing recommended practices
	CMC	Sanctioning LBCs
	Local private sector	Quality control, drying cocoa, training farmers
	Farmer groups	
	KKFU	Purple bean seminars, small bonus for dried cocoa
	Individual farmers	Traditional fermentation and drying practices, pre-selection of good pods/beans, pest management, selling remnant beans
Sub-strategy 1.2 Increase in productivity and higher volumes of production	International institutions	
	Research institutes (for example CIRAD)	Research
	International buyers	Research on pests and diseases, integrated pest management, new varieties, etc.
	Ghanaian government	Increase producer-price, bonuses (compensation)
	CODAPEC	Mass-spraying programme, High-tech programme (fertilizer on credit)
	Cocobod	Rehabilitation of (abandoned) cocoa farms, Infrastructure
	MoFA	Extension services
	CSSVD	Swollen shoot programme
	CRIG	Research and development of new varieties
	Local private sector	
	Wienco (input provider)	Provision of fertilizer on credit to farmer groups, combined with extension services
	LBCs/PCs	Provision of credit
	Banks	Provision of credit
	Multi-stakeholder initiatives/PPP	
	STCP	Farmer Field Schools: farmer-based extension services and training (IPM)
	CI, KKFU, MoFA and CRIG	FFSs in conservation areas
	Farmer groups	
	Informal (nnoboa)	Exchange labour and knowledge
	KKFU	Credit unions
	Ad hoc organisation	Get advice/training/access to products
	Individual farmers	Planting new varieties, Applying good farm practices, Pest management, Using fallow land, Hire more labour, Savings and apply for credit, Participation in training, Apply higher levels of technology, On-farm investment

Sub-strategy 1.3 Producing under more remunerative contracts	International institutions	
	International banks	Financing forward sales
	International buyers	Forward sales premium
	Ghanaian government	Forward sales
	Local private sector LBCs/PCs	Investment in (selection of) PCs, Prompt payment, provision of services, credit, subsidized inputs etc.
	Farmer groups Nnoboa	Informal exchange labour contracts (nnoboa)
	KKFU	Fair trade contract
	Individual farmers	Sharecontracts, Membership of KKFU, Loyalty to LBC
STRATEGY 2: PRODUCING NEW FORMS OF EXISTING COMMODITIES		
Sub-strategy 2.1 Producing for specialty/ niche markets	International institutions (for example Fair Trade Movement)	Opening of alternative marketing channels
	International buyers ADM	Installing processing facility for 'origin cocoa' from Ghana
	Barry Callebaut	Processing small amounts of organic/fair trade cocoa products
	Traders/processors/manu-facturers/retailers/consumers	Paying premium for organic/fair trade cocoa
	Multi-stakeholder initiatives/PPP	Certification schemes (labour, sustainable trade)
	NGOs (AgroEco) and Cocobod	Organic cocoa
	Farmer groups KKFU	KKFU produces for fair trade market
	Individual farmers	Become member of KKFU, Participation in organic cocoa projects
Sub-strategy 2.2 Development of non-traditional uses of cocoa	Ghanaian government CRIG	Development and marketing of cocoa by-products
Sub-strategy 2.3 Diversification into non-traditional products and other (non-farm) income-generating activities	Multi-stakeholder initiatives/PPP	In FFS attention is focused on diversification
	Farmer groups KKFU	Income generating projects for women
	Individual farmers	Inter-cropping/shade management/production of other cash crops, other activities (cf. teaching)
STRATEGY 3: LOCALIZING COMMODITY PROCESSING AND MARKETING		
Sub-strategy 3.1 Processing cocoa waste	International buyers Resigha	Buying inferior cocoa
	Ghanaian government	Research, development and marketing of cocoa by-products (soap, fertilizer, liquor, food for poultry, etc.)
	Farmer groups	Soap making
	Individual farmers	Soap making
Sub-strategy 3.2 Processing cocoa beans	International buyers	Outsourcing of processing to Ghana
	Ghanaian government	20 % discount on light-crop beans
Sub-strategy 3.3 Marketing cocoa beans	Farmer groups KKFU	KKFU is shareholder in Divine Chocolate, a chocolate marketing company based in UK
	Individual farmers	Become a PC

Source: composed by author.

selection of four interventions which are discussed in-depth in this chapter. The main selection criteria were a) number and type of farmers being reached; b) type of impact; and 3) available data. First, I will discuss two large-scale public interventions; one aimed at the production of high quality cocoa and the other at increasing the volumes of cocoa production. These interventions differ both in the type of impact and type of farmers they reach. While the quality control system reaches all farmers and helps them to be competitive on the world market, the mass spraying programme is not equally accessible for all farmers. This is a problem, because access to the spraying programme can result in higher yields and higher incomes. Second, I discuss one medium-scale multi-stakeholder initiative (which includes public, private and civil actors) namely the only formal farmer union, the Kuapa Kokoo Farmer Union (KKFU). This Farmer Union, which encompasses around 50,000 farmers and their families, produces a small share of its beans for the fair trade market. In addition to opening up an alternative marketing channel, membership in the union also empowers farmers. Third, I will discuss briefly an intervention by international processing companies, which outsourced part of their processing capacity to Ghana. This intervention has no direct impact on farmers but does contribute to the long-term demand for Ghanaian cocoa by consolidating relations between Cocobod and international processing companies.

An analysis of these interventions will provide insight in: how the interests of the different players in the cocoa chain are manifested locally, who dominates the upgrading agenda and which upgrading issues are prioritised? Furthermore, it makes it possible to highlight the strengths and weaknesses of the interventions. I will also discuss the farmers' strategies and their responses to the different interventions. Table 7.3 gives an overview of the selected strategies.

Table 7.3   Selection of upgrading strategies

Strategy	Sub-strategy	Type of intervention		Actor involved in intervention	
Scope	Main impact on farm-level				
Strategy 1	Sub-strategy 1.1: Quality	Quality control	Public intervention	Large scale (all farmers)	Impact 1: Competitiveness
	Sub-strategy 1.2: Productivity and volume	Mass spraying exercise	Public intervention CODAPEC	Large-scale (the majority of farmers)	Impact 2: Remunerative income
Strategy 2	Sub-strategy 2.1: Production for niche market	Fair trade cocoa	Multi-stakeholder initiative Kuapa Kokoo Farmer Union, Twin trading, Fair trade organisation	Medium scale (50,000 farmers)	Impact 3: Empowerment
Strategy 3	Sub-strategy 3.2: Processing cocoa beans	Local cocoa processing	International buyers	Large scale (all farmers)	No impact

Source: composed by author.

For each strategy, I will first explain what the specific upgrading strategy entails for Ghanaian cocoa producers. Second, I will provide a farmer's perspective on different sub-strategies and explain the role of the producer in their implementation. Consequently, I will discuss the selected interventions in terms of their impact on producers. In the analysis, several key questions will be raised: Who intervenes? How do farmers benefit from these interventions? Through what kind of mechanisms do they benefit? Who exactly is targeted by the interventions and who is excluded? Why are some interventions not successful or not implemented at all? How do the different interventions interact? Are there constraints and/or unexpected side-effects (economic, social and environmental trade-offs)?

In order to measure the effectiveness of public interventions (the quality control system and the mass spraying programme) and the responses of farmers, I will mainly use qualitative data (based on in-depth interviews with different actors, group discussions with farmers and 'grey' literature), combined with quantitative data, based mainly on the farmer survey held in 2005. I will mainly use qualitative data to measure the effectiveness of private interventions and of multi-stakeholder initiatives

7.2.1 Strategy 1: Capturing higher margins for unprocessed cocoa

High(er) margins for unprocessed cocoa (cocoa beans ready for storage, i.e. already fermented and dried) can be captured by producing premium quality cocoa, increasing productivity and offering a reliable supply of high volumes of cocoa, or by securing more remunerative (informal) contracts with buyers of cocoa (Figure 7.5).

Sub-strategy 1.1: Capturing higher margins by producing better quality cocoa
For a long time capturing higher margins through the production of better quality cocoa has not been considered an option as Ghana already grows the finest cocoa in the world, and is awarded a premium price (for producers there is no price-differentiation in Ghana for different quality beans).[147] However, recent developments put 'product upgrading' back on the agenda. First, there were some problems as lower-grade cocoa beans got mixed in with premium cocoa beans. Second, a small part of Ghanaian cocoa was rejected on the world market due to excess levels of chemical residues. Third, Cocobod officials have been debating the lack of price-differentiation for some time now, exploring the possibility to introduce some kind of price-differentiation for producers. This would provide the incentive to produce the so-called 'Ghana Super beans'[148] (or grade 93A).[149]

In 1963 the international standard for cocoa was forged with Ghanaian cocoa as its base (at that time Ghana was the dominant producing country) (Daviron and Ponte, 2005: 9). The Ghanaian standard has not been adopted by other cocoa producing countries, but there have been attempts to adopt the Ghanaian system of quality control, for example in Vietnam (PSOM, 2004). It is generally assumed that the main positive distinguishing characteristic of Ghanaian premium cocoa is its post-harvest quality performance, where fermentation is particularly important as

Figure 7.5   Strategy 1: Capturing higher margins for unprocessed cocoa

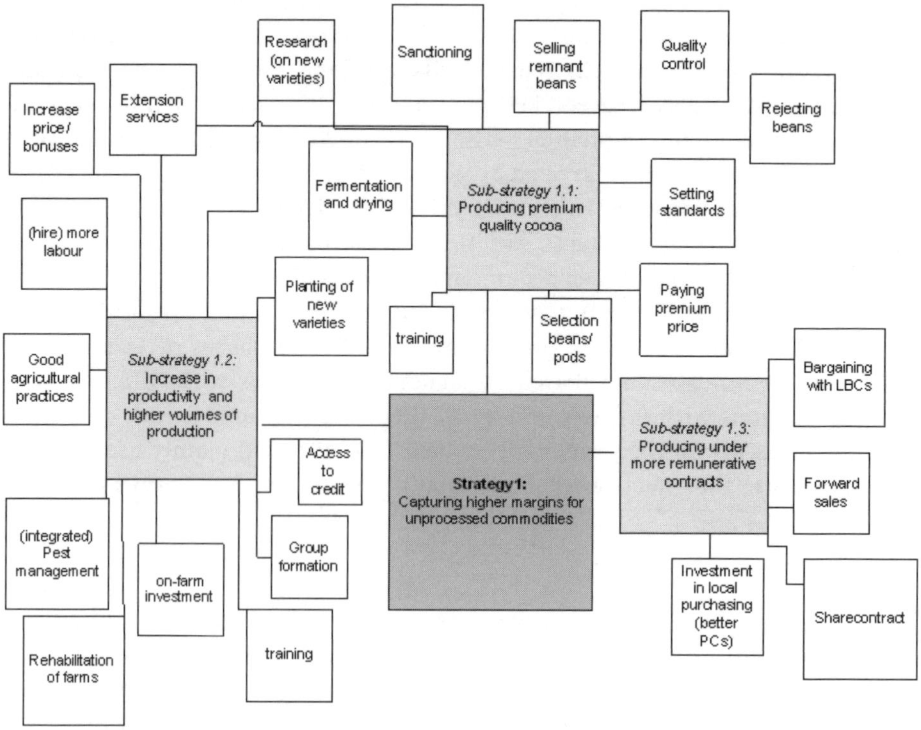

[Figure: diagram of Strategy 1 with three sub-strategies — Sub-strategy 1.1: Producing premium quality cocoa; Sub-strategy 1.2: Increase in productivity and higher volumes of production; Sub-strategy 1.3: Producing under more remunerative contracts — surrounded by related elements including Research (on new varieties), Sanctioning, Selling remnant beans, Quality control, Extension services, Increase price/bonuses, Rejecting beans, Fermentation and drying, Setting standards, (hire) more labour, Planting of new varieties, training, Paying premium price, Good agricultural practices, Selection beans/pods, Bargaining with LBCs, Access to credit, Forward sales, (integrated) Pest management, on-farm investment, Group formation, Investment in local purchasing (better PCs), Sharecontract, Rehabilitation of farms, training.]

Source: composed by author.

the cocoa flavour develops during this process.[150] Actually, there are more elements that distinguish the Ghanaian standard:

- *Higher fat content* – Ghanaian cocoa has a higher fat content, which results in higher butter yield;
- *Lower levels of defects* – Ghana cocoa has a lower level of 'moulty' (not being dried well) and 'slaty' (not being fermented well) beans than other origins;
- *Preferred flavour* – as a result of better fermentation practices and the lower level of defective beans, Ghana cocoa produces liquor with a favour preferred by some end-users;
- *Shipping weight basis* – the CMC sells cocoa on a shipping weight basis;
- *Contract performance* – the CMC has the reputation of honouring its contracts with global buyers (or of alerting buyers to problems well in advance) (Cocobod, 2003).

Recently there have been some problems with Ghana's quality performance. The 'purple beans' are threatening Ghana's reputation as *the* producer of the world's finest cocoa.[151] Even though, formally none of the 'purple beans' were rejected on the world market, nevertheless international buyers did officially warn Ghana about the increasing quantities of infected beans (personal communication Cocobod Research Department, May 2007) and reduced the price offered for Ghanaian cocoa beans for the 2004/05 season. Generally LBCs and farmers are blamed for the drop in quality; but, as already illustrated in Chapter 5, there are different views on the exact cause of the decline. The next quotes illustrate some of the farmers' perceptions of this issue and summarise their views:

- Companies are competing with each other to buy the cocoa and don't really care if the cocoa is dry enough. That also explains the decline in the quality of cocoa. I believe if the companies are strict with the rules and regulations the quality of the cocoa will improve again.
- Farmers should not be blamed [for the purple beans], we are adhering strictly to the teachings of the Quality Control Board's fermentation and drying of the beans but we still have the purple beans. (...) We don't know the cause of this. It is creating a lot of problems for farmers.
- Some years back, when our forefathers were cultivating cocoa, the quality of cocoa was one of the best. The introduction of new farming techniques has resulted in the decline of quality cocoa.
- From the teachings we had on the fermentation and drying of the cocoa we have to cover the beans with plastic sacks. We have realised that this method is not very good for fermentation. We were also advised to leave the pods for a week before we break them but that is also not working, so we are confused now because when we send our cocoa to the buying companies they refuse to buy because the beans, they say, are not dried well. This has created a lot of hardship for farmers.
- The crisis in Côte d'Ivoire has brought hardships to farmers in Ghana, some of the cocoa farmers in Côte d'Ivoire brought their cocoa down to Ghana to sell and as a result their cocoa got mixed up with cocoa from Ghana and this affected the quality of the cocoa from Ghana. (...) that is why we have not received payment for our produce.
- The decline of the quality of the cocoa could also be attributed to the method used to dry it. Some new companies advise to dry the beans for three days instead of the six days the government proposes. I think the government should institute measures to make sure the beans are well dried for six to seven days. Six years ago farmers were drying the beans for six days and there was no problem with that, so I will suggest we stick to the six days (farmer profiles 2005; group discussions 2005).

These views illustrate not only the different perceptions of farmers and their confusion about the exact causes of the decline, but also their knowledge on developments in the sector and their responses.

Another critical issue with quality performance is meeting international standards on excess levels of pesticide residues. These violations occur due to inadequate extension services and the 'widespread, excessive and abusive use of unapproved (not recommended) pesticides to protect the cocoa crop from insects, pests and diseases'.[152] This takes place both during the growth cycle on farms as well as during cocoa storage. In Ghana the Cocoa Research Institute Ghana (CRIG) (a Cocobod subsidiary) is responsible for informing the farmers on the use of chemicals, for recommending appropriate remedies and for assuring that Ghanaian cocoa beans are accepted on the world market. In some cases this is not enough, as legislation can change overtime. For example, after Japan introduced more restrictive legislation on maximum residue levels it consequently rejected Ghanaian beans.[153]

In response to these problems with quality performance, in 2005 Cocobod issued sanctions against LBCs. Some LBCs temporarily stopped buying cocoa from farmers, which directly affected the farmers' income. This (partial) shifting of risk from the government to the private sector and farmers contributed to growing tensions and mistrust between Cocobod, LBCs and farmers. Still, compared to other cocoa producing countries in the region and their decline in quality, Ghana's cocoa is still considered as (relatively) good quality cocoa by buyers and they still offer a premium price.

A farmer perspective on quality aspects and strategies to upgrade their beans
In discussions on quality, the Ghanaian government and international buyers of cocoa mainly look at the quality standards of the beans. Farmers have a different perspective on quality and worry more about crop losses that take place earlier in the chain. The next Figure (7.6) indicates the main reasons for the quality decline, comparing season 2003/04 with season 2002/03, as perceived by farmers that participated in the survey held in 2005[154] (FS 2005).

Figure 7.6   Main reasons for the quality decline in 2003/04 – a farmer perspective

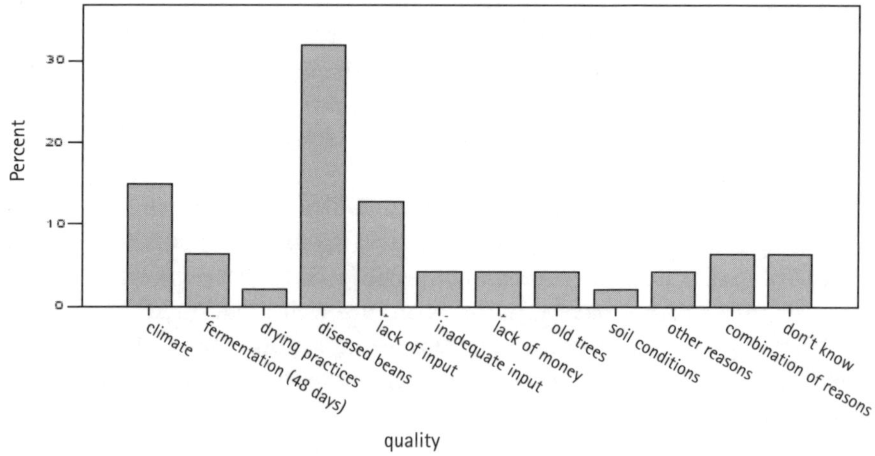

Source: FS 2005.

According to farmers 'diseased beans' was the main cause for quality decline, followed by unfavourable weather conditions and the lack of input. From the perspective of farmers, the best way to avoid further losses is through more effective pest management. Many cocoa-producing countries face this problem of considerable 'crop losses'. It is estimated that worldwide at least one-third of cocoa production is lost to pests and/or diseases every year.[155] The largest part of this loss is discovered and dealt with at the farm-level. Farmers interviewed in Ghana indicated that for cocoa seasons 2002/03 and 2003/04 a considerable part of the cocoa pods was affected by diseases. They pre-selected the bad pods and beans from the good ones. Of the good pods, around one fifth of the beans were of inferior quality, mainly due to pests and diseases (FS 2005). This pre-selection, which takes place prior to official procedures, obviously contributes to the high uniform quality of Ghanaian cocoa. However it implies additional costs for farmers, which are higher for larger farmer who have to hire labourers to perform this task.

The different perspectives on good quality performance show that 'product quality' is not always easy to separate from 'process quality'. The farmer perspective informs that there are various options for 'good quality performance'. There are alternatives to focusing only on the quality of the selected beans (by making sure adequate fermentation and drying takes place) and the excess levels of residues (such as reducing the number of discarded cocoa pods and beans through more effective pest management). These contrasting views can cause tensions. From a farmer perspective, pest management is the way forward, while for international buyers and their governments excess levels of pest residues are problematic and result in their rejection of Ghanaian cocoa.

Other interventions that contribute to the production of premium quality cocoa beans

Different interventions, taking place at different levels, aim to advance the production of premium quality cocoa. In the international market place, international institutions, such as the FAO and EU, are involved in standard setting. International buyers determine the premium price paid for good quality cocoa. Moreover, international buyers can reject beans if they do not meet the prescribed quality standards. Nationally, Cocobod and its subsidiaries control the quality of cocoa, provide information and extension services. This enables farmers to produce increasing quantities of premium quality cocoa. The previous chapters made it clear that extension services worsened since the introduction of reforms. In response to recent problems with quality performance, Cocobod started using sanctioning in order to avoid further down-grading of the product and in order to protect its good reputation. Local buyers of cocoa have also become involved in quality control procedures. Purchasing Clerks (PCs), hired by Licensed Buying Companies (LBCs) to buy cocoa in the communities, have been given the responsibility by QCD to conduct the first official quality check. Because of the problems with the quality of the beans, local buyers have been accused of 'downgrading' Ghanaian quality standards, by buying cocoa which is not adequately dried and fermented. PBC, the largest buying company, responded to this accusation by providing training to a number of their suppliers.[156] In addition, LBCs intervene in quality control procedures by involvement in drying the cocoa they buy (normally farmers dry their cocoa themselves). For LBCs this is a good way to compete with other LBCs, as well as a way to overcome problems of 'mouldiness'. For farmers it is also beneficial as they save the time and labour they would normally expend on drying the beans. Because LBCs compete on volume, they pay farmers the same price for the 'wet' cocoa as they would pay for well dried cocoa. This service provided by LBCs is especially lucrative for larger farms with higher labour costs.

In countries that have fully liberalised cocoa marketing and pricing systems (such as Côte d'Ivoire) the organisation of cocoa farmers in cooperatives is regarded as a way to safeguard quality standards. In Ghana, this was one of the main reasons to set-up farmer cooperatives in the 1920s (see Chapter 5). But, nowadays, formal organisation among Ghanaian cocoa farmers is scarce and has other goals. Only the Kuapa Kokoo Farmer Union (KKFU), the only major cocoa farmers' union in Ghana that at the same time functions as a LBC, paid some attention to the issue by organising 'purple beans seminars' in season 2003/04. In addition, some of its members received a small extra bonus for thoroughly dried cocoa.[157] Individually, farmers do play a key role in the production of premium quality cocoa. They contribute to good quality performance by applying traditional drying and fermentation techniques, by pre-selecting beans and by applying pest-management measures to reduce crop losses.

Missing interventions

Surprisingly, in Ghana there is no price-differentiation for the different categories of 'accepted' beans. At present, for 'premium cocoa' farmers receive an annually fixed percentage of the producer price; they can sell inferior cocoa at a very low price, equalling around 2 per cent of the producer-price, to the processing company Resigha[158] (see next section). Another intervention that can safeguard quality but is not yet implemented is to assist farmers in their effort to set-up producer organisations and to provide capacity building training to existing farmer groups. While in other cocoa producing countries there is already supportive legislation in place; the Ghanaian government is very hesitant to take similar steps (see Chapter 5 and 6).

In the next paragraph I will analyse one of the main governmental strategies that secures the production of premium cocoa – a good quality control system.

Intervention – the public quality control system

The choice for gradual liberalisation was linked to the lessons learned from the negative experiences of other cocoa producing countries. There the privatisation of the system of quality control, together with a lack of professionalism among new local buyers, resulted in considerable losses in quality. Consequently, in Ghana Cocobod held on to its public quality control system, although some changes were introduced. In the past QCD held five quality inspections before exporting the beans; now the number is reduced to three: inspections at the up-country store (Picture 7.1), at the take-over point from the LBCs to the CMC, and at the point of export (Cocobod, 2003). As part of this last check QCD carries out 'fumigation and disinfection' of beans to ensure that only cocoa beans free of insects are exported. Additionally, to prevent damage to the stored beans rodent control takes place in all cocoa storage facilities (GAIN Report, 2005: 5).[159]

Prior to these inspections, at the community level PCs used to check the quality at their buying stations: 'they are to buy cocoa which is thoroughly dry, of uniform bean sizes, not slaty, not germinated or broken, and no evidence of adulteration'.[160]

Picture 7.1   QCD personnel is checking cocoa quality at a buying depot in Dunkwa (Central Region)

Picture 7.2   QCD officials take samples out of cocoa bags

After this step, in the buying-depot at the district level, a representative of QCD checked the quality, by taking a sample of hundred beans out of each bag (see Picture 7.2).

The cocoa beans are graded according to bean size and strict international quality standards, established by the FAO. These standards are mainly based on adequate fermentation and drying of the beans. There are three Grades: Grade 1, Grade 2 and sub-standard cocoa. The rest is waste. A distinction is made between 'main crop' and 'light crop' beans, showing both distinctive patterns of bean size. Most main crop beans are large and receive a premium price on the world market, while most light crop beans are medium and sold at a 20 per cent discount to local processing industries (Ministry of Finance, 1999: 66-7). Small and remnant beans can be sold directly to Resigha. In terms of capturing higher margins for unprocessed cocoa this option is a very marginal way of upgrading.

Evidently, the cocoa producers are the main actors who determine cocoa quality and perform the first (unofficial) quality check. Local buyers and QCD officials, the Ministry of Food and Agriculture (MoFA) and the Cocoa Research Institute Ghana (CRIG) are all involved in the provision of extension services to make sure only recommended chemicals are applied. In addition, CRIG conducts cocoa research on quality related issues (Figure 7.7).

Figure 7.7   Local quality aspects in the production and control of premium quality cocoa

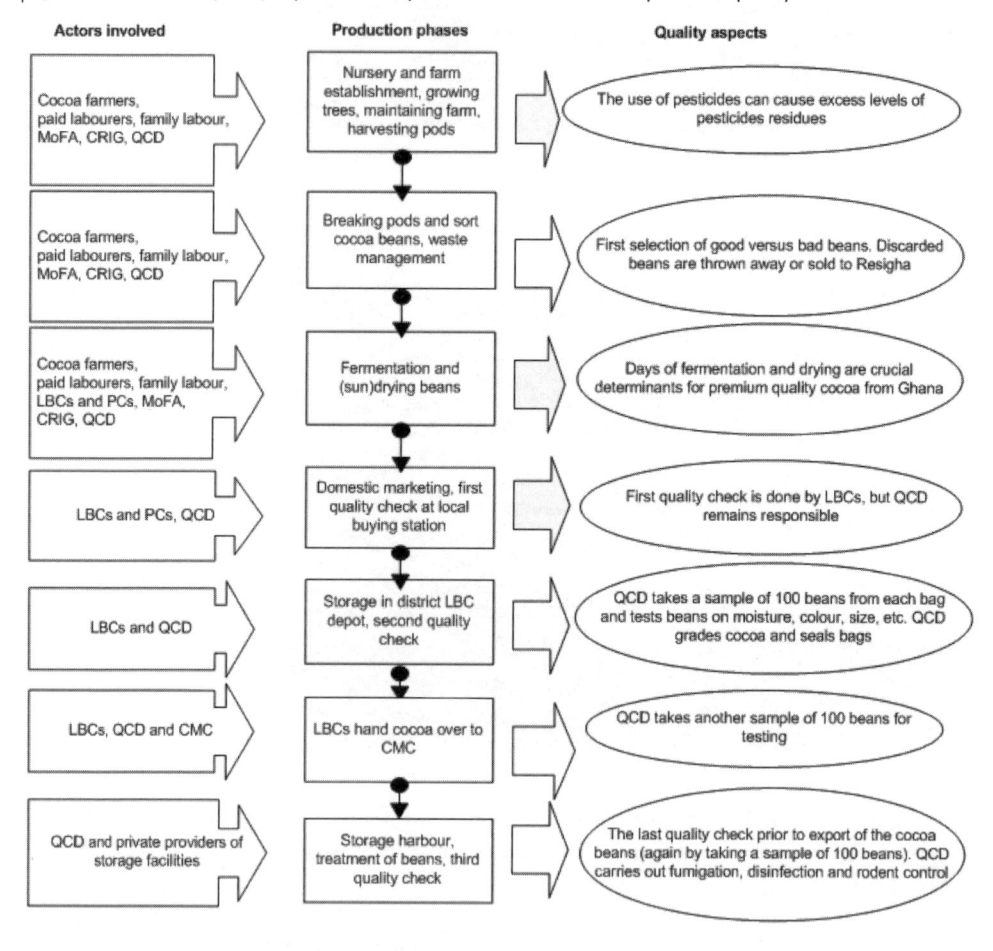

Actors involved	Production phases	Quality aspects
Cocoa farmers, paid labourers, family labour, MoFA, CRIG, QCD	Nursery and farm establishment, growing trees, maintaining farm, harvesting pods	The use of pesticides can cause excess levels of pesticides residues
Cocoa farmers, paid labourers, family labour, MoFA, CRIG, QCD	Breaking pods and sort cocoa beans, waste management	First selection of good versus bad beans. Discarded beans are thrown away or sold to Resigha
Cocoa farmers, paid labourers, family labour, LBCs and PCs, MoFA, CRIG, QCD	Fermentation and (sun)drying beans	Days of fermentation and drying are crucial determinants for premium quality cocoa from Ghana
LBCs and PCs, QCD	Domestic marketing, first quality check at local buying station	First quality check is done by LBCs, but QCD remains responsible
LBCs and QCD	Storage in district LBC depot, second quality check	QCD takes a sample of 100 beans from each bag and tests beans on moisture, colour, size, etc. QCD grades cocoa and seals bags
LBCs, QCD and CMC	LBCs hand cocoa over to CMC	QCD takes another sample of 100 beans for testing
QCD and private providers of storage facilities	Storage harbour, treatment of beans, third quality check	The last quality check prior to export of the cocoa beans (again by taking a sample of 100 beans). QCD carries out fumigation, disinfection and rodent control

Source: composed by author.

The quality control system in Ghana ensures the export of traceable high quality cocoa. After the first quality check of the QCD, the bags are sealed and stamped with a seal that indicates the grade, size, the buying station of the respective LBC and the person responsible for the quality control (interview with quality control examiners, 2003). In other cocoa producing countries quality control occurs much higher up in the chain, making it difficult to trace back the cocoa all the way to the source (Cocobod, 2003).

How do farmers benefit from this public intervention?
Cocobod sees the existing institutional framework for quality control as its main tool for securing the production of premium quality cocoa. Even if the costs for (maintaining) the production of premium quality cocoa are high, it has been recommended that the final quality control should remain in the hands of the QCD (Ministry of Finance 1999: 70). The international buyers of cocoa support this stance. Farmers also benefit from the export of premium quality cocoa. Ghana's ability to sell quantities of consistently good quality cocoa contributed to its good reputation and favourable contracts with overseas buyers and international banks.[161] As a result, Ghanaian cocoa farmers have no difficulty to sell their produce for a stable annually fixed price.

For the production (and consistent delivery) of uniform premium quality cocoa beans, Cocobod receives an extra high price, ranging from an extra €40-60 per metric tonne[162] (between four and six per cent of the producer-price) (GAIN Report, 2005). The cost of grading and certifying a tonne of cocoa is estimated by QCD officials at around 10€ per metric ton (6.000 cedis per bag).[163] It is not clear what part of the premium (if any) reaches the farmers. NGOs and researchers working in Ghana have stressed the importance of conducting further research on this topic.[164] We have seen that the farmers bear the costs of pre-selecting pods and beans. Furthermore, the risks involved in the production of premium quality cocoa are unevenly distributed among the different actors in the chain. When the quality deteriorates (e.g. the purple beans problem) and the quality control system fails to intercept inferior beans, the risks are partly shifted towards the buying companies and the farmers (hitting farmers the hardest). Fragmentation of extension services also produced conflicting advice on quality issues.

The farmers always have the option to sell their remnant beans to a private buyer. Although the price for this type of beans is very low, it is an attractive opportunity for some additional income. In season 2002/03 less than 10 per cent of the farmers reported that they sold their remnant beans, while in 2003/04 this percentage multiplied manifold to around 43 per cent for this same group (FS 2005; FS 2003). Farmers that participated in the survey in 2005 did not equally take advantage of this option to sell to Resigha. Almost half of the farmers with a 'very strong position/status' in the community/chain (such as chief farmers)[165] sold inferior beans to Resigha, while for farmers who held no position this percentage was only 31 per cent.[166] Location also turned out to be significant. Farmers from the Central and Western Region made more use of this option than the respondents living in Brong Ahafo and the Ashanti Region. A possible explanation for this is that Resigha's central depot is in Winneba (located in the Central Region, but also easily accessible from the Western Region).[167]

In summary, the quality control system is central in helping the Ghanaian cocoa farmers to remain competitive on the world market with this premium quality product and to distinguish themselves from other cocoa producing countries. This intervention reaches all farmers, but because there is no price-differentiation for different qualities farmers do not have the choice to produce lower quality cocoa (for example by selling unfermented beans) and save time for other activities. Due to a lack of transparency in recording the exact costs and benefits, it is not possible to determine whether the quality control system contributes directly to increasing farmer incomes. It is clear that it contributes to the stability of their income.

In terms of empowerment this intervention has no impact. Farmers have no control; they are not involved in setting quality standards; and the quality control system does not affect their negotiating power. The quality control system does not change the position of the farmer in the chain.

Constraints and trade-offs of the quality control system
A positive side-effect of the quality control system is that it makes it possible to trace the Ghanaian bean (prior to bulk transport) back to the local buying centre. Another positive side-effect is that the QCD system contributes not only to Ghanaian cocoa obtaining a premium price but also reduces cost for international buyers. Both aspects contribute to the long-term demand for Ghanaian cocoa.

However, there are also some constraints and negative trade-offs. First, the fact that competition is based on volume provides little incentives for local buyers to be very strict on the quality control of the beans they buy locally. This also reduces the incentives for farmers to follow labour-intensive traditional fermentation and drying practices. A lack of incentives is also problematic for quality control officials. On several occasions (mainly during informal discussions with local buyers) it was mentioned that these officials are sometimes corrupt and allow inferior cocoa to enter the market. Farmers are not only lacking substantial incentives to produce high quality cocoa beans but in addition they are also confused regarding the mandated farming practices. The fragmentation of extension services is primarily to blame for the contradictory advice on how to produce good quality cocoa. A final negative aspect of the quality control system geared at delivering only premium cocoa is that farmers shoulder a large share of the costs and risks (Table 7.4).

The central question remains: why does Cocobod want so strongly to maintain control over the quality of cocoa and is reluctant to contract this out to the private sector? The main reason is the negative experiences in other countries where quality control was privatised, which convinced Cocobod that quality control should remain a public affair. There are also underlying reasons that are more difficult to identify. For Cocobod additional reforms would pose a threat to their powerful position. It is likely that privatisation of the quality control system, especially if successful, will be a catalyst for change and result in Cocobod losing control over its other subsidiaries, such as the Cocoa Marketing Company. This is likely to threaten the privileged position of many officials involved in the cocoa sector.

Sub-strategy 1.2: Increase in productivity ane higher volumes of production
Next to 'product upgrading', another way to capture higher margins for unprocessed cocoa is by upgrading the production process. In Ghana, cocoa is

Table 7.4 Measuring the inclusiveness of the public quality control system

Strategy 1: Capturing higher margins for unprocessed cocoa	Intervention	Activity	Mechanism	Number of farmers reached	Expected impact	Constraints	Trade-offs
Sub-strategy 1.1. Capturing higher margins for unprocessed cocoa by producing better quality cocoa	Ghanaian government QCD	Quality control	Control (compulsory)	All farmers	Impact 1: Meeting standards creates access to world market. Impact 2: Due to a lack of transparency it is not sure if farmers benefit from the premium. The quality control system does contribute to more remunerative contracts with international buyers. Impact 3: This intervention does not contribute to empowering farmers.	Little incentives for local buyers (PCs), who are responsible for first quality check to be strict on quality control, which also reduces incentives for farmers to follow traditional fermentation and drying practices. Corruption among QCD officials. Fragmentation of extension services: farmers receive little and sometimes conflicting advice. Quality control system is not directed towards prevention of inferior cocoa.	+ QCD system contributes to traceability of cocoa. QCD system contributes to receiving a premium price for Ghanaian cocoa. Cost saving for international buyers. − Quality control system is expensive. There is no transparency on distribution of costs (and benefits). Farmers shoulder a large share of the costs and risks for delivering only premium cocoa to the market.
Measuring inclusiveness	The public quality control system involves all Ghanaian cocoa farmers and helps them in acquiring access to the world market. Nevertheless, in terms of inclusiveness the system is not optimal. The quality control system is compulsory and offers no stimulating incentives; farmers are not involved in standard-setting; farmers have no choice (there is no price differentiation) and the system does not enhance the farmers' capacity. The lack of transparency regarding the distribution of costs and benefits undermines to some extent the effectiveness of the system. In addition, the incentives for different local actors to support the system seem to be diminishing, this results in quality losses.						

cultivated on some 1.6 million hectares in Southern and Central Ghana. About three quarters of output is produced by around 700,000 small-scale farms (Ton et al., 2008: 5). As the political and economic value of cocoa is very high for the Ghanaian government, it is one of the main initiators of programmes that aim to increase the productivity and increase the volumes of cocoa produced.

Cocoa has been dominating the political economy of Ghana since 1920s. There was one serious blow in the late 1960s when cocoa production plummeted (Figure 6.7), partly due to the outbreak of Cocoa Swollen Shoot Virus Disease (CSSVD), which was 'cured' by cutting down the sick trees. In the 1970s the government attempted to again increase scale of production by growing cocoa in plantations. This was not successful because of problems with land acquisition and the scarcity of labour (Ministry of Finance, 1999: 9). The government also attempted to rehabilitate cocoa areas in the Eastern Region and in the Ashanti Region, with loans from the World Bank. The first two attempts were unsuccessful. But, when the producer price almost doubled in 1987 (as part of Ghana's Economic Recovery Programme [ERP]),[168] farmers seemed to have enough incentive to return to their abandoned cocoa farms. For Ghana as a whole it is estimated that for the period 1990/91 to 1997/98 the harvested area increased by 73 per cent (from 707,000 ha to 1,220,000 ha) (Teal and Vigneri, 2004: 8).

In Chapter 5, I already pointed out that fluctuations in producer price have an impact on the volume of production of cocoa. But the way in which cocoa farmers respond to prices is generally complex. According to Anim-Kwapong and Frimpong (2004) farmers respond to price by changing the intensity with which they tend their farm (for example they stop with maintenance and with new planting activities). Conversely, if prices cover or exceed variable costs farmers will intensify farm management (for example by investing in harvesting, weeding and the use of inputs). Price responsiveness to price fluctuations is usually delayed (Anim-Kwapong and Frimpong, 2004).[169] Figure 7.8 illustrates the causality between price fluctuations and volume of production and the delay in response.

This causality was the main reason behind the introduction of gradual reforms in the cocoa sector in the 1990s (Ministry of Finance, 1999; Fold, 2002). The government's ambitious production targets were met (Chapter 5), however not only due to the price increase (see also Takane, 2002).[170] There were also other explanations. A key-factor that boosted cocoa production, leading to record outputs in seasons 2003/04 and 2005/06, was the product life-cycle of cocoa (increasing and later decreasing cocoa yield after establishing a plantation) (Ruf, 2007a). As already mentioned in the previous chapter, smuggling from Côte d'Ivoire (and the end of smuggling from Ghana into neighbouring countries) also contributed to the increase in the volume of cocoa beans exported.

Although the volume of production increased, productivity remained relatively low. In the period 1990/91 to 1997/98, production increased only with 37 per cent, indicating a reduction in productivity of 21 per cent (FAOSTAT Database in Teal and Vigneri, 2004: 12). This was already pointed out in Chapter 6. According to their research, there is no evidence that reforms brought innovation in techniques.

Figure 7.8   Cocoa production and prices in Ghana

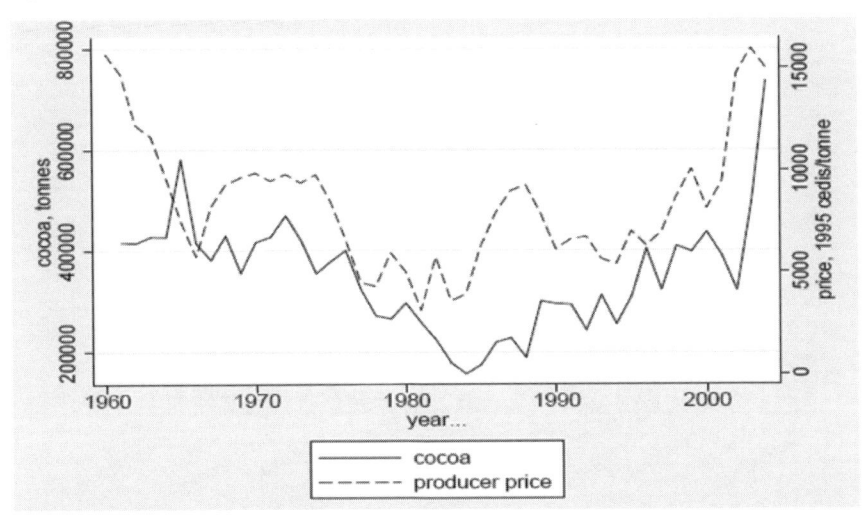

Source: ICCO 2003/04 (in Teal et al 2004).

Low productivity is also linked to the high costs involved in cocoa production. Pest management is relatively expensive and farmers have to pay world-market prices for chemicals. Farmers complained that the domestic cocoa price does not even covering all their production costs (interview MoFA, 2005; Mehra and Weise, 2007; FS 2005). Furthermore, farmers lack access to credit and have little savings, key constrains on investments. Low productivity is also linked to poor extension services and high levels of illiteracy, simply farmers do not know how to adequately apply the chemicals. Other causes include old trees, low tree density, reduced soil fertility, the type of varieties that dominate the tree stock, and the small scale of operations.

Farmer perspective on process upgrading
As explanation for the production increase in season 2003/04, farmers reported good farming practices as the main reason, followed by the additional use of chemicals, and third, the implementation of public mass spraying programme (Figure 7.9).

Examples of 'good farm practices' include regular weeding, pruning, cutting mistletoes, managing direct sunlight, adequate planting of trees, and others. Good farming practices and the use of farming input require access to equipment and labour. Farmers perceived the extra use of pesticides, fungicides, herbicides and fertilizer as one of the main reasons for increases in production. Despite the increase in the price of chemicals, the application of chemicals is growing in Ghana, partially due to their improved availability and due to interventions of Cocobod (mass spraying programme). In case farmers bought the inputs themselves it has been argued that the resulting increases in production-costs have led to 'self-exploitation' among farmers (interview MofA, 2005; Blowfield, 2003).

Sadly, the majority of farmers have had regular exposure to harmful chemicals because they neglect to use protective clothing. More than 35 per cent of the farmers

Figure 7.9   Main reasons for improved yield in cocoa season 2003/2004

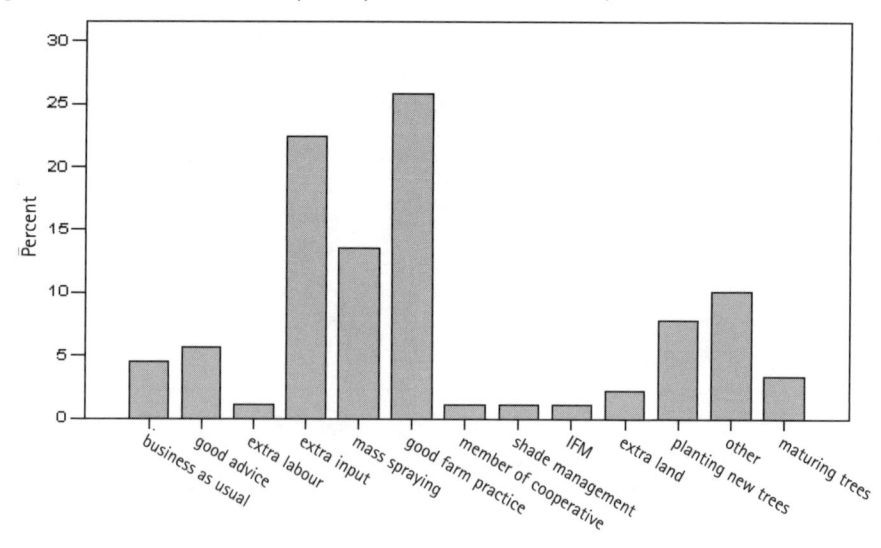

[Bar chart showing percent for categories: business as usual, good advice, extra labour, extra input, mass spraying, good farm practice, member of cooperative, shade management, IFM, extra land, planting new trees, other, maturing trees]

Source: FS 2005.

that participated in the farmer survey held in 2005 reported that they sprayed Gammalin for capsid and termite control; one of Gammalin's active ingredients is Lindane 'which is on the forbidden list of Persistent Organic Pollutants (POPs) (CREM, 2002: 32).[171] Almost 55 per cent used Kocide, which is a copper fungicide with acute toxic effects. Ridomil, used by 7 of the respondents, is only slightly toxic (FS 2005).

In Ghana the fragmentation of extension services and the privatisation of input distribution resulted in more providers of services and inputs. Although there is no firm data on changes in farmers' input use over the years, a number of farmers complained that they receive conflicting advice.[172] While the importance of following the advice of the national cocoa research institute is recognised by cocoa producers (FS 2005), the (informal) input providers can pose a problem as they mislead farmers by selling them the inappropriate (or even forbidden) chemicals (black market) (personal observations and *Daily Graphic* 15 December 2005). In these cases, the illiteracy of the farmer greatly contributes to inefficient application.

Farmers indicated that the use of toxic chemicals caused health problems and affected the environment. Health problems are exacerbated due to the limited use of protective clothing. In 2005 only twenty per cent of the interviewed farmers protected themselves while sprayings on their farm.[173] Inadequate spraying makes the negative impact of spraying even worse. Almost all the respondents sprayed their entire farm (FS 2005), even though fungicides can be locally applied. Also the timing of the sprayings is not always optimal, which affects the productivity of the cocoa farm. Inadequate spraying is particularly problematic if the chemicals are bought on credit. If the spraying did not result in higher yields, it may be difficult for farmers to pay back the loan. 'Mass spraying' was indicated as another important reason for the increase in production. The public mass spaying programme will be analysed later in this section.

Producers who experienced a decline in production in season 2003/04 (44 per cent of the respondents) reported pests and diseases (primarily Black Pod) as the main causes (Figure 7.10).

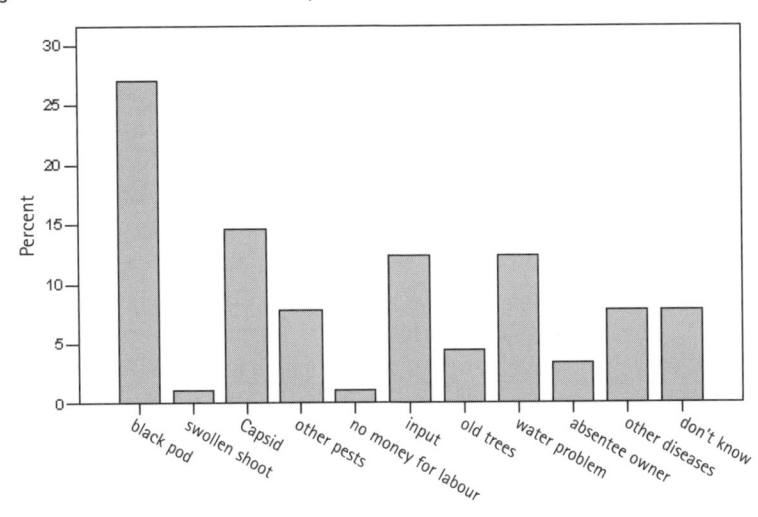

Figure 7.10   Main reasons for the production decrease in season 2003/4

Source: FS 2005.

Farmers are not idle, they take (both individual and collective) actions to remedy the challenges and to improve their production process. Almost 70 per cent of the respondents replied that they help each other with weeding, breaking pods and carrying fermented cocoa. The perceived positive effects of working together are primary time-efficiency and lower production costs; also the farmers report knowledge sharing as a positive side-effect (FS 2005). I already indicated that for my respondents the location, type of contract and gender influence whether they work together or not. Farmers who did not work together in these informal groups generally did not provide a concrete justification. Nevertheless, of the reasons provided two were mentioned most often: lack of trust and a lack of incentives to cooperate (FS 2005).

Other interventions that aim at increasing productivity and the volume of production

In Ghana, the majority of the interventions in the cocoa sector aim at increasing the volume of production and productivity levels. International (and national) research institutes and international buyers are both involved in research, for example researching new ways of combating pests and diseases or developing new (for example more resistant) crop varieties. However, the Ghanaian government is the main intervener; it set increased production volume and improved quality as the key priorities for the future development of the sector. I already mentioned some of the public interventions, such as the increase in producer price and the rehabilitation of abandoned cocoa farms. During the first two attempts at rehabilitation, in total

28,000 hectares of cocoa with high yielding varieties were replanted; however, the farmers were not interested to tend to their farms (Amoah, 1998). In 1987, the government was more successful with its Cocoa Rehabilitation Project and reached its objective, namely to increase cocoa production and yield, stabilising output at an annual level of about 300,000 tonnes. I have no information on which type of farmer was targeted by this intervention.

Another incentive given to farmers is government bonuses. This unique institutional arrangement supports the access of producers to rising world-market cocoa prices. Cocobod reinvests part of its marketing margin back into the cocoa sector, giving farmers incentives to remain involved in cocoa production and to increase their volume of production. This bonus is an outcome of the yearly recalculation of margins and prices by Cocobod (Ministry of Finance, 1999). This bonus is distributed by the Cocobod through the LBC's and Purchasing Clerks. The total payments made to farmers between cocoa season 2000/2001 and 2005/2006 are presented in Table 7.5. The individual bonuses paid to farmers for cocoa season 2002/2003 and 2003/2004 was between 1 and 2 dollars per bag (FS 2005). Generally, for farmers this amount seemed to be disappointing, especially in light of the high expectations that the government raised, stressing the important contribution of cocoa farmers to the Ghanaian economy.[174]

Table 7.5   Total bonus payments to farmers 2000-2006

Season	Amount in cedis	Amount in US$
2000/2001	¢70.1 billion	6.79 million
2001/2002	¢41.5 billion	4.12 million
2002/2003	¢157.9 billion	15.30 million
2003/2004	¢161.2 billion	15.62 million
2004/2005	None	17.26 million
2005/2006	¢178.2 billion	58.99 million
Total	¢608.9 billion	118.18 million

Source: Ministry of Finance, 2007.[175]

Access to this bonus is not without costs; the associated cost with fulfilling the administrative regulations make it less feasible in more remote areas. Furthermore, some farmers seem to have more difficulty in getting the bonus, e.g. respondents in the more remote Brong Ahafo region, the farmers without any social status and the caretakers[176] (FS 2005; Figure 7.11). At the moment of conducting the interviews (winter 2005) around 12 per cent of the farmers had not received a bonus for seasons 2002/2003 and 2003/2004, while almost half had received only one of the bonuses in these two cocoa seasons. These findings reflect that some of Cocobod's reinvestments into the cocoa economy clearly prove ineffective.

The establishment of the Cocoa National Cocoa Disease and Pest Control Committee (CODAPEC) in June 2001 made another important contribution to the recent increase in the volume of cocoa production. It consisted of two main programmes of

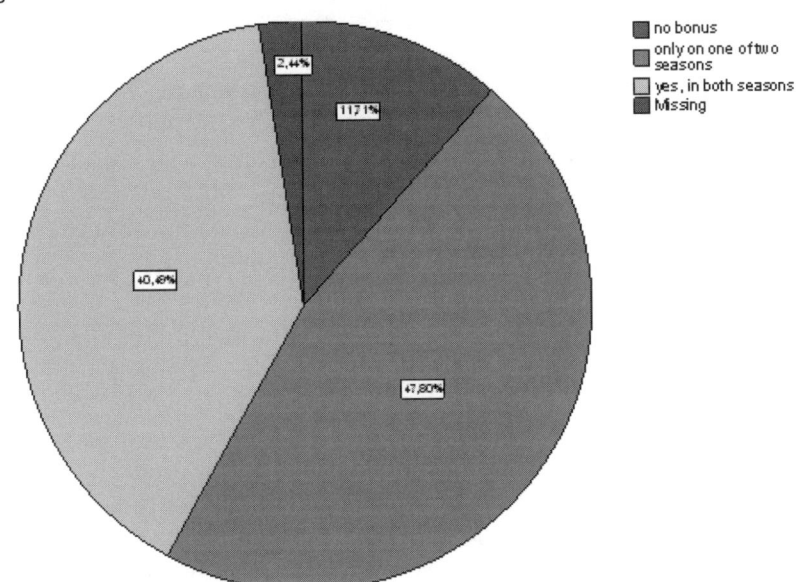

Figure 7.11   Bonuses in seasons 2002/03 and 2003/04

- no bonus: 47.80%
- only on one of two seasons: 40.6%
- yes, in both seasons: 11.71%
- Missing: 2.44%

Source: FS 2005.

combined Capsid (insecticides) and Black Pod (fungicides) control,[177] known as the 'mass spraying programme',[178] and the provision of fertilizer on credit, known as the 'high-tech programme'. The high-tech programme did not survive the pilot phase due to problems with loan repayment. The private sector (Wienco) has successfully taken over this initiative, albeit on a smaller scale (see Box 7.1).

Box 7.1   The Cocoa Abrabo-pa package

The private sector also got involved in extension services. Wienco for example, one of the major input providers, is active in educating farmers on the effective use of inputs and the importance of good farming practices. Although their main objective is to sell chemicals to farmers, they also train farmers on general farming practices and efficient pest management (FS 2005).

The Cocoa Abrabo-pa package brings together inputs (agro chemicals) and services (training and credit). The inputs contain fertilizer (for the soil), fungicides (Ridomil for spraying at the beginning of the season and Nordox at the end of the season to combat black pods) and a chemical named Confidor (to combat capsid). This combination of inputs and know-how on spraying techniques (adequate input and timing of spraying) will increase productivity considerably (interview Wienco, 2005).

The introduction of the package has been remarkably successful:
- average farm production increased by 20 %
- the increase in production was worth nearly three times the value of the loan (10 % had difficulty repaying the loan).

What contributed to its success was that farmers were grouped and jointly responsible for paying back the loan. A problem that occurred was a nearly 40% drop-out, mainly due to inconsistent use of the inputs.[179]

The Ministry of Food and Agriculture (MoFA) also actively intervenes by informing farmers on good agricultural practices and pest management. It took over the responsibility from to provide extension services the Cocoa Services Division. The CSSVD Division intervenes through the swollen shoot programme, removing sick cocoa trees in order to stop the spread of the disease. The Cocoa Research Institute is involved in research and in bringing the research results to the farmers. Finally, a last identified intervention by the government concerns road rehabilitation in cocoa growing areas. This has improved the movement of high volumes of dry beans to the ports in good time for shipment. This measure directly targeted at farmers but has contributed to increasing the volume of exported beans.

Private input suppliers provide fertilizer on credit to farmer groups and provide farmers with advice on how to apply input in an adequate way. Also local buyers occasionally provide farmers with inputs. Local buyers of cocoa also share the farmers' interest to increase production levels and productivity. If farmers produce more, LBCs can buy more and thus increase their income. Cocoa production can be stimulated by providing micro-credits that can enable the farmers to buy input or hire labour. Also some banks provide credit to farmers; still it should be emphasised that formal credit services are very limited. Besides purely private initiatives, there have been some public-private initiatives, such as the farmer field schools. They trained farmers in good agricultural practices, in the use of new types, more environmentally friendly and more efficient ways of pest management (for more information on FFSs in Ghana see Box 7.2).

Box 7.2   FARMER FIELD SCHOOLS

Conservation International (CI) Ghana initiated the Farmer Field Schools (FFSs). In 2000 CI was awarded a grant by USAID to develop a pilot program in agroforestry. This pilot provided the opportunity to test the linkages between agroforestry and biodiversity. Corridor strategies were chosen (linking protected areas) in order to provide 'a greater potential for biodiversity conservation that is consistent with the needs of the local residents and preserve ecological services vital to the well being of those residents' (CI, no date: 5). In Ghana the pilot is implemented together with existing organisations. CI formed a partnership with CRIG and MoFA's Integrated Crop Pest Management Unit (ICPM) to assist the Kuapa Kokoo Farmer Union (KKFU) in developing its own extension service by using the FFS approach. The location of the project was the corridor surrounding the Kakum Park, in the Central Region.

The system relies on farmers themselves learning by doing and passing the information on to their children, neighbours and association members; in short CI:
- promotes agro-ecological approaches to farming adapted to local conditions;
- supports the development of sustainable marketing approaches that bring more revenue to farmers and thus provides an incentive to maintain these systems;
- develops markets for cocoa that provide benefits to farmers and their environment;
- researches and monitors the landscape where cocoa is grown in order to better understand the links between this cropping approach and biodiversity (Adapted from CI, no date: 3).

The goal of FFS is to 'foster farmer learning by experimentation with known practices to identify those that work best under local circumstances', by making use of a farmer-driven mechanism (CI, no date: 6). According to CI this approach towards extension has been successful around the world and is highly recommended (see also KIT et al., 2006: 31). In Ghana one of the main successes was the bringing together of two governmental agencies. This initiated a process of collective learning, which for years had been prevented by politics (CI, no date.: 6).

There is also some criticism of FFSs. A key limitation is cost, it is very expensive to establish the schools, mainly due to the high number of expensive experts who are involved (interview CI, 2005). Another critique is that evaluations on the effectiveness and impact of FFSs are not always made public, which makes it difficult for farmers, researchers and policy-makers to learn from these experiences.

The pilot program in agroforestry in Ghana introduced improved cocoa varieties (hybrids) through five demonstration agroforestry nurseries in four communities surrounding the Kakum Park. Between 30 and 50 farmers per community participated in the pilot phase of FFS. Lead farmers were trained as Trainer of Trainers (ToTs). The first ToTs included people from the Kuapa Research and Development Office, the MoFA extension unit and exceptional Kuapa and non-Kuapa farmers. Then, these 19 trainers each further trained 25 to 30 farmers (jointly in groups of three). In the pilot phase between 120 and 150 farmers were trained. Preliminary findings of the socio-economic study showed that the farmers are adopting many of the practices being tested under the validation phase, noting for example an increase in the numbers of farmers who set up their own cocoa nurseries (CI, no date: 7-8). The pilot project ended in 2003.

Perspectives of farmers participating in FFS
In general, the small number of farmers that I interviewed who participated in FFS reported that they were very pleased with the FFS initiative, and the attention they received. They indicated that their yield increased because of their participation in FFS. This was mainly the result of increased weeding and better prevention against insect/pest infestations. According to the interviewed participants, the FFS training mainly focused on Integrated Pest Management (IPM) where they learned to make a distinction between insects that destroy and those that help the crop. Other frequently mentioned topics included good planting methods and good farming practices (such as weeding and pruning). The ToT aspect of FFS was evaluated positively; most of the participants indicated that they still meet and exchange knowledge regularly, even after the ending of the pilot project (farmer profiles 2005).

Only a very small part of the cocoa farmers in Ghana participated in FFS. Although FFS claims to be open for everyone, its obvious link with the KKFU (which accepts caretakers as members only under very stringent conditions) has resulted in limited caretaker participation. Also, the farmers who participated in FFS outlined several additional constraints:
- Financial constraints make it impossible to weed and/or prune as much as required;
- Financial constraints make it difficult to buy the necessary inputs;
- Although FFS were open to everyone, only few farmers were willing to cooperate because a lot of they face time constraints;
- There is limited availability of chemicals and spraying machines (farmer profiles, 2005).

CI would like to continue its activities in the Western region, where forests are being rapidly converted for agricultural uses.180. Recently, the STCP followed up the work initiated by CI and started a pilot with FFS in the Ashanti region. In 2007 around 15,000 farmers were trained (as ToTs) by the STCP (personal communication Mars, 2007).

Individually farmers follow different strategies that aim at increasing productivity and the volume of production: planting new varieties, applying good farming practices, pest management, using fallow land, hiring more labour, saving more and applying for credit to make on-farm investments, and participating in training programmes. Many farmers work together in labour exchange groups; this saves them time that that they can invest in their farm or in other activities. During this collective work farmers also share knowledge on farming practices, which helps them improve their yields.

Missing interventions

Although international buyers have an interest in increasing volume of production and productivity, most intervene only indirectly, through research and participation in multi-stakeholder initiatives, such as the STCP.[181] On the one hand, this seems logical; there is no apparent need to intervene as the volume of production is continually improving. On the other hand, it can be argued that the Ghanaian system, where the government plays a central role, does not allow international buyers to intervene directly, for example by trading directly with farmers and paying them higher (or lower) prices. In other cocoa producing countries that fully liberalised their cocoa sector, direct relations between international buyers and local suppliers are being established, whereby international buyers increasingly tend to buy directly from farmer organisations. Also, an increasing number of farmers are trained by multi-stakeholder initiatives in these countries. I have no data that could demonstrate the effect these activities have on production levels and productivity in these countries.

Another missing intervention is process upgrading through regional supply management. Cocoa production is concentrated in West-Africa. Together, Côte d'Ivoire, Ghana, Cameroon and Nigeria supply more than 70 per cent of the world's cocoa. However, this geographically concentrated supply has not resulted (yet) in the formation of an effective production-cartel. According to Bass (2006: 259) an explanation might be 'that the largest producer country (Côte d'Ivoire) did not have sufficient incentives to join the agreements [to limit production or to set up physical buffer stocks]'. In Ghana, the government seem to have closer alliances with international buyers of cocoa than with other producing countries. Some farmers gave their own reason. They argued that currently there are no options to manage supply; the high costs of living make it impossible for farmers to delay shipments (Group discussions 2005).

In the next paragraph I will analyse one of the main governmental strategies that has successfully contributed to increasing production, namely the mass spraying programme.

Public intervention: Mass spraying programme

CODAPEC, a subsidiary of Cocobod, is responsible for the mass-spraying programme, which is principally open for everybody. In order to combat Capsid and Black Pod, it envisages the spraying of every maintained farm four times a season with insecticides and fungicides. The national headquarters of CODAPEC is stationed in Accra, within the Cocobod office; where also the national coordination committee and the technical head office group are located. There are six regional committees with regional coordinators. At district level, there is a 'district Task Force Coordinator', who monitors the work in the district. At the village level there is also a 'village task force'. This unit supervises actual spraying by the 'spraying gangs', checks all the inputs and logistics and makes sure that the spraying is implemented well. In these village tasks force the following actors participate: a spraying gang leader (chosen by the 'gang'), village chief farmer, one representative of a LBC, and a farmer (interview CODAPEC, 2005). I do not have data on the selection criteria for farmers to be part of this gang.

How do farmers benefit from this public intervention?

According to CODAPEC the success of mass spraying is obvious: the government provides all the inputs and logistics, 'farmers pay nothing', 'the yield almost doubled, [and] farmers receive higher incomes' (interview CODAPEC, 2005). The success of the mass spraying exercise is partly explained by the explicit link between farm maintenance and spraying; good farm maintenance (which in itself contributes to increasing farm productivity) is a precondition for participating in the programme.

Farmers generally confirm this success; in 2003, 93 per cent of the farmers observed an improvement and indicated that for the season 2001/2002 the mass spraying programme was the main reason for increased productivity (FS 2003). In 2005 the response of this same group of farmers (n=103) was still positive, but their enthusiasm was somewhat tempered; less than 65 per cent of these producers (fully) agreed with the statement that the mass-spraying programme had helped them. Of the farmers who actually experienced a production increase in 2003/2004, around 65 per cent indicated that they received a minimum of two sessions of mass spraying (FS 2005).

One of the most striking results of the 2005 survey was that only 6.3 per cent of farm were sprayed the planned four times (see Figure 7.12) (similar findings are shared by GCFS[182], 2002, 2004 in Teal et al., 2006: 15).

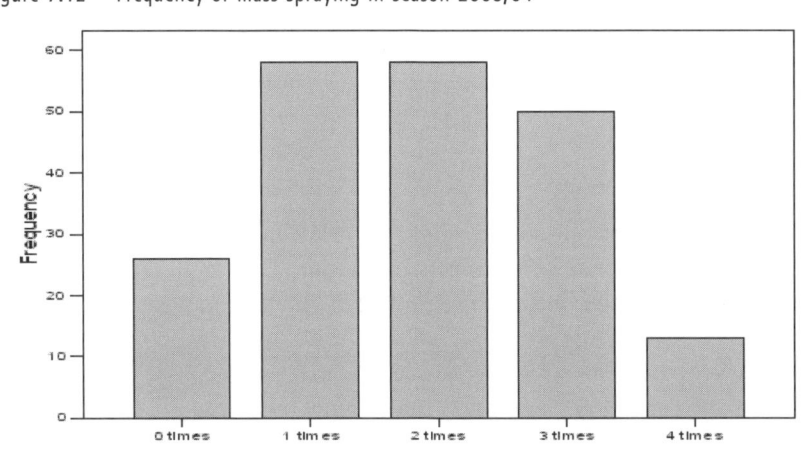

Figure 7.12   Frequency of mass spraying in season 2003/04

Source: FS 2005.

Although compared to other interventions the public mass spraying programme is a large scale intervention that reaches the majority of cocoa farmers, farmers do not have equal access to this programme. For example, in Brong Ahafo almost 90 per cent of the respondents received a minimum of 2 sprayings (n = 44), while in the Western region this percentage was lower than 50 (n = 113) and in the Central region less than 35 per cent received a minimum of 2 sprayings (n = 31) (FS 2005).[183] This is quite remarkable as the density of the population in Brong Ahafo is generally low, communities are more remote and the infrastructure is weak. Research of Teal et al (2006), who gathered data on the mean number of visits of a government spray gang to the three main producing regions, confirms my findings for Brong Ahafo. Their explanation for the remarkable outcome is that Brong Ahafo received most visits of these spraying gangs.[184] Earlier I already showed that use of technologies, such as the knapsack sprayer and mistblower is also relatively high in this region.

A farmer's position in the community or chain also mattered as it enhanced access to sprayings. Around half of the respondents with a strong position (e.g. chief farmer), received 3 or 4 sprayings, while for the farmers with a weak position this was only true in 30 per cent of the cases and for the farmers with no social status this was around 22 per cent (Table 7.6). The survey (FS 2005) also showed that the interviewed farmers who worked together had slightly more chances of receiving a higher number of sprayings.[185] This can be explained by the mutual assistance of farmers in clearing each others farms, which was a condition to get your farm sprayed by fungicides and pesticides.

Next to location and working together, a significant variable is the farmer's position in the community (Table 7.6). The correlation of this relationship is rather weak (FS 2005).

Table 7.6 Cross-tabulation between position in community or chain and frequency of mass sprayings; horizontal percentages

		mass spraying		n
		0-2 spraying	3-4 spraying	
position in community	no (significant) position/status	76,6%	23,4%	107
	moderate-strong position/status	61,5%	38,5%	13
	very strong position/status	61,3%	38,7%	75
Total		195 (100%)		

Gamma 0,174* (FS 2005).

The qualitative data gathered shows a stronger correlation between social network and the benefits from the mass-spraying programme. This is illustrated by the complaints that farmers raised on the functioning of spraying gangs:

> ...they are very greedy. They only spray their own farms and their relations ... The sprayers are not reliable and they don't spray it properly because it is not their farm.

and

> My farm is on a hill and the sprayers were very reluctant to go there, I wasn't the only one, even those who had their farms on lower grounds could not get their farms sprayed. It is politics whom you know (farmer profiles 2005).

However, a high position or a good network does not always help. One of the regional chief farmers, because of his position, did not want his (very large) farm to be sprayed more than one time:

> I am the regional chief farmer; if I insisted they spray my farm, people would start questioning why that was so. I could not push them to spray my farm. Moreover I have my own machines to spray my farm (farmer profiles 2005).

When we asked a farmer who did not receive the four times mass spraying, if he knew who was responsible for spraying his farm, he said: 'I know the leader he is even married to my daughter, I complained to him but he did nothing about it' (farmer profiles 2005). Farmers also stressed more logistical problems, such as insufficient quantities of chemicals and fuel for the spraying machines: 'When we enquired they told us there was no fuel to put in the spraying machines, they also told us the chemicals were insufficient'; and 'When the sprayers got to my farm they told me to buy the fuel for the machine after I had done that they told me the chemical for spraying was finished' (farmer profiles 2005)

CODAPEC recognised some of these difficulties and argues that they are mainly due to the large scale of the programme. According to CODAPEC delays in spraying mostly happen because of logistical problems and bad infrastructure. There are also indications that supervision is not optimal and that cheating is a problem (stolen chemicals are sold on the black market).

In the media complaints have been raised about the over-politicisation of the mass-sprayings; '(...) too many NPP [New Patriotic Party] party activists were serving as chairmen of the task forces'. In addition, several cases were mentioned where spraying gangs had been put in place without the presence of a task force and without the farmer's knowledge.[186] But in general, farmers do know who is responsible for spraying and most of them did complain about the insufficient number of sprayings, unfortunately without results. Several times during the discussions and in-depth interviews, the suggestion was made to give farmers the chemicals and to let them do the spraying themselves, in a response to the problems with cheating or clientelism in these programmes.

Impact of the mass spraying programme

The mass spraying programme contributed to an increase in the volume of production of Ghanaian cocoa beans. For some farmers this directly resulted in higher incomes. Others reported different reasons for the increase in production (such as weeding prior to spraying their farms), but did acknowledge that mass spraying also played a role. For some farmers, the mass spraying replaced the spraying they normally did. For others facing a decline in production, the mass spraying programme limited their loss of income. Some farmers could not gauge the impact, as so many factors played a role.

In terms of empowerment, the idea behind the mass spraying programme was to show farmers the benefits from spraying and clearing their farm. Besides this 'training' element there is no impact in terms of empowerment; farmers were not involved in the organisation of the exercise or other management issues, nor did the intervention change the type of activities farmers normally take or their management skills.

Constraints, trade-offs and flexibility of the mass spraying programme
Although the spraying exercise is presented as free of charge, in reality farmers do pay for spraying their farms. According to a member of the Ghanaian Parliament, 564.9 billion Cedis was spent on the exercise in the 2005/06 cocoa season and 479.91 billion Cedis in the following season.[187] This money has been set aside in the current 2006/2007 season from the export value (gross FOB value) of cocoa (see Chapter 5, Table 5.3) and was reinvested in the cocoa sector. Thus all farmers pay for this programme, regardless whether they receive the promised number of sprayings, less than the promised or no sprayings at all.

Another negative trade-off is the environmental cost of the exercise. As long as the mass-spraying exercise continues, it will be a serious obstacle for the introduction of organic cocoa in Ghana. In addition, it obstructs current attempts to introduce more environmentally friendly pest management, such as Integrated Pest Management (IPM), in the FFSs. Some recipients of the IPM training refused the spraying of their farm. A positive social trade-off of the program was that it provided 'white collar jobs' for rural young people (interview CODAPEC, 2005).

Sub-strategy 1.3: Producing under more remunerative contracts
In the beginning of this chapter I made a selection of different interventions that fall under different sub-strategies. For this sub-strategy (1.3) I will not analyse any intervention in particular and will limit myself to providing a description of the different types of remunerative contracts.

The literature sees contract farming as a central feature of the restructuring of agro-food systems and perceives it as 'an alternative to parastatal marketing boards [...] to avoid government-regulated markets and price controls' (Little, 1994: 219). In this context, contract farming or contract production has been defined as 'arrangements between a farmer and firms (for example local buyers, exporter, processor, etc.) in which non-transferable contracts specify one or more conditions of marketing and production' (based on definition Glover and Kusterer, 1990: 4 in Little and Watts, 1994: 4). These arrangements are extremely varied. It can involve several small individual farmers under contract from a foreign-owned export company, or for example it may involve a contract between a large state-owned estate with thousands of highly differentiated outgrowers (Little and Watts, 1994: 5).

In Ghana the parastatal marketing board is still in place and the partial reforms did not give much room for the private sector to get involved in these types of arrangements. I already elaborated in Chapter 5 on the inability of LBCs to compete on prices. The partially liberalised system did not give LBCs many incentives to invest in building relationships with their suppliers. Nevertheless, LBCs do make

Table 7.7 Measuring the inclusiveness of the public mass-spraying programme

Strategy 1: Capturing higher margins for unprocessed cocoa

Intervention	Activity	Mechanism	Farmers reached	Expected impact	Constraints	Trade-offs
Sub-strategy 1.2 Capturing higher margins for unprocessed cocoa by increasing productivity and volume of cocoa						
CODAPEC	Mass-spraying	Economic incentive and learning (stimulating)	Exclusive: Farmer survey in 2005 indicates that only 6 per cent received 4 sprayings. Frequency of spraying depended significantly on region and social position	Impact 1: Mass spraying exercise does not lead to competitiveness and gives no added value to the beans. Impact 2: Higher yields. Impact 3: Farmers experience the impact of weeding and application of chemicals	Spraying gangs favour their relations and these farmers obtain a stronger position. Part of the chemicals is transferred to the black market. Logistical problems (fuel, timing, etc.)	+ A condition for receiving spraying was weeding. This in itself contributed to higher yields. – The mass spraying programme is supposed to be 'free'. But costs are paid from difference between net and gross FoB price. The mass spraying programme obstructs the introduction of more friendly methods of pest management and makes it more difficult to introduce the production of organic cocoa.

Measuring inclusiveness The mass-spraying programme is a large scale programme that reaches many farmers. However, while indirectly all farmers pay for the programme they do not benefit accordingly and also not equally. The programme is not transparent on its effectiveness. Among the respondents, especially the farmers working in the Western and Central Region had difficulty in getting the full number of sprayings. Access was also relatively difficult for farmers who have no or a weak position in the chain. Nevertheless, the programme reached most farmers, and the majority appreciates the government's support. The difficulties with distribution of the chemicals have raised questions on alternative ways of getting farmers to use the chemicals.

small investments in trust building and social capital (Chapter 5); these are relatively novel efforts, mainly informal and still in an experimental stage. The farmer survey showed that prompt payment and social relations were most important reasons for the farmers to select a buyer. It also showed that there is no formalised arrangement. Loyalty between farmers and buyers is not guaranteed. Farmers can decide to sell to another buyer. But also local buyers are not always trustworthy; the survey showed that despite promises only a small number of the farmers received any services or bonuses from LBCs. The ones who did were mainly farm-owners and farmers living in the Western and Central region (where competition between LBCs is more intense) (FS 2003).

There is one example of a more institutionalised arrangement between farmers and their local buyer, namely the farmer-owned LBC, Kuapa Kokoo Farmer Union. This union produces a small part of its cocoa for the fair trade market and has to meet specific demands with respect to process quality.

There are different reasons why contract farming in Ghana did not succeed. The still predominant role of the government in the coordination of economic activities is one clear reason. Another reason is that farmers are not organised, which would make contract farming very difficult due to the large number of smallholders involved. Another constraint is more general and has to do with problems of land acquisition; contract farming becomes more lucrative for buyers if they have some economies of scale.[188] The Ghanaian government experimented with growing cocoa on a plantation basis but this did not work out well.

There is a type of 'contract farming' that does take place between farm-owners and their caretakers (shareholders). There are two types of share contracts, Nhwesoo (Abusa) and Yemayenkye (Abunu) (Chapter 6). Even though, almost all caretakers claimed to be satisfied with their contract, working under a Yemayenkye contract is generally more favourable. In this respect, shifting from a Nhewsoo contract to a Yemayenkye contract would be considered 'upgrading'. The opportunity to become a landowner has significant advantages. In addition to providing income, farms are perceived as a form of social security/inheritance and can be used as collateral to take out a loan (FS 2005). In terms of these kinds of remunerative contracts, also working for a number of different farmers can be considered upgrading. My survey showed that as a risk management tool, part of the caretakers work under different contracts and they also have other sources of income or some land available for other activities.

In addition to 'contract farming' there are also other kinds of arrangements between different actors in the chain. On the international level there was a cocoa commodity agreement, but it is now abandoned (Chapter 4). Another type of contract relevant for producers is the forward sales contract between international buyers and the Ghanaian government. Forward sales enable different chain actors to plan their economic activities according to the agreement and contributed to favourable contracts between international buyers and the Ghanaian government (Ministry of Finance, 1999). This secured an effective marketing channel for Ghanaian producers and the export of premium quality cocoa.

Informally, the labour exchange groups function under similar conditions, based on mutual trust. Every involved farmer agrees on a rotating scheme where each farmer provides an equal amount of labour on someone else's farm. The importance of trust is illustrated by the next quote by a farmer who is not participating in *nnoboa*: 'It [nnoboa] is a good idea but some of the farmers are cheats when it gets to their turn to work on their farm they usually give you the most difficult places to weed' (farmer profiles 2005).

In summary, producing under more remunerative contracts or arrangements can have an impact not only on the farmer's income, but can also result in additional benefits (for example access to know-how, inputs etc.). There are different types of contracts, which influence how farmers benefit from cocoa production. Some of these contracts are formal and some informal; some are between actors further up in the chain; some arrangements are made between the private sector and farmers; and some are made among the farmers themselves, for example between farm-owners and their caretakers (sharecroppers). What hinders 'contract farming' in Ghana is the dominant role of the state in coordinating cocoa activities; it leaves little space for direct relationships between farmers and actors further up in the chain.

Conclusions on Strategy 1
Looking at the different sub-strategies that aim to increase the margins for unprocessed cocoa and at the interventions, I observed that the government plays a dominant role and leaves little room for other actors, especially for international buyers, to intervene.

Analysing effective interventions in terms of inclusiveness shows that both governmental interventions that were discussed are large-scale and reached the majority of farmers. However, their large scale and the weak institutional environment made these types of interventions also vulnerable, for example to corruption or clientalism. The lack of transparency on (the distribution of) costs and the benefits of these interventions makes it difficult to discuss their effectiveness. The interventions of the government are major reinvestments in the sector, paid mainly from the export value. Other actors cannot duplicate these governmental efforts on the same scale.

Farmers have little to say about these kinds of government intervention. The public quality control system involves all farmers and also here, similarly to the mass-spraying programme farmers have little choice. Although all farmers are not obliged to have their farms sprayed, as I demonstrated all of them do pay for this exercise. The idea is that the mass-spraying programme works as an incentive for farmers to start spraying their farms, but in reality some farmers take government spraying more as a time-saving exercise – they don't have to spray by themselves. The end result is that it does not stimulate entrepreneurial behaviour automatically.

Looking at impact, both interventions have a positive impact in terms of providing access to international markets and remunerative contracts; however, they do not contribute to empowerment. The farmers remain chain actors and do not move (vertically or horizontally) in the empowerment matrix (see Figure 7.13).

In terms of inclusiveness both interventions are sub-optimal. Although the quality control system reaches all farmers, it is compulsory and does not stimulate entrepreneurial behaviour: farmers are not involved in standard-setting; farmers

Figure 7.13   Changes in the empowerment matrix due to public interventions

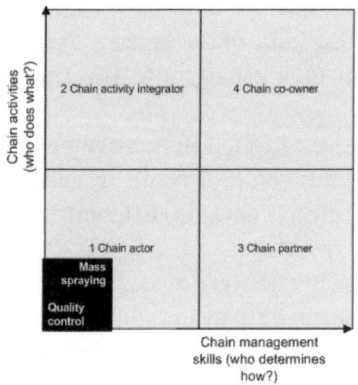

Source: author.

have no choice (there is no price differentiation); and the system does not support farmers in building their capacity. The lack of transparency on distribution of costs and benefits to some extent undermines the effectiveness of the system. In addition the incentives for different local actors to support the system seem to be diminishing. The second intervention, the mass-spraying programme, is a large scale programme that reaches the majority of farmers. However, while indirectly all farmers pay for the programme they do not benefit accordingly and also not equally. Especially intereviwed farmers working in the Western and Central region had difficulty in obtaining the total number of sprayings. Farmers who had no position or a very weak position in the chain also face challenges with access. Nevertheless, the programme did reach most farmers, and the majority is appreciative of government support. The difficulties with distribution of the chemicals raised question: Why not give farmers the money so they could do the spraying themselves?

Conditions under which farmers are included
When discussing the different sub-strategies (under Strategy 1) it becomes clear that the conditions under which different groups of farmers are inserted in the cocoa chain have changed over time. For example, in the early 1920s quality was safeguarded through the formation of farmer groups, later these farmer groups disappeared. Another example is the fragmentation of Cocobod's extension services, which used to support farmers in the production of premium quality cocoa. There is an observed lack of transparency regarding the costs, benefits and risks involved in 'product-upgrading' and how these are distributed among the different actors involved. The weakened institutional environment and the lack of transparency reduced the incentives for LBCs and farmers to invest in quality, and mounted tensions between Cocobod, LBCs and farmers.

7.2.2   Strategy 2: *Producing new forms of existing commodities*
There are different ways to produce new forms of existing commodities. For example cocoa can be produced for other types of markets, such as specialty or organic

markets. Some authors perceive this as product upgrading (producing a more sophisticated product) and others as a way of process upgrading (producing according to more responsibility demanding practices). Alternatively, new forms of existing products can refer to the use of cocoa beans as ingredients for other types of products, such as cosmetic or health products. A third way is to look for diversification opportunities for income and production of a diversity of cash crops. This can result in a shift towards non-traditional products or investments in non-farm activities.

Figure 7.14   Strategy 2: Producing new forms of existing commodities

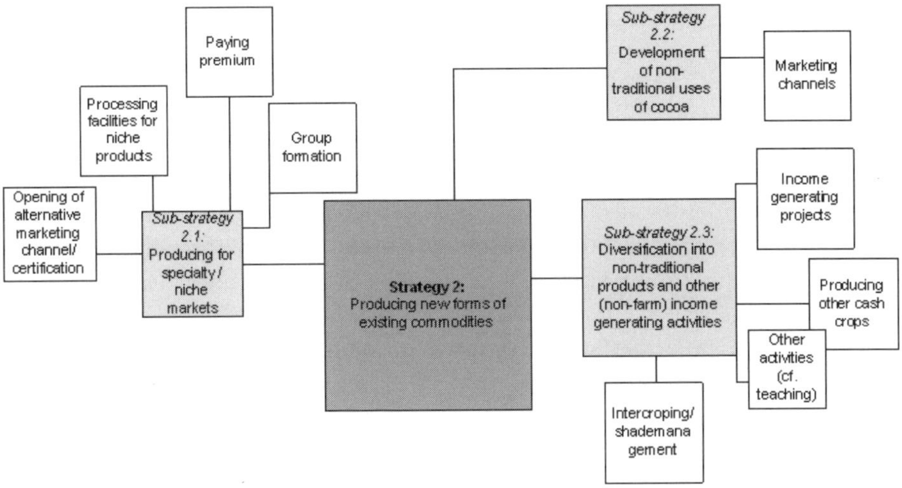

Source: composed by author.

Sub-strategy 2.1: Producing new forms of existing commodities by producing for niche markets

Production for niche markets is considered as upgrading because producers of these more sophisticated products generally receive a higher price. Traceability is one of the key-conditions for selling on niche-markets. Ghana is considered to be the only cocoa producing country where cocoa is still traceable back to the community where the cocoa was sold to a local PC. However, the increase in bulk transport of cocoa (where bags are removed prior to shipping) is threatening the current state of full traceability.

The main niche markets for cocoa are organic cocoa and fair trade cocoa. Both products receive premium prices on the world market. If we look at the origin of organic certified cocoa, this cocoa is mainly produced in Latin American countries. The Dominican Republic is by far the largest supplier. According to estimates by a large trader in organic products (Tradin Organic Agriculture),[189] the Dominican Republic produced around 30,000 tonnes of organic cocoa in 2008, around two-thirds of the world's total organic cocoa production. African countries produced 3,000 tons, mainly coming from Tanzania, Uganda and Sao Tome. In Latin America around 10,000 tonnes were produced (with Peru and Ecuador as main contributors). Taken all together, the organic market represents only a very small share of the total cocoa market, estimated at less than 0.5 per cent (ICCO, 2007). Between 2003 and 2005 the annual growth rate of organic cocoa was 38 per cent, with a total production of almost 21,865 tons in 2006, mainly in Latin American countries. In 2008, total organic cocoa production was around 40,000 tons (see Table 7.8).

Table 7.8 Growth of the global organic market

Year	1998	2000	2002	2004	2006	2008
Production in Tonnes	8,390	13,050	18,065	21,865	28,575	40,000
Growth rate	-	56%	38%	21%	31%	39%

Source: KIT et al., forthcoming.

According to Ayenor (et al., 2004: 263), in Ghana the concept of organic might be new, but the production of organic cocoa has a long tradition: 'some do not use inorganic pesticides because they cannot afford to use them and others because they consider them poisonous and hazardous to human health'. There have been several attempts in Ghana to introduce the production of organic cocoa (see further down in this chapter).

Organic cocoa production has an intrinsic environmental value, promoting and enhancing the health of agro-ecosystem. It is also interesting from an economic perspective as organic cocoa commands a higher price than conventional cocoa and attracts premiums (ICCO, 2007). This premium should cover both the cost of fulfilling organic cocoa production requirements and the fees paid to certification bodies. The costs of certification can be perceived as a problem, especially when the production volumes are low. Considering the lack of adequate organic inputs to combat pests and diseases in Africa, producing organic cocoa does not automatically translate into increased incomes for the farmers (Koning and Steenhuijsen Piters, 2009; Laan, 2007).

Cocoa is a very suitable product for organic trade. It is consumed in large quantities, has structured trade channels, and is processed into a luxury item that has a high perceived value and few substitutes. For this reason, large traditional chocolate processors and manufacturers have moved into organic cocoa by-products and chocolate, making the organic sector increasingly mainstream. The only issue is limited processing opportunities, organic cocoa needs to be processed separately from regular cocoa and not all factories have this capacity.

The price for organic cocoa beans is formed by adding a price premium on top of the spot market price of mainstream cocoa. There are no official numbers on the development of the premium paid for organic cocoa, as this depends on the specific buyer and seller, as well as the negotiations between them. Organic premiums fell sharply in 2001 to USD 100–200 per tonne above conventional cocoa prices. Premiums began to recover in 2003, reaching USD 200–300 per tonne at the end of the year. In April 2007 the premium for organic cocoa varied between USD 500 and 1,500 per tonne. This maximum of 1,500 USD was also paid in mid 2008. Since then the level of the premium drastically declined to a value of a few hundred dollars and even to zero when there was a situation of oversupply (February 2009). Based on experiences with organic cocoa production, it is estimated that a premium of USD 200 per ton is the minimum needed to sustain organic production (KIT et al, forthcoming).

As the worldwide demand for organic cocoa greatly outstrips supply capacity, it is expected that market prices for organic cocoa beans will remain high in the coming years. However, the world prices for conventional cocoa beans have also increased, which has put traders, processors and manufacturers under pressure of high commodity prices. More importantly, the supply of organic cocoa increased steadily. These two trends affected the premiums paid for organic cocoa. Especially countries that produce relatively high quality cocoa (and sell over the London future market) and are not subject to discounts, seem to out-price themselves when the premiums are too high. Recently this was the case with organic cocoa production in Ghana.[190]

More and more organic certified commodities also apply for a fair trade certificate from the Fair Trade Labelling Organisation (FLO). The most essential characteristic of fair-trade is that producer organisations receive a higher price for their cocoa beans. Even though, worldwide the demand for fair trade cocoa is growing, in season 2003/04 the cocoa beans sold with the fair-trade label captured a very small share of the cocoa market (0.1 percent or 2 687 tonnes) (ICCO, 2005). In 2005, fifteen producer organisations were FLO certified, of which twelve were located in Latin America and the Caribbean. The largest fair trade producer organisation is based in Ghana (Kuapa Kokoo Farmer Union); together with Conacado (based in the Dominican Republic) they are responsible for around 90 per cent of the fair trade sales. In Africa two other FLO producer organisations were set up in Cameroon and Côte d'Ivoire. In season 2003/04 they were not actively selling on the fair trade market (ICCO, 2005: 9). In recent years also the fair trade market has become more 'mainstream', just like the organic cocoa market. For example, the Dutch company Verkade is 100% fair trade and Barry Callebaut is processing fair trade cocoa beans for the Dutch 'media-bar' of Tony Chocolonely.[191] Also Cadbury and Mars have committed themselves to sustainable cocoa sourcing.

In contrast to organic premiums, fair trade premiums are not decided at moment of purchase, but are rather fixed beforehand. Fair trade prices are calculated on the basis of world market prices, plus fair-trade premiums of US$150 per tonne of cocoa beans. The minimum price for fair-trade standard quality cocoa, including the premium, is US$1,750 per tonne. If the world market price of the standard qualities rises above US$1600 per tonne, the fair-trade price will correspond to the sum of the world market price and the US$150 premium per tonne (ICCO, 2005: 3). For fair trade cocoa that also has an organic certification, there is an additional organic premium of USD 200 per tonne. Fair trade organic cocoa beans cost a minimum of USD 1,950 per tonne.[192]

In Ghana an additional premium for quality and consistent delivery is added to the fair trade price. Generally a large part of the social premium is allocated to a social fund; farmers receive only a small part of it. The producer organisations decide on the desired use for the social fund. Most producer organisations sell only part of their cocoa under the fair-trade arrangement; the benefits are distributed among all of their members. Certified producer organisations pay part of the certification costs. The financial benefits and the additional costs for co-operatives associated with fair-trade, compared to the conventional market are summarised below (adapted from ICCO, 2005: 5):

Sources of additional benefits
- Fair-trade price – the FOB price paid to the co-operative is higher than the conventional price and, by definition, more stable;
- Direct sales – the fair-trade supply chain does not usually involve as many intermediaries as the conventional one;

Sources of additional costs
- Cost of participation in the FLO system – certification fees, documentation costs, and other associated costs;
- Production costs of meeting FLO standards – possible additional labour, social and environmental costs.

Another niche is specialty or 'single origin' commodities. This trend is already clearly visible in the coffee market (the coffee sector is a front runner in more aspects, for example in organic production, fair trade and mainstream certification) (e.g. Daviron and Ponte, 2005). Although cocoa has a different marketing process than coffee (it is only one of the ingredients of the end-product) this trend is increasingly noticeable also in the cocoa sector. 'Single origin' cocoa is being launched by major chocolate manufacturers, such as Barry Callebaut and ADM. Cadbury, which has sourced its most important raw material from Ghana since 1908, always had a focus on 'origin' cocoa-products.

Farmer perspective on producing for alternative markets

The perspectives of farmers on niche markets and their benefits are discussed in the analysis of this multi-stakeholder intervention. A general comment is that most farmers are not familiar with concepts such as organic and fair trade. This is likely to change as a result of mainstream initiatives of large cocoa buyers, such as Cadbury and Mars, that aim at sustainable sourcing of (part of) their cocoa supply.

Interventions that aim to open alternative marketing channels

There were several attempts to launch the production of Ghanaian organic cocoa. The first came from the NGO Conservation International Ghana (CI Ghana) in 1998. The Ghanaian government obstructed the joint CI Ghana and Kuapa Kokoo Farmer Union pilot project to produce organic cocoa in the Central Region. According to CI Ghana the government opposed the promotion of organic cocoa. So far, more recent attempts of the Dutch Rabobank Foundation to create a marketing channel for organic cocoa also failed due to Cocobod opposition.[193] It is suspected that Cocobod fears loosing their grip on state-controlled marketing and pricing systems if they allow new foreign buyers in the country. A recent attempt by the international company AgroEco was more successful. The likely reason for this success is the partnership with Cocobod. In 2008, around 250 cocoa farmers from the Eastern Region produce 8,000 ton of organic cocoa for export under this project.

A multi-stakeholder partnership (involving Kuapa Kokoo Farmer Union and Twin trading) introduced the production of fair trade cocoa in Ghana (discussed in-depth below in this section). Farmers can become involved in fair trade cocoa production through membership of the farmer union. I already mentioned that ADM constructed a new processing facility for 'single origin cocoa-products'.

Missing interventions

The government in Ghana prioritises the production of conventional cocoa and focuses clearly on product quality and not on process quality. Although recently the Ghanaian government was rather proactive by supporting events such as the Round Table "Towards a Sustainable Cocoa Economy" and worked together with AgroEco on the introduction of organic cocoa, it is not proactively engaging in the production for niche markets. It will be exciting to watch how the government responds to more mainstream activities in this field. These initiatives demand support from Cocobod for opening separate marketing channels, but it is not yet clear how these initiatives will develop. In the next paragraph I will discuss the third selected intervention – fair trade cocoa production.

Intervention fair trade cocoa production

The involvement of Ghanaian cocoa farmers in fair trade cocoa production started with an influential cocoa farmer (see also Chapter 5). In the mid 90s he set up, together with support of two foreign NGOs a farmer-owned LBC, Kuapa Kokoo Ltd (KKL) together with a farmer union, the Kuapa Kokoo Farmer Union (KKFU). When KKL received its license to trade, it simultaneously negotiated a special agreement with Cocobod for some of its members' cocoa to be set apart and exported under fair trade terms. This small proportion of Kuapa Kokoo's cocoa receives the minimum fair trade price and an additional fair trade social premium.

The farmer's union is a democratically elected union of primary societies with a National Executive Council of local leaders. It has grown quickly from the original 22 farmer groups or village based 'societies' with 2,200 members, to a very wide expansive net in 2004 (48,854 registered members who hailed from 1,124 societies located in nineteen areas in six cocoa regions) (Kuapa Kokoo Annual Report, 2004: 38). Membership is open to farm-owners who sell their entire cocoa harvest to KKL. If a caretaker wants to become a full member he/she needs certification of the owner of the farm (interview Research and Development Officer KKFU, 2003).

Throughout the years, the Kuapa Kokoo group has developed into a complex organisation, with a number of different bodies and committees that manage key aspects of its operations and mandate (Figure 7.15).

The KKFU's mission is to 'empower farmers in their efforts to gain a dignified livelihood, to increase women's participation in all of its activities, and to develop environmentally friendly cultivation' (Kuapa Kokoo Annual Report, 2001/2002). KKFU produces only a small part of its cocoa as 'fair trade cocoa' for the Max

Figure 7.15   The organisation of the Kuapa Kokoo Group

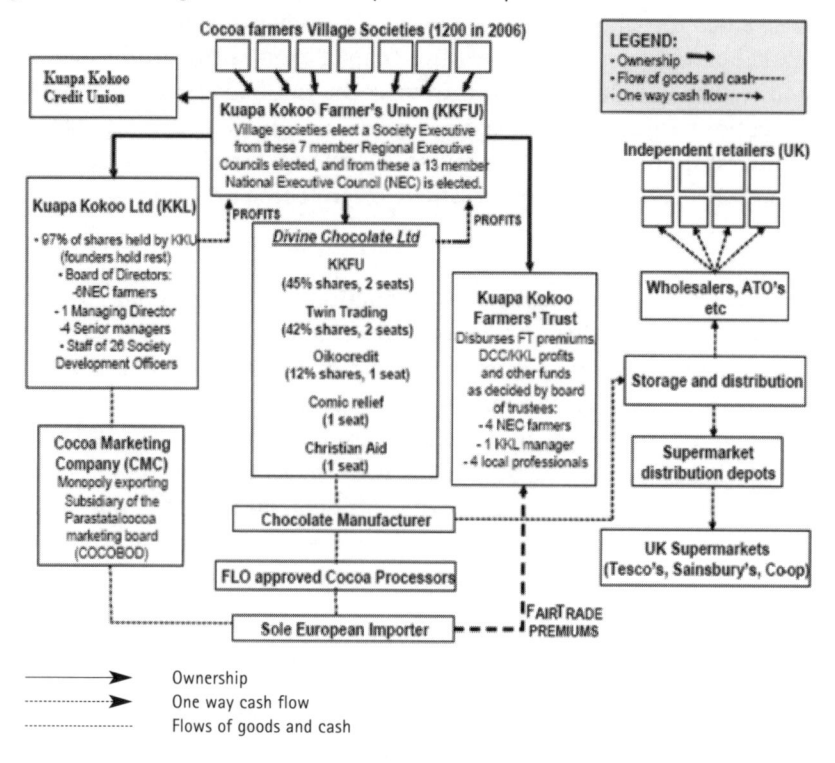

Source: Doherty and Tranchell, 2005.

Havelaar Foundation. In the 2003/2004 season, KKL purchased a total of 64,975 tonnes of cocoa, with a portion of fair trade sales totalling 1,800 tonnes (2,7 %) (Kuapa Kokoo Annual Report 2004: 38). In season 2005/2006 the purchases of KKL reduced to 42,676 tonnes (see Chapter 5, table 5.2). The portion of fair trade sales has increased somewhat over the last years but still remains fairly low. Divine Chocolate Ltd puchases around half of total fair trade sales.

How do farmers benefit from their multi-stakeholder partnerships?
KKFU receives a fair trade minimum price of US$1,600 per tonne and the social premium is US$150 per tonne. In Ghana farmers are assured a fixed price. Kuapa Kokoo farmers received part of the fair trade premium in terms of small cash bonuses. While the number of Kuapa Kokoo members has greatly increased in the past decade, the increase in fair trade sales is marginal, mainly due to low consumer demand.

The social premium was allocated to the Kuapa Kokoo Farmers Trust (KKFT) which used this income to fund a range of activities, including the construction of water wells, schools, women's income generating projects, medical facilities, etc. (see also Vuure, 2007). For the past four years, Cadbury Ghana Limited (a major chocolate manufacturer) has been channelling some of its social projects to the Trust Fund and donated an average of forty wells per year to farmer communities where KKFU is active (Annual Report 2004: 10). In 2004 the three primary expenses of the Trust 2004 were to provide additional farmer income (55 per cent); to support the Farmers Union (e.g. the educational programme of the Union) (16 per cent); to purchase Union President Vehicle (27 per cent)/Office Renovation (2 per cent). Remarkably, almost one-third of the fund was allocated to the purchase of a vehicle for the president.[194]

The farmers also benefit from the two credit schemes (Kuapa Kokoo Women's Revolving Scheme and Kuapa Kokoo Farmers Credit Scheme), where farmers can save money and apply for a loan. In 2004, almost 20,000 Kuapa Kokoo members (around 41 per cent of total members) were also members of the Credit Union (an increase of almost 17 per cent over the previous year). The members were 77 per cent men, 21 per cent women and 2 per cent farmer groups. Compared to the previous year savings increased by 41 per cent (Kuapa Kokoo Annual report 2004: 27).

In addition to its activities in cocoa production and internal marketing, KKFU is an official shareholder of the Day Chocolate Company that distributes and sells Divine fair trade milk chocolate in the UK (the Divine company is not involved in cocoa processing). Recently KKFU became the major shareholder of Day Chocolate Company; it owns 45 per cent of the company and has two seats on its Board.[195] As a shareholder in the Divine Company, KKFU has a voice in strategic decision-making. Moreover, ownership contributes to a sense of pride, trust and commitment among the cocoa farmers involved. Being a shareholder also contributes to empowering the farmers through training and by involving them in a democratic decision-making processes.

As shareholders, the farmers also receive a share of the profits from the Divine Chocolate Company. In 2007, it was the first year that ownership brought some direct (although marginal) economic benefits to the farmers. The 'success' of Divine chocolate can be explained by the fruitful partnership between a consolidated producer organisation, an experienced NGO (Twin Trading has a lot of experience in the coffee sector with 'Cafedirect'), NGOs with a large network of consumers and celebrities (such as Christain Aid and Comic Relief) and the Divine office, with a managing director with excellent marketing skills.[196] The branding strategy centred on farmer ownership played a key part in this success. (Doherty and Tranchell, 2005; Koning and Steenhuijsen-Piters, 2009)

Impact of the intervention
The introduction of fair trade cocoa had several impacts at the level of the farmer and upgraded their position from a chain-actor to an official chain-owner. Through membership of KKFU, farmers became the official owners of the farmer union and recently they also became shareholders of the Divine Chocolate Company. Moreover, members of Kuapa Kokoo had easier access to training programmes, such as the farmer field schools and benefited from the support of community development projects. In terms of direct economic results, the benefits are only marginal: they receive small bonuses and have received the first dividend of the chocolate company; each member received a direct payment of USD1 (personal communication KKFU research division, 2007).

Constraints, trade-offs and flexibility of the scheme
In general the achievements of the KKFU are evaluated positively (see evaluations by Mayoux, no date; Tiffen, no date; Vuure, 2006). Besides a (marginal) extra income

that the farmers appreciate, through their membership of Kuapa Kokoo farmers also had access to new marketing channels, the opportunity to become involved in chain management, had ownership, had access to training (such as FFSs) and had a chance to have a funded community project in their village. Other main benefits mentioned by Kuapa Kokoo members in 2005 concern the financial advantages, such as access to credit (from the credit unions) and extra bonuses.

Despite the successes of the Kuapa Kokoo Farmer Union and the positive evaluations, I find it rather difficult to assess their achievements. Not only because I interviewed only a small number of registered Kuapa Kokoo members, but also due to the absence of critical self-reflection among Kuapa Kokoo's staff-members. Fieldwork indicated that not all farmers who sell of their cocoa to KKL are aware of its semi-cooperative status and consider KKL similar to any other LBC (FS 2003; interview Nana Osafo-Ansong, Senior Advisor SNV, 2003). Despite that selling all your cocoa to KKL is a pre-condition for membership of the farmer union, it is common practice for farmers to sell a part of their cocoa also to other LBCs.

Moreover, also farmers who are not members of Kuapa Kokoo can sell to the PC of the farmer union; the result is that 'fair trade cocoa' is mixed with conventional cocoa. Also the empowerment of farmers is not optimal. Some farmers complained about promises that were never fulfilled and about being cheated on weighing scales. Especially with regard to the credit union several constraints were identified. According to some of the farmers the amount of credit was too low and the union lacked transparency. Sometimes the union was located too far away. The next quote illustrates some of these complaints:

> When Kuapa arrived they did not paint any picture that it is a farmer's organisation. All they told us is to bring our produce to them to sell, so that if farmers needed any financial support

Picture 7.3   A Kuapa Kokoo buying station

they will also help them. I sent my produce to them and when I needed financial assistance they told me it is a new company so they are facing teething problems so they could not assist us. There is a saying in Akan that onyeame boa onnea woaboa no ho literally meaning "heaven helps those who help themselves". I did honour my promise by selling my produce to them but they failed to give me a loan (farmer profiles, 2005).

Ownership of KKFU and being a shareholder of Divine was not mentioned by the farmers as a reason to join this union, and the direct benefits of it were not directly clear. While officially farmers own KKFU's LBC, this seemed a rather abstract idea for most of the farmers. But over time, it is possible that the benefits resulting from KKFU membership will become more apparent to farmers.

One of the conclusions is that both KKFU and promotions of its fair trade principles depend to a high extent on the benevolence and skills of the PC who buys the cocoa beans from the farmer.

Sub-strategy 2.2 and Sub-strategy 2.3

In the introduction of this chapter, which discussed the link between upgrading and exclusion, I raised the question on the extent to which vulnerable groups are prepared to leave the cocoa farm and their realistic alternatives. Although in conventional upgrading debates 'diversification' is not considered as an upgrading strategy, it seems vital to acknowledge diversification as a strategy, especially for countries that depend heavily on only a few export commodities with little added value, and for farmers who depend heavily on cocoa. More of the players in the cocoa sector have come to appreciate diversification and stimulate farmers to invest in multi-cropping, shade management and other practices. Also, 'diversification' is part of the curriculum of the farmer field schools (public-private partnerships).

There are different types of diversification. In the section above I already demonstrated that production for niche markets is a good way to open alternative marketing channels. Another strategy (Sub-strategy 2.2) is to use cocoa beans as an ingredient for non-chocolate products, such as cosmetics (there is a variety of cocoa products available in the Body Shop) and health products. This strategy is however not directly accessible for farmers; farmers sell their beans to CMC and have no influence on the further allocation of their beans. In Ghana only the cocoa research body of Cocobod (CRIG) is involved in investigating and developing alternative uses of cocoa. For farmers diversification mainly involves producing other crops or becoming involved in other income-generating activities (off-farm).

For almost all cocoa farmers that participated in the farmer survey (FS 2005) cocoa-farming is perceived as a life-fulfilling occupation. A farm does not only generate immediate income but is also regarded as a way of advancing social security and inheritance. For most respondents cocoa provided a considerable part or almost all of their income (FS 2005). As already indicated, diversification in terms of producing other crops is common, for example: cassava, palm oil, plantain and cocoyam. These other crops are planted both on separate farms and directly on the cocoa farms. Almost 80 per cent of the respondents owned and/or cultivated other land. Caretakers often worked under more than one contract and/or took care of

Table 7.9 Measuring inclusiveness of membership in the Kuapa Kokoo Farmer Union

Strategy 2: Producing Intervention new forms of existing commodities

Sub-strategy 2.1	Intervention	Activity	Mechanism	Farmers reached	Expected impact	Constraints	Trade-offs
Opening of niche markets	Farmer groups KKFU	KKFU produces for fair trade market	Economic incentive (stimulating)	Exclusive. Only members of KKFU are reached. For caretakers it is difficult to become a member. KKFU sells only small part of its cocoa to fair trade market; the rest is conventional cocoa.	Impact 2: remunerative contract and attention for diversification Impact 3: empowerment of farmers, through training, participation in extension programmes, alternative income generating activities for women	Little demand for fair trade Extra bonus is not being given to all farmers → dependence on good will of PC Community development projects do not reach all farmers Also non-Kuapa members can sell to KKFU → doubts on status of fair trade cocoa (is all Ghanaian cocoa fair trade?)	+ income generating activities for women Credit union Community development projects

Measuring inclusiveness

This multi-stakeholder intervention is not fully inclusive in numbers (excluding caretakers); however, in terms of impact it is more inclusive than public interventions. One identified problem is low demand for fair trade cocoa. Another difficulty is that the status of ownership of the union and of the chocolate company is primarily symbolic; it takes a long time before the farmers feel direct benefits and ownership becomes a tangible variable. The success of Kuapa Kokoo can be largely explained as a fruitful partnership that links producer organisation to investors, strong marketers and a large network of consumers.

several farms (under the same contract). Table 7.10 illustrates how farmers use this extra land.

Table 7.10   Main uses of extra land

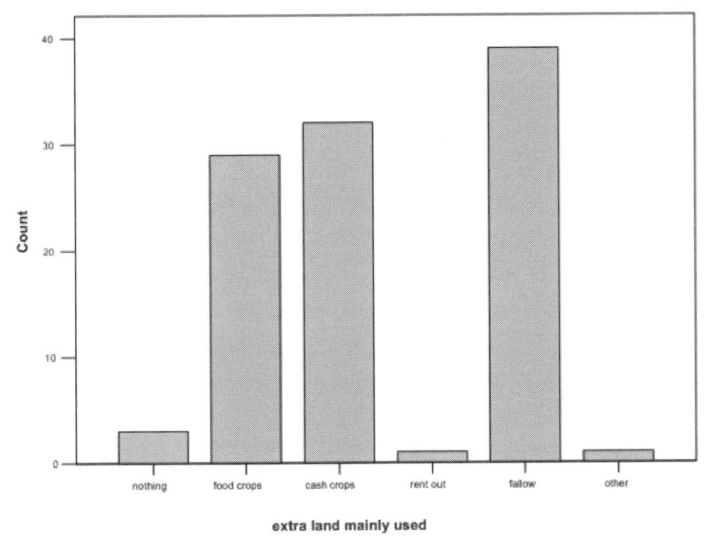

[Bar chart showing counts for categories on x-axis "extra land mainly used": nothing (~2), food crops (~29), cash crops (~32), rent out (~2), fallow (~39), other (~2). Y-axis "Count" ranges 0–40.]

Source: FS 2005.

The main reasons behind diversification include generating extra food and money and also spreading out the risks of depending on a seasonal crop such as cocoa (FS 2003). In case of inter-cropping and shade management this contributes to the establishment of a 'natural eco-system'. However, in the Western Region, marked by the most rapid cocoa production increases in the country, there is another trend. In this region, almost 30 per cent of the cocoa is cultivated without any form of shade-management. This is worrisome as it causes rapid soil depletion (Gockowski, 2007).

Diversification does not necessarily involve on-farm activities. Almost 20 per cent of farmers interviewed in 2005 obtained additional income from non-farm activities, for example teaching (FS 2005). Different sources of income help cocoa farmers to be more flexible and even negotiate; farmers can choose to dedicate more/less time to cocoa in case of price-fluctuations. Clearly, diversification directly contributes to empowering farmers.

Conclusions on Strategy 2

There are many actors involved in producing new forms of existing commodities. In the development of niche markets, NGOs take the leading role and work together with farmer groups and governmental bodies. The government plays a facilitating role in the organic and fair trade market, but has also hindered a number of earlier attempts by NGOs and international banks to introduce organic cocoa in Ghana. Without governmental support the opening of alternative marketing channels is not possible in Ghana. Niche markets tend to work positively for farmers, but only affect a very small portion of farmers. For example, in the case of fair trade cocoa caretakers have difficulty to become a member of a farmer union. In another example with organic cocoa, it was only cultivated by a very small number of farmers in the Eastern region. In addition, the assumed benefits are not always realised, for example fair trade cocoa does not automatically result in improved income for farmers.

In terms of impact, a difference with state interventions is that the main impact of these types of public-private partnerships is empowerment. Farmers who are reached by these interventions receive training, participate in decision-making processes and are involved in activities higher up in the value chain (in the case of fair trade the farmers are involved in internal marketing of their produce) (Figure 7.16).

Figure 7.16   Changes in the empowerment matrix due to a multi-stakeholder initiative

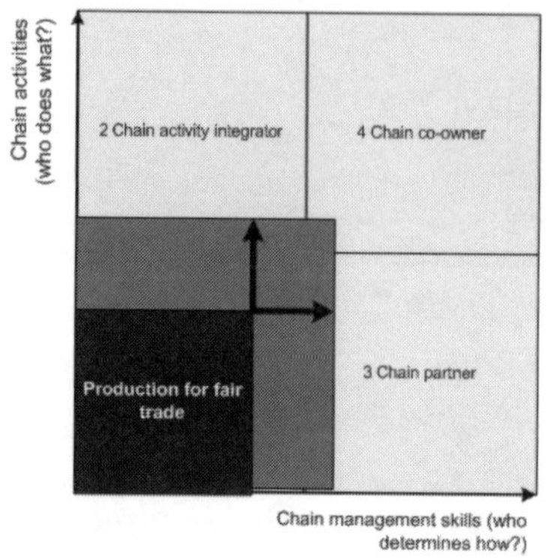

Source: author.

In terms of inclusiveness the production of fair trade cocoa and membership in the farmer union are more optimal than the interventions that aim at increasing margins for unprocessed cocoa (Strategy 1). Membership in the Kuapa Kokoo union is voluntary and stimulating. There are different incentives for farmers to become a member of this group, membership rewards include access to training, credit schemes and community development. Members are owners of the union and they are also owners of a chocolate company. Their involvement in trade and in decision-making processes significantly contributed to enhancing their empowerment. Still, this intervention is not optimal as it reaches only a small number of farmers and caretakers have difficulty becoming members. The economic benefits from membership of KKFU are marginal. A main cause is that the fair trade share of the total cocoa production by Kuapa's members is very small. The benefits of selling part of the cocoa as fair trade cocoa are divided among all members.

Conditions under which farmers are included

The opening of alternative markets in Ghana cannot take place without the involvement of Cocobod. The government is mainly interested in product quality and increasing volumes of production. As a result only a few initiatives have been realised so far. Another requirement for the opening of alternative markets is the set-up or formalisation of farmer groups. This is not an easy task as trust is lacking and farmers need to be convinced that farmer organisation has tangible benefits. As I illustrated, in Ghana the introduction of reforms did not go hand-in-hand with incentives for farmers to organise themselves.

Compared to farmers who continued with the production of only conventional cocoa the conditions for farmers who produce niche cocoa are generally favourable. However, it can also be a risky affair. For example, producing certified organic cocoa requires an effort from the farmers; organic cocoa production demands more intensive farm management. In theory, the premium paid for organic cocoa more than compensates for these costs; but, premiums fluctuate. It is possible that sometimes the premium does not cover the increased costs of production.

7.2.3 Strategy 3: Localising commodity processing and marketing

Localising commodity processing and marketing involves different sub-strategies. For example, farmers can add value to their cocoa by processing cocoa waste. Involvement in marketing of cocoa is another option, although this option is limited to involvement in internal marketing (external marketing is still under state control). Another sub-strategy is the local processing of cocoa (by-)products (Figure 7.17).

Figure 7.17   Strategy 3: Localising commodity processing and marketing

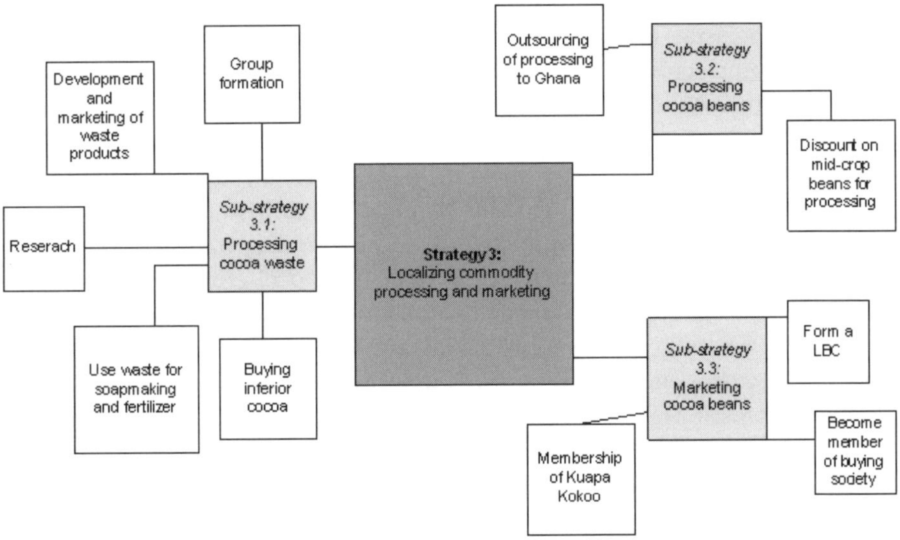

Source: composed by author.

Sub-strategy 3.1: Processing of cocoa waste

Material from the cocoa pod and from the cocoa beans (which is normally discarded [ICCO, 2003 in Bass, 2006: 260 and personal observation]) can be used as an ingredient for other commodities. Processing cocoa waste adds value to cocoa-bean production. For example, cocoa processing companies separate the shell from the

177

cocoa beans they buy from CMC and export these as fertilizer, mainly suitable for application in gardens.[197] Recently there were some experiments with using shells as bio fuel. The national cocoa research institute (CRIG) is also exploring ways of using cocoa waste as input for fertilizer, soap making, cocoa liquor, cocoa jam and animal food. In 2005 they organised a workshop where they tried to find over-sea marketing channels for these products. The primary aim of CRIG was to generate new sources of income for the research institute itself; it did not aim to improve the income generating opportunities for farmers (interview CRIG, 2003). On the level of the farmers there is little involvement in the processing of cocoa waste into by-products. Only 'soap making' occurs on a regular basis; in 2003 around 28% of the interviewed farmers (mainly women) used cocoa waste for the production of soap (FS 2003; Norde and van Duursen, 2003). This soap is mainly used for domestic consumption, with a small portion being sold at the local market. During my fieldwork in 2005 I met several women's groups involved in soap making. There is also a farmers' cooperative that was established in Asikuma (Central Region) to concentrate on soap making. The government promised to give assistance for scaling-up soap production and commercialising soap making activities in Asikuma, but in 2005 farmers complained that the government did not fulfil its promise. It would be worthwhile to explore new opportunities for farmers to utilise cocoa waste, especially as fuel and as fertilizer (which is an expensive input for farmers).

Sub-strategy 3.2: Localising processing of cocoa products
Localising processing and marketing of cocoa products, i.e. functional upgrading, are in theory presented as the most promising fields of adding value to cocoa. Gibbon and Ponte (2005: 153) put it more firmly by arguing that in the case of cocoa second-tier suppliers (in the form of local exporters or smallholder cooperatives) 'can upgrade only by taking on first-tier supplier roles – that is, by engaging in international trading/or grinding'. Similar to coffee, serious physical and financial obstacles constrain the development of a local grinding industry. In Ghana the government has been actively involved in setting up local grinding facilities. It also actively stimulated foreign processors to outsource part of their processing facilities to Ghana, by offering processing companies a 20 per cent discount on light-crop beans. Without this subsidy cocoa processing in Ghana is not a profitable business (informal discussions with industry, 2007). The Ghanaian beans are relatively expensive, building a factory is a tremendous investment and the other ingredients for making chocolate have to be bought at world-market prices.[198] Although cocoa processing entails a cost for both international processors and the Ghanaian government, it is also a strategic interest of both parties. For the government it is a way of securing the future demand for their product, while for international buyers outsourcing of their processing facilities to producing countries is a way of 'physically' moving closer to the Ghanaian farmers. This will become especially important if the process of liberalisation proceeds any further. Moreover, the political crisis and the relatively low farm gate price of cocoa in Cote d'Ivoire have encouraged the movement of cocoa into Ghana. In the analysis on the intervention I will provide more information on processing that is currently taking place in Ghana.

The farmers' perspective on processing cocoa products

The processing of cocoa into cocoa-products, such as liquor, cocoa powder, cocoa paste etc., which are used in chocolate manufacturing and confectionary stage, requires large financial investments and a completely different type of knowledge than is needed in the production phase. For individual farmers direct involvement in local processing of cocoa is not a goal. There are no interventions that aim to involve farmers in cocoa processing. In the next section I will discuss the fourth intervention – private grinding activities in Ghana.

Private intervention outsourcing of grinding activities in Ghana

In the period 1985-1995 parastatals and public/private joint ventures in Ghana and especially Côte d'Ivoire established local grinding operations. Today, most of these ventures are in foreign hands (and can no longer be seen as a way to upgrade). In 2004 in Ghana four processing factories were operational (Table 7.11).

Table 7.11   Cocoa processing in Ghana

Cocoa processing companies	Owner	Type of production	Installed processing capacity in 2004 (tonnes)
PORTEM (Tema)	Privatised in 2002, but Cocobod is major stakeholder	Consumer products (Golden Tree Chocolate)	65,000
WAM and WAMCO II (Takoradi)	Joint venture Cocobod/ Schroeder of German Hosta Group	Semi-finished products	70,000
Barry Callebaut (Tema)	Barry Callebaut	Semi-finished products	75,000
Cargill	Cargill	Semi-finished products	Not yet installed[199]
ADM	ADM	Semi-finished original products	Not yet installed[200]
Total			210,000

Source: Adapted from Bass, 2006: 251 and completed (based on interviews with CPC, Barry Callebaut and Cargill in 2005).

The Cocoa Processing Company Limited PORTEM (CPC) at the harbour city Tema (Greater Accra) used to be a subsidiary of Cocobod but now is privatised with Cocobod as its major shareholder (picture 7.4). Its products, the Golden Tree Chocolate, are consumed in Ghana and surrounding countries. The Ghanaian government is actively promoting domestic consumption of cocoa. In these campaigns there is a strong emphasis on the consumption of chocolate and its positive effects on health.

Two other processing factories in Takoradi (harbour city in the Western Region) are run by WAM and WAMCO II, a joint venture between Cocobod and a small German processor. They process semi-finished products. In 1999, all of these processors used between 18 and 22 per cent of Ghana's total bean production (Ministry of Finance, 1999). A fourth processing company was installed in 2004 in Tema, by the Swiss 'giant' Barry Callebaut, with an installed capacity 75,000 tonnes per year. In 2006

Picture 7.4   Visiting the Cocoa Processing Company in Tema

Cargill also opened a processing facility in Tema, with a processing capacpity of 65,000 tonne per year[201]. Very recently, in October 2009, ADM opened a new processing plant in Kumasi for 'single source origin' cocoa, with a processing capacity of 65,000 tonne per year.[202] In ADM's press release on this issue (June 7, 2007), Mark Bemis, the president of ADM Cocoa, stated that this investment:

> also represents Ghana's growing importance in the cocoa processing value chain. By locating the plant in Kumasi, we will be processing cocoa closer to the farmers and providing local jobs to the community, [...] In addition, it fits securely within the Ghanaian government's strategic and economic objectives of adding value to its cocoa production.

Benefits for farmers

These cocoa processing companies use small (high quality) beans from the low season, which they buy at a 20 per cent discount. Even though the farmers receive the same price for light crop beans as for mid crop beans, indirectly they actually subsidise industry because the discount is paid from the export value (gross FOB).

In the long-run the outsourcing of grinding facilities to Ghana does help to guarantee the future demand for Ghanaian cocoa. Moreover, the export of cocoa beans and cocoa-products makes significant contributions to the country's budget; cocoa bean exports account for about 40 percent of the country's foreign exchange earnings. Cocoa provides the second largest source of export dollars to Ghana bringing in almost $500 million yearly to the Ghanaian government (Bass, 2006). A portion from the profit from cocoa is (partly) reinvested in the economy.

It is interesting that the Ghanaian government (in its cocoa strategy published in 1999) questioned the fairness of the discount provided to processing industries: 'there arises a fundamental question of fairness as to whether removing subsidies at farm level [on inputs] but providing them at the level of processing industries at the expense of reduced prices to farmers is fair' (Ministry of Finance, 1999: 77).

Constraints, trade-offs of intervention
Ghana just like other West African countries enjoys tariff-exempt status for exporting to the European Union through the Lomé/Cotonou Treaties. However this did not lead to a favourable situation for Ghana to export semi-processed and processed cocoa beans. The grinding facilities are almost exclusively owned by the same multinationals that dominate the international cocoa business, thus reducing possibilities for technology transfer (Bass, 2006: 251). There is also a constraint for international processing companies – there are not enough light crop beans to meet the expansion of cocoa processing capacities. This means that the government cannot fulfil its promises to all the processors of cocoa, which affects adversely the relations between Cocobod and the foreign processors (informal discussions with industry, 2007). A positive trade-off for processors is that their presence in Ghana helps to consolidate their relationship with Cocobod and it also helps them gain insight into local dynamics. If changes do occur (in terms of further liberalisation or in demand) they will be better able to respond.

For farmers there is a negative trade-off. The discount given to multinational processors is actually paid by the cocoa farmers themselves. Farmers have no idea how reinvestments in the sector take place and who benefits from these interventions.

Sub-strategy 3.3: Marketing of cocoa beans
The marketing of cocoa beans is still controlled by Cocobod. Despite the introduction of gradual reforms, external marketing is not liberalised. Internal marketing is liberalised and in hands of Licensed Buying Companies. Currently there is one farmer organisation that also functions as a LBC. The establishment of a private buying company is a way of moving up in the value chain: if farmers become owners of a buying company they become involved in activities higher up in the chain, which adds value to their cocoa beans. LBCs are paid a fixed margin of the FoB for their marketing activities. Becoming a LBC also involves costs as the cocoa has to be collected, stored and there is also a process of quality control to administer. Also, internal trading demands time and skilled local purchasers with a good reputation in the community where they buy the cocoa. It also requires a license, which is obtained from Cocobod (see Chapter 5). In short, obtaining a LBC licence requires experience, investment, and must be grounded in a long-term approach. Foremost, obtaining a license requires good contacts with Cocobod (interviews with different LBCs, 2005). So far only one farmer organisation (Kuapa Kokoo Farmer Union) established a buying company (Kuapa Kokoo Ltd.) (recently the same organisation established a second LBC).

Table 7.12 Measuring the inclusiveness of cocoa processing in Ghana

Strategy 3: Localising processing and marketing	Intervention	Activity	Mechanism	Farmers reached	Expected impact	Constraints	Trade-offs
Sub-strategy 3.2 Processing of cocoa	International buyers	Outsourcing cocoa processing in Ghana	None reached	Impact 1: Contribute to long-term demand for Ghanaian cocoa	This is no upgrading strategy involving farmers	Discount is indirectly paid by farmers (as part of gross FOB)	

Effectiveness This type of intervention is not inclusive. Farmers are not involved, and do not directly benefit. However, it does secure a future outlet for their products. In addition, the presence of international buyers will stimulate the government to adapt more pro-actively to changes on the world market. It is likely that some of the processors that have outsourced part of their activities to Ghana will ask for ways of certification of their mainstream product. Their presence gives some weight to the alliance between processors and Cocobod. Furthermore, if liberalisation proceeds it will be easier for international buyers to establish direct relationships with farmers, as they would be already present on the ground.

Farmers can become involved in marketing cocoa beans through membership in Kuapa Kokoo, or in a more direct way by applying for the job of local purchaser of cocoa (PC). Generally LBCs look for literate clerks with a good social network. The purchasers of cocoa are paid on commission basis and most PCs are also cocoa farmers.

Conclusions on Strategy 3
The private sector is the main player involved in the local processing of cocoa. There are some joint ventures where Cocobod is shareholder. Local processing within Ghana is beneficial for both multinational processors as well as for the Ghanaian government. The Ghanaian government has a strategic interest in attracting this type of industry to their country; it helps to secure long-term demand for their product and to consolidate their relationship with this industry. Multinational processors are very interest in Ghanaian cocoa because of its high quality and dependable delivery. In addition the political instability in Cote d'Ivoire and a growing interest in traceability made Ghana an even more attractive source country. This alliance between Cocobod and international buyers does have an impact on the farmer, but this impact is indirect. Farmers are not directly reached by these types of interventions and are not involved in processing activities.

Figure 7.18 Changes in the empowerment matrix due to outsourcing of local processing of cocoa to Ghana

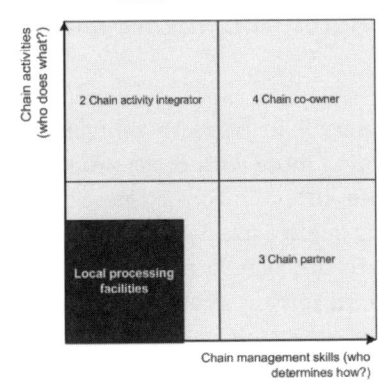

Source: author.

In terms of inclusiveness this intervention is not effective. Farmers are not involved and do not directly benefit from the presence of foreign processors.

Conditions under which farmers are included
Since the introduction of reforms the number of foreign processors in Ghana gradually increased. These processors process cocoa-beans into cocoa-products which they then export. Moving 'physically' closer to the farmers has not altered the conditions under which cocoa farmers run their businesses. The planned activities of individual buyers of cocoa, for example Cargill, give some indication that this might change. There is a large number of training activities for farmers that are

planned to take place in 2012 (TCC, 2009) (see also Chapter 4). The involvement of buyers in certification schemes and their commitment to sustainable cocoa sourcing are other indications that the conditions under which farmers are included might change. It is not yet clear exactly which groups of farmers will be targeted by these kind of programmes, and which groups will be excluded.

7.3 Discussion on more inclusive upgrading for cocoa farmers in Ghana

In this chapter I analysed a small number of interventions. In order to understand the rationale behind these interventions and their impact, I placed them in a broader context, providing some information on other interventions and some background information on the corresponding sub-strategies. Earlier in this chapter I made some concluding remarks on the three strategies and discussed the main drivers of the selected interventions, their impact and the (changing) institutional environment that facilitates the interventions. In this section I will look at all the identified interventions (presented at the beginning of this chapter) and unravel some patterns of upgrading, with an emphasis on different their different impacts. A full summary of the analysis will be presented in the appendix (7.1).

7.3.1 Upgrading patterns in Strategy 1: Capturing higher margins for unprocessed cocoa

This strategy defines the objective of most farmers – to increase the volume of production and to improve productivity (thus earning more with cocoa production). This first strategy is dominated by large-scale interventions. The Ghanaian government is the main intervener in safeguarding quality standards and increasing volumes of production. International buyers share the agenda of the government but play a rather passive role. Control and standard setting affect all farmers, but interventions that provide services are not easily accessible to everyone. For my respondents, the main determining factors were ownership, social position in the community and location (region). The farmers themselves are responsible for producing high quality cocoa, but toil under diminished incentives. In terms of volumes of production and productivity, the farmers are actively involved and have developed different ways of increasing their volume of production, for example through effective pest management, planting new varieties of cocoa and working together in labour exchange groups.

Competitiveness
Interventions that aim at the production of premium quality beans (Sub-strategy 1.1) contribute to the good reputation of Ghanaian cocoa and its competitive advantage on the world market. In addition, the premium that Ghana receives for its beans adds value to the produce and offers farmers a stable income. Generally these are

compulsory measures; but (at the same time), they are also beneficial for all farmers. Due to a lack of transparency in the distribution of costs, benefits and risks it is not possible to measure whether the production of premium quality beans ultimately results in higher incomes for farmers. These interventions do not empower cocoa farmers.

The incentives for the local private sector, quality control officials and farmers to invest in quality management are reducing. Farmers shoulder part of the risks in case quality problems result in lower prices or rejected produce.

The farmers' knowledge on producing high quality cocoa beans mainly comes from their families. Traditionally, also extension services play an important role. Access to advice on quality issues depends significantly on ownership, kind of contract, yield, position in the community and region.

Remunerative income
Interventions that aim at the production of increased quantities of cocoa and higher levels of productivity (Sub-strategy 1.2) generally contribute to farmers obtaining higher returns. The scale of interventions varies; governmental interventions are large-scale. Generally, the measures taken to increase production levels are stimulating and exclusive, with three important determinant variables: the farmers' position in the community, his/her position on the farm and the location of the farm. Among the interviewed caretakers also the type of contract mattered.

The cocoa farmers' main objective is to invest in higher volumes of production; although many of them are constrained by the high production costs and the lack of credit. The opportunities for farmers to invest (time or money) in their farm are not equal and farmers make different choices. Some farmers for example chose not to work together in exchange labour groups or were constrained (for example by gender or type of contract). In some cases the region where farmers work determined their investment (for example planting new varieties of trees).

Interventions in the field of more remunerative contracts (Sub-strategy 1.3) are generally stimulating. Contract farming between farmers and actors higher up in the value chain does not exist (due to the gatekeeper role played by Cocobod). The type of share contracts between owners and caretakers is determined by the owner. Caretakers do have a choice in working under more than one contract.

Empowerment
Some of the interventions made in order to increase production levels contributed the empowerment of farmers (mainly through training and extension services). These types of stimulating measurements are exclusive. It is simply too costly to provide training to all cocoa farmers. Another reason is that since the merger of Cocobod's extension services with MoFA's services the quality of the service declined and less cocoa farmers have been reached.

Generally, caretakers have more difficulty accessing training programmes than farm owners. The position in the community also influences the opportunity to receive public extension services. Location was another significant variable. Farmers in (the more remote) Brong Ahafo and Ashanti Region had less access to services than farmers working in the Western or Central Region.[203] Unfortunately, training and extension do not always yield positive effects. Adoption rates are low (related to high costs of input), inputs are often not applied adequately and services are generally top-down.

There are two multi-stakeholder initiatives that use the concept of farmer field schools (FFSs). These schools provide farmer-based extension services. Both initiatives are stimulating, small-scale and exclusive (implemented as pilots in specific regions); both contributed to empowering farmers.

7.3.2 Upgrading patterns in Strategy 2: Producing new forms of existing commodities

The second strategy concerns mainly multi-stakeholder initiatives and NGOs, which are generally small-scale and exclusive. Among the respondents ownership and region played a decisive role in access to this type of strategy.

Competitiveness
Interventions that aim at the production of cocoa for niche markets (Sub-strategy 2.1) open alternative markets that generally pay higher prices for unprocessed beans (although this premium not automatically benefits the involved producer). These types of interventions are still in a pilot phase or small-scale; initiated by international buyers and/or NGOs, they are heavily dependent on good collaboration with Cocobod. These interventions involve small groups of cocoa farmers in specific locations. Some of these interventions are stimulating, paying farmers a bonus or offering other types of benefits. An exception is a recent initiative to produce 'single origin cocoa-products'. The production of 'specialty cocoa' however does not benefit Ghanaian farmers directly. Because the beans (and cocoa products) are still marketed by Cocobod, farmers do not get a higher price. They still do obtain some financial benefits as this has a positive effect on long-term demand.

Cocobod has been reluctant to open up separate marketing channels for niche cocoa and obstructed several attempts to introduce organic cocoa production in the country. This consistent resistance resulted in absence of product differentiation in Ghana.

Remunerative income
There are more interventions that aim at fostering non-traditional uses of cocoa (health and cosmetics) and local consumption (Sub-strategy 2.2), however many of them are still in an initial (research) stage. So far, these types of interventions do not directly benefit farmers; there are no separate marketing channels and there is no product or price differentiation.

The development and marketing of cocoa by-products (Sub-strategy 2.3) do not involve cocoa farmers directly (they do not have the technologies, knowledge and marketing opportunities). CRIG does invest in cocoa by-product research, but the initial aim is to increase its own budget.

Producing other cash crops and food crops in order to diversify sources of income is a common strategy used by farmers. Around 20 per cent of the farmers generate some income from non-farm activities. Generally diversification is a risk management tool and the means to obtain some additional income throughout the entire year.

Empowerment
The production for niche markets can have additional benefits for farmers in terms of empowerment. The Kuapa Kokoo Farmer Union is an example in Ghana. It produces a small amount of cocoa for the fair trade market and is a dominant shareholder in a small chocolate marketing company based in the UK. The members of the union (around 50.000) also formally own the union, which is at the same time a LBC. The farmers' involvement in decision-making processes and in activities higher up in the value chain, in combination with access to training programmes (such as the FFSs) and income-generating activities contributed to their empowerment.

Caretakers are generally excluded from membership of KKFU (as they need permission of the farm-owner). Women, on the other hand, are a specific target-group of KKFU; women are stimulated to become members and receive training on (other) income-generating activities. So far, the economic benefits resulting from KKFU membership have been marginal.

7.3.3 Upgrading patterns in Strategy 3: Localising commodity processing and marketing

The third strategy is exclusive or does not reach farmers at all. The multinational buyers are the main interveners, they share an interest with the government in outsourcing part of their processing capacity to Ghana. The interventions aimed at marketing reach farmers indirectly.

Competitiveness
The government intervened in localised processing (Sub-strategy 3.2) by providing cocoa processors with a 20 per cent discount on mid-crop beans and by offering other incentives. As a result, over the last years an increasing number of processors have opened local processing facilities. The impact on farmers is indirect. Farmers still market their cocoa through Cocobod and have no direct relations with international buyers. But it is likely that the establishment of cocoa processing facilities within the country will contribute to the long-term demand for Ghanaian beans, which benefits all farmers.

The incentives given to international processors are financed from the gross FOB price; thus, indirectly farmers do pay for the subsidies given to these multinationals.

Remunerative income
Processing cocoa waste (Sub-strategy 3.1) takes different forms. Cocoa processing companies are involved in this activity on a large-scale, but this does not have an impact on farmers. Soap making by (groups of) farmers does take place on a regular

basis, mainly for domestic consumption and for small scale trading on local markets. Women are the main producers of soap. Whether or not they benefit from this depends on the arrangements within the household. Generally, the economic benefits resulting from soap production are small. The government is minimally involved and provides no substantial benefits.

Empowerment
Farmers are involved in cocoa marketing activities (Sub-strategy 3.3) in different ways. In discussing Strategy 2, I already indicated that through membership in the KKFU farmers get involved in internal and external marketing activities. The economic benefits are marginal, but involvement in these kinds of activities can potentially have an empowering effect. Another way for farmers to become involved in marketing is by becoming a society member or by becoming a Purchasing Clerk.

7.4 Capturing dynamics, thinking in scenarios

The Ghanaian case, often presented as best practice, embodies two important dimensions: first, it is unique due to its partially liberalised economy; and second, it is exceptional for its production of large quantities of premium quality cocoa. The partially liberalised system reflects partly the strong role of the Ghanaian state and partly the global buyers' interest to maintain or only slightly to modify the Ghanaian system. The production of premium quality cocoa reflects both the capacity of the national government to coordinate the supply chain as well as the existing high demand for premium quality cocoa.

The upgrading strategies taking place in Ghana, which focus on quality and volume of production, reflect these dimensions. But the conditions underlyin g these dimensions are not fixed. First, there is a trend in the global cocoa chain that product requirements become less important. Gibbon and Ponte (2005: 200) warned of the potential for exclusion and marginalisation if African farms fail to meet the new expectations concerning: quality, lead times, volumes, and prices. A risk is that they will fail to capitalise and actively participate in shaping new standards to their advantage – including those that are related to social and environmental concerns raised by Northern NGOs. There is also a risk that achieving/maintaining high quality standards may not attract higher prices (or add value) for producers. Second, it is not sure if a partially liberalised system is the end-stage of the reforms. So far, global buyers of cocoa support the Ghanaian government and Cocobod in continuing its coordinating role and are prepared to pay a premium price for Ghanaian cocoa. But preferences of global buyers can change or pressure for change can come from the World Bank or can come from within.

Understanding the position of Ghanaian cocoa farmers in the chain and the kind of upgrading strategies that are beneficial for farmers require a dynamic perspective, not only by drawing lessons from the past and making comparisons with experiences in fully liberalised countries, but also by taking into account possible future scenarios. Ghanaian farmers are better off now, but what if the main pillars that underpin their strong position disappear?

Ongoing liberalisation and/or a growing demand for process standards (such as environmental and social standards) require other types of upgrading strategies and interventions. I use the 'scenario matrix', as introduced at the beginning of this chapter, to understand better the vulnerability of the current system. Already in Figure 7.3 (Section 7.1) I distinguished four possible scenarios: 1) Status quo with passive private sector; 2) Opening up; 3) Loosing control; and 4) Status quo with active private sector. The first scenario reflects the current situation. For each scenario I will discuss briefly its main features, I will reflect on the kind of interventions that are needed to guarantee benefits for farmers under these changed conditions. Furthermore, I will indicate what developments possibly trigger moving from one scenario into another. This exercise of 'thinking in scenarios' is based on my findings presented in the previous chapters, particularly on the chapter analysing the comparison with other fully liberalised cocoa producing countries and on the trend among global buyers to source cocoa sustainably.

7.4.1 Scenario 1: Status quo with passive role of private sector

In this scenario, the cocoa sector in Ghana is partly liberalised and there is high demand for premium quality cocoa beans (product quality). This scenario reflects the current situation. Currently, the state controls the supply chain and the demand for premium quality cocoa is high. In order to remain competitive farmers will have to continue to produce high quantities of premium quality cocoa. For this purpose, the Ghanaian government intervenes actively in the sector, for example through its quality control system and the mass spraying programme. These large-scale interventions have been mainly contributing to competitiveness of the cocoa sector as a whole and also to increased and stable incomes for farmers. Still, this scenario, which is supported by the alliance between international buyers and the national state, disregards some of the interests of the farmer and of the local private sector and the system in place does not automatically provide sufficient incentives for farmers and for actors higher up in the chain to continue with the production of high volumes of high quality cocoa. Without such incentives and without more transparency regarding the distribution of costs, benefits and risks Ghana's cocoa sector might become locked in a negative quality performance spiral. Without better extension services and the provision of (input on) credit farmers will have difficulty to prevent or overcome diseases, which hinder farmers continuing to produce high volumes of cocoa and to increase their productivity.

Still, comparing to cocoa farmers in neighbouring countries and comparing to other type of farmers in Ghana, Ghanaian cocoa farmers are relatively better of (World Bank, 2007b). But how is this if (one of) the main pillars that underpin Ghana's relatively favourable position disappear? There are a few developments that might trigger change, such as the global trend of product requirements becoming less important in favour of process requirements. Bu as demand for Ghana's premium quality cocoa (product quality) remains high, Ghana has been rather slow in broadening its focus and has obstructed for example cocoa production for niche markets (such as organic cocoa) (process quality). While Ghana has the capacity to develop new domains of rent, they put little effort and continue to focus on current activities. Considering the increasing attempts of global buyers to look for sustainable sourcing of cocoa, there is a risk that the Ghanaian government may start to become an obstacle instead of an enabler in facilitating 'inclusive' upgrading strategies.

The risk of inertia is also put forward by the IMF (2009); it argues that governments 'should do well to pay attention to structural dynamics in global trade such as new trends in market information processing, logistics, customer analysis etc. in order to explore emerging and niche markets' (IMF, 2009).

7.4.2 Scenario 2: Opening up

This scenario reflects a partially liberalised system and increasing demand for process requirement. Currently Ghana is the only country where cocoa beans are still consistently separated by national origin for grinding purposes (Gibbon and Ponte, 2005: 136). Product quality is getting even more important as processors, such as ADM, started to build processing facilities for 'origin cocoa products' from Ghana. Nevertheless, globally there is an increasing demand for process requirements (or performance requirements). Consumer behaviour is not only determined by 'price-quality' decisions; consumers (and supermarkets) are also increasingly interested in conditions under which the cocoa is produced. Evidence for practices of child labour and slavery in cocoa supplying countries contributed to this demand, but it is also a reflection of a growing global demand for organic and healthier products. The attention given to corporate social responsibility is (partly) a response to this. Global buyers of cocoa have become increasingly involved in public-private partnerships that aim at securing sustainable sourcing for part of their cocoa, at strengthening farmer organisations and at exploring niche markets.

Inclusive upgrading: moving from scenario 1→2

If Ghana moves from scenario 1 to scenario 2, other kinds of strategies will be required (Figure 7.19 illustrates this shift).

The current interventions that aim at more sustainable practices are small-scale and mainly initiated by public-private partnerships. It is necessary to up-scale these initiatives. The high cost of these types of programmes (such as the farmer field schools) is a clear constraint. More inclusive upgrading in this situation would also call for more transparency with respect to the state's re-investments in the sector. Ghana is doing relatively well but a lack of transparency and information makes it impossible for other actors in the chain to evaluate the Ghanaian situation and to act accordingly. Research and more information on, for example, child labour in the sector can help to demonstrate Ghana's level of sustainability.

In order to motivate farmers to produce cocoa differently, extension services and credit services have to be improved. In addition, farmers should also be given price-incentives, which would require the introduction of a system of price and product-differentiation. Sustainability is not only about producing cocoa differently but also about improving the method of cocoa production, enabling higher levels of

Figure 7.19   Moving from scenario 1 ⟶ 2

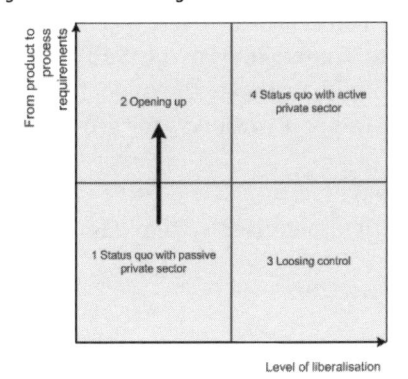

[Figure: 2x2 matrix with axes "From product to process requirements" (vertical) and "Level of liberalisation" (horizontal). Quadrants: 1 Status quo with passive private sector; 2 Opening up; 3 Loosing control; 4 Status quo with active private sector. Arrow pointing from quadrant 1 to quadrant 2.]

Source: composed by author.

productivity, and about strengthening farmer organisations. Lastly, it is about environmental objectives. In all areas Ghana is currently lacking behind.

7.4.3   Scenario 3: Loosing control

This scenario reflects the continuation of focusing primarily on product quality. But instead of an active role of the Ghanaian government, it assigns the coordination of the supply to the private sector. Cocobod can play a supportive (or hindering) role, or its subsidiaries can be privatised.

Ghana is an exceptional case because it is only partially liberalised. The current status quo reflects partly the interests of multinational buyers that are currently benefiting from the Ghanaian mixed system, and partly the strength of the Ghanaian government in resisting World Bank pressure to fully liberalise its cocoa economy. Officially, the reasoning behind the gradual introduction of reforms is that Ghanaian government wants to give the private sector more time to build its capacity to to become successfully engaged in external marketing. But, so far, local buyers have not been given the license to export part of their cocoa directly on the open world market. Even though this persistent resistance has frustrated some of the larger buyers, it is understandable from the point of view of the government. They are looking to the neighbouring countries and their negative experiences with fully liberalised cocoa producing systems. The result is that LBCs can only compete on volume and receive little incentive to invest in quality control and in building relationships with farmers. This has also frustrated farmers as purchasing clerks cheat farmers on scales and do not honour their promises.

Ghana is the only country that produces premium quality cocoa, a necessary ingredient for making good quality cocoa products. The growing outsourcing of processing activities to Ghana (for example by Cargill and ADM) seems to indicate that future demand for Ghanaian cocoa is secured. Nevertheless, a push for further reforms could come from international buyers. Global buyers choose to intervene in cocoa production or processing only if it helps them mitigate their risks. These risks can be global and in particular it should be emphasised that the solutions they propose are not necessarily beneficial for individual source countries (See Chapter 4). A push for further liberalisation could also come (again) from international institutions. Looking at the impact of the recent financial crisis on Sub-Saharan Africa, the IMF (2009) suggested that 'countries should also seize the opportunity to advance their structural reform agendas in order to boost prospects for growth'.

Inclusive upgrading, moving from scenario 1→ 3
Moving from a partially liberalised system to a fully liberalised system (Figure 7.20) has some implications for upgrading.

Figure 7.20   Moving from scenario 1 → 3

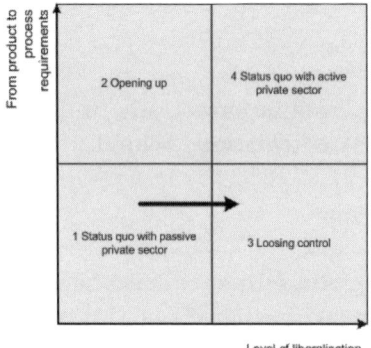

Source: composed by author.

Looking at experiences in other fully liberalised cocoa producing countries, it already became clear (Chapter 5) that in such a setting prices tend to fluctuate, the costs of cocoa production generally increase, and the quality of the cocoa declines. All together this causes loss of premiums and losses in demand. Full liberalisation without offering incentives to farmers to produce premium quality cocoa (for which in this scenario there is still demand) and without offering incentives to local buyers and private quality controllers in order to make sure product quality standards are met is likely to cause problems. In addition to strategies that help secure the production of high volumes of quality cocoa, in this situation there is a need for strategies that support farmer organisation. The partially liberalised system does not provide the incentives for farmers to organise themselves. In a fully liberalised setting this neglect could be disastrous. Individual farmers cannot deal directly with large buyers and have no bargaining position. A fully liberalised setting would also require investments in the relations between smallholders and traders. Moreover, in order to secure tangible benefits for the farmers, a fully liberalised setting demands an effective information system and price-differentiation.

7.4.4   Scenario 4: Status quo with an active role of the private sector
This scenario, where the sector is fully liberalised and the demand for process-requirements increases, is not likely to occur over-night. However, increasing

demand for process requirements can exert pressure on the Ghanaian government to open up alternative marketing channels for 'sustainable cocoa'. It can also result in elevated demands for increasing the transparency and traceability of the Ghanaian cocoa sector or it could push for an alternative 'process quality control system'. These different steps can accumulate and push Ghana into the direction of introducing additional reforms. In a situation where the strong Ghanaian government becomes a hindrance instead of an ally of international buyers, the status quo of a partial liberalised system is at risk. This could be the case if for example the Ghanaian government was reluctant to participate in mainstream certification programmes.

Inclusive upgrading: moving from scenario 1 ➤ 4

This scenario would demand a variety of interventions in order to contribute to developing inclusive approaches for upgrading, combining the interventions already mentioned in Strategies 2 and 3.

Figure 7.21   Moving from scenario 1 ➤ 4

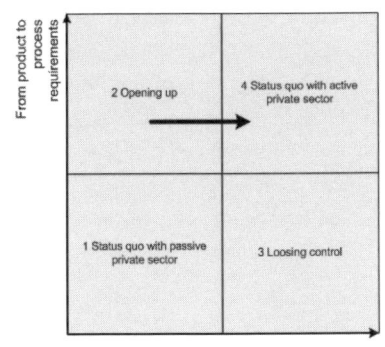

From product to process requirements

2 Opening up	4 Status quo with active private sector
1 Status quo with passive private sector	3 Loosing control

Level of liberalisation

Source: composed by author.

This scenario would require the upscaling of sustainable practices, more transparency, better services for the farmers and the strengthening of their organization capacity. In addition, this shift requires capacity building of other actors involved in the cocoa sector that are taking over public tasks. An option is to reorganise Cocobod and its subsidiaries, thus enabling them to continue to play a meaningful role. It should be avoided to create a situation similar to Côte d'Ivoire where even though the sector is liberalised, the government still collects a large share of the margin but does not reinvest money back into the cocoa sector.

Final reflections

Ghana is not well-prepared for change. Farmers and the private sector are particularly vulnerable in the current system. Looking at experiences in other cocoa growing countries in the region that fully liberalised their cocoa sector it has become clear that weak farmer organisations and a weak private sector are severe bottlenecks for farmers to benefit from further reforms. Nevertheless, the Ghanaian government is not investing in capacity development of private buyers of cocoa and farmer organisations. The lack of investment in farmer organisation also makes it increasingly difficult to meet (changes in) demand, for which being organised becomes more and more a prerequisite. Moreover, this has contributed to a lack of agency among farmers to change their position and to benefit more from the current partially liberalised system. More inclusive upgrading requires more emphasis to be placed on empowering farmers and local private actors, it also requires more awareness (beforehand) of whom interventions are likely to include and whether they intensify unequal social structures in the Ghanaian society or contribute to transforming them.

8
CONCLUSIONS

8.1 Introduction

The objective of this study was to develop a thorough understanding of the opportunities and constraints that producers of primary commodities face in their effort to improve their position in the global value chain. This research is particularly relevant now when farmers are expected to behave more like 'firms'. However, as my results show, agents higher up in the chain and the institutional environment that surround farmers still largely determine scope for change and the direction of 'progress' available to farmers.

The central question in this study was how global chain governance, national governance processes and social structures interact in making more inclusive upgrading strategies for small-scale cocoa farmers in Ghana (thus linking upgrading explicitly to development). In my analysis, I took a multi-level and dynamic approach, by analysing shifting power relations, structures and human agency (on the global, national and local level) and also by looking at the ongoing interactions between these different levels. Employing several research tools, I looked at different dimensions to understand the upgrading opportunities and constraints that farmers face. The value chain approach was used as a tool to identify upgrading opportunities and constraints for cocoa farmers in Ghana. The concept of state governance was used to identify (additional) upgrading opportunities and constraints by looking specifically at the changes in cocoa farmers' national institutional environment, i.e. the introduction of reforms in the cocoa sector in Ghana and the changing role of the state. At the level of the farmer, the concept of 'embeddedness' was used to identify social relations that constrain or facilitate upgrading strategies and affect the way farmers benefit from these strategies.

8.2 Global Value Chain approach and upgrading

I employed the global value chain (GVC) approach to answer the following question: How are the main interests/risks of global actors currently governing the cocoa chain being manifested locally, both through their involvement in local upgrading strategies in Ghana, and through establishing more direct relations with cocoa suppliers and the formation of new public-private partnerships?

The main hypothesis, underlying this question, is that actors higher up in the chain than the producers are the ones who (have agency and) determine the producers' room for manoeuvring and the direction of upgrading. This conflicts with one of the key assumptions of the GVC literature; namely, that a value chain is considered a dynamic open system where producers in developing countries can act as active agents and upgrade their 'business'.

Another assumption of GVC literature is the presumed existence of open market conditions. I illustrated that in Ghana this is not the case. The government, through Cocobod, curtails the role of international buyers (and other players in the chain). Consequently international buyer interventions are small-scale and limited to extension services and training. Despite the fact that international traders/processors (and to a lesser extent also chocolate manufacturers) are increasingly driving the global chain, in Ghana the state remains a powerful entity in control of the supply chain. With respect to the GVC theory the dominant role of the state in Ghana implies a need to discuss other more hybrid types of governance based on various forms of coordination and public-private-civil partnerships, and to move beyond the usual distinction between buyer, trader and producer-driven chains.

Do international buyers want to become more actively involved in Ghana? In other fully liberalised countries international buyers established direct trade relations with farmer groups. Besides extension and training, buyers invest in information systems, quality control and marketing skills. Buyers execute large scale interventions in other countries primarily due to the problems in these countries (for example poor quality of cocoa in Cameroon and Nigeria, political instability in Côte d'Ivoire and the involvement of children in cocoa production) and the resulting higher risks of supplier failure. In addition, the reduced role of the state in providing marketing channels, extension services, and other services makes trade relations between buyers and suppliers more direct and interventions more common. However, this shift in chain coordination does not give much information about the well-being of the smallholders involved: were these direct relationships beneficial for the individual farmer? In Ghana, the state guarantees a consistent supply of premium quality cocoa and one obvious argument is that interventions in Ghana are simply less needed. Consequently, international buyers have limited their activities in Ghana to initiating/supporting small-scale programmes, primarily aimed at enhancing the use of more environmentally friendly practices, at addressing the problem of child labour and at financing community development. Physically international traders did move 'closer' to the farmers, but establishing grinding operations in sourcing countries did not result in stronger direct relationships between buyers and cocoa farmers.

The involvement of global buyers in social and environmental programmes is not only 'window dressing'. Since 2003, an important change is taking place among multinational traders, processors and manufacturers. Round table meetings, national cocoa platforms and multi-stakeholder initiatives have contributed to a continuous dialogue with (new) partners and, as a result, to the recognition that a sustainable cocoa economy requires the active participation of the public sector, members of civil society and farmer groups and some level of cooperation. This trend however involves high transaction costs, which explains why the private sector still favours a state intervention that is able to arrange everything (where costs remain partly invisible), such as in Ghana. Both international buyers and the Ghanaian state have (still) an interest in maintaining or only slightly changing the partially liberalised system. However, this consensus does not necessarily lead to the best outcomes for farmers.

From a development perspective it is important to recognise that the upgrading strategies initiated by actors further up in the chain can be 'sub-optimal' for (groups of) farmers. This observation connects to more recent debates on value chains and social inclusion and exclusion, where it is argued that insertion in a GVC does not automatically lead to upgrading and development for all the involved producers. This research confirmed that upgrading is a selective process. It also demonstrated the importance of looking beyond the risk of being 'excluded' from a chain (or from upgrading opportunities within a chain) to the risk of being 'included'. It is common practice to transfer risks down the chain to the producer level.

GVC analysis is limited because it abstracts too far away from the local and national socio-political context in which the producers and some other (local) chain actors are embedded. This research shows that, in models of GVC analysis, the extent and type of upgrading are determined not only by the international character of the GVC itself and inter-firm relations. Upgrading strategies and their outcomes for individual producers are also substantially determined by their heterogeneity and the character of national institutional arrangements – notably, the strength of producer organisation and the marketing sector within producer countries.

8.3 State governance and upgrading

The concept of state governance is used to answer the question: How does the changing role of the state in the Ghanaian cocoa sector affect the conditions under which cocoa smallholders operate and their opportunities and constraints for upgrading?

In the sector, there is a lively discussion on the role of the state in promoting development in the light of international competition. The international process of liberalisation reduced the direct involvement of the state in economic activities in developing countries; however, this did not automatically lead to development. This failure to provide the expected social and economic progress gave impetus to new discussions on the role of the state in sectors in transition.

Compared to other cocoa producing countries, such as Côte d'Ivoire, Cameroon and Nigeria, the partially liberalised system in Ghana has some clear-cut advantages for smallholders. The system of price stabilisation is still intact and provides Ghanaian cocoa farmers with a stable income. The Quality Control Division, still controlled by the government, contributes to the production of good quality cocoa. Ghanaian cocoa beans receive a premium price on the world market. The Ghanaian government acts as a kind of 'lead agent'; it assures consistent supply of high volumes of premium quality cocoa, resulting in remunerative contracts between Ghanaian government and international buyers and international banks. In addition the incentives for foreign processors' outsourcing of their processing capacities to Ghana more or less guarantees strong future demand for Ghanaian cocoa and cocoa products.

Contrasting the cocoa sector in Ghana with the sector in other West African countries, it becomes clear that the Ghanaian state seeks to maintain its crucial role as 'chain actor'. Its interventions are mainly directed towards economic development of the sector. Government interventions are generally large-scale and reach the majority of farmers. This cannot be easily duplicated by other actors. In 'softer' issues, such as sustainable development and pro-poor growth, the state intervenes less actively. Instead of functioning as an enabler, in some cases the Ghanaian government 'disables' (or constraints) actors from taking initiatives on these issues. In other countries that fully liberalised the cocoa sector, such as Nigeria and Côte d'Ivoire, the trend is to bring coordination of the supply chain back to the state.

In Ghana, the 'upgrading agenda' still prioritises economic sustainability, only recently giving serious attention to social and environmental issues. As a result, the level of farmer organisation remains weak. The gradual reforms also constrained the performance of public and private players in the cocoa sector. The fragmentation of extension services and the lack of bargaining power of local buyers and farmers make farmers vulnerable in the transition process. As a result farmers are not able to enjoy the full benefits of the reforms. There were different attempts at indentifying winners and losers of the liberalisation (e.g. Akuyama, 2001; Teal and Vigneri, 2004), but these studies mainly looked at price, volume and quality aspects. Understanding the full impact of reforms on the producer level requires a wider set of indicators, including remunerative farmer income, changes in the farmers' institutional environment and the agency of farmers (e.g. Haque, 2004; Long, 2001). It also requires a longer time line, as outcomes of liberalisation are not immediately visible and quite dynamic over time. My findings also suggest that farmers are not benefiting fully nor equally from the partially liberalised system; ownership, social position and locality to a significant extent determine who has access to services and who benefits from local buyer-supplier relationships. Also, the new reforms exposed farmers to new risks, such as increased production costs and living costs.

The findings show that the theoretical assumption of the state playing an enabling in creating more favourable condition for agricultural development only holds true in a fully liberalised setting. But this does not mean that in these cases the state is always a 'neutral player'; economic interests and personal gains are a powerful driving force. Nor does the state automatically possess the capacity to successfully manage interventions and to successfully contribute to the capacity-building of other actors.

In a partially liberalised economy the state still intervenes directly in sectoral development. By taking this role the state functions both as 'balancer' as well as 'bottleneck'. As a balancer it mainly protects farmers from price-fluctuations on the world market, providing them with a stable income and reinvesting part of its income back into the cocoa sector. It also guarantees international buyers the supply of premium quality beans. So, the risks for suppliers as well as for international buyers are mediated by the state, assuring smooth trade relations. In a partially liberalised system the state can also function as 'bottleneck', preventing other public, private and civil actors from taking a more active role and contributing to accomplishing development goals. The state as hinderer of development is linked to the lack of transparency on how the state calculates and distributes the costs, benefits and risks involved in cocoa production and marketing.

8.4 Social structures and upgrading

The notion that the economic behaviour of firms, markets or economic institutions is embedded in wider social relations implies that in order to identify 'inclusive upgrading strategies' it is necessary to gain insight in the differences in farmers' social relations. The concept of 'embeddedness' is used to answer the following question: How do upgrading strategies and the resulting interventions benefit different groups of cocoa farmers and to what extent do these groups participate actively in the activities and have initiated 'their own' strategies?

The theoretical assumption that local upgrading is initiated by farmers themselves (individually or collectively) and that other (global, national and local) factors enable farmers to act as active agents, is only partly true. Due to severe disempowerment among farmers, scarce few bottom-up initiatives actually improve the farmers' position in the chain. For example, the lack of formal organisation (and solidarity) among farmers reduces their capacity to negotiate for services and to benefit more from existing and alternative marketing channels. This study demonstrated that 'upgrading strategies' and the interventions that facilitate upgrading very much reflect the agenda of the dominant drivers of the chain; they are not 'neutral'. It is within this restricted space that farmers can create changes, individually or collectively.

The main strategies that farmers follow are directed towards maintaining the quality of their produce and increasing the volume of production. The data collected through farmer surveys shows that farmers faced several institutional constraints and that access to institutional support is unequal. This contrasts to another theoretical assumption that asserted that upgrading is accessible to all producers inserted in a chain. Informal institutions and social networks play a significant role in the cocoa sector and determine the choices farmers make in relation to selling their produce, their involvement in upgrading strategies and the extent to which they benefit from this participation. Therefore in this study, in evaluating upgrading outcomes, I took the heterogeneity among farmers into account.

It turns out that both 'exclusion' and 'inclusion under unfavourable conditions' are not natural processes but an outcome of social structures, land and shareholding systems and interventions. Whether or not farmers own the farm they tend, their social position in the farming community and the location of the farm can be significant in determining how farmers respond to reforms in the cocoa sector and the associated interventions. For example, my survey showed that farmers that play the role of leader in their community have better access to quality services and have higher yields (sometimes this is the reason for becoming a leader). There are other relevant indicators. For example female farmers have more difficulty to secure access to high technologies, work less with other farmers and have lower yields. Collective action was not a significant determinant in a partially liberalised system such as Ghana but is more important in other (fully liberalised) contexts.

Another important outcome of the study is that differences between farmers are inter-related. The inter-relation between farmer characteristics makes it even more urgent to understand the different impact of interventions on farmers. Understanding this process also helps policy-makers and NGOs to develop more effective interventions. Moreover, it also helps the private sector to think more strategically for realising some key goals, such as securing delivery of large quantities of good quality beans from Ghana and the strategic targeting of specific groups for assistance in developing sustainable sourcing of cocoa.

What became clear was that conventional approaches towards upgrading (GVC approach and clustering approach) have a rather narrow view on what upgrading actually entails. Conventional theory assumes that competitiveness is the goal of upgrading. Cocoa farmers in Ghana do not try 'to do things differently compared to competitors'; on the contrary, the objective for all farmers is to produce premium quality cocoa, which has a secure market. My study also showed that although learning is an important mechanism that makes upgrading possible there are other mechanisms that deserve more attention such as the more adequate distribution and application of already existing technologies. In other words, in a partially liberalised system other upgrading goals, upgrading means and upgrading impacts have to be identified. For farmers the main incentive to invest in their farm is to obtain higher levels of income and to secure a kind of social security. Another upgrading 'goal' missing from the theory is empowerment. Maybe not surprisingly, the interventions I identified are embedded within existing structures without aiming to transform these structures. If the aim is to narrow down existing inequalities in a chain and on a producer level, one has to move beyond sectoral interventions and focus on empowering specific (more vulnerable) groups of farmers. For a full understanding of the relationship between upgrading and development, also the extent of 'self-exclusion' from a GVC and diversification needs to be included in the analysis. The analysis revealed that the public-private intervention, involving the fair trade movement, was in terms of impact more 'inclusive', than for example the larger-scale public interventions.

A final conclusion in this section is that a thorough analysis of upgrading strategies requires the unravelling of upgrading concept. Besides different types of strategies, the focus should be on the different ways that contribute to upgrading as an outcome. The typology, proposed in earlier work by Gibbon (2001), is useful to unravel upgrading patterns and the extent to which upgrading strategies are 'inclusive'. In order to be able to say something about the outcomes of upgrading more information is required on issues of control and the impact of upgrading.

8.5 Interactions between governance structures and upgrading

In response to the central question (How do different governance structures interact in creating opportunities and constraints for more inclusive upgrading among different groups of farmers?) several authors emphasised the importance of studying the interactions between vertical and horizontal relations (e.g. Bolwig et al., 2008; Guiliani et al., 2005; Humphrey and Schmitz, 2000; Lambooy, 2002; Palpacuer, 2000; Westen, 2002). Also, they looked more closely at whether governance processes reinforce or block each other and under what conditions.

Such an approach is necessary to understand the link between power structures, upgrading and poverty. In this study, I looked at different levels of interaction, focusing on interventions. For example, I analysed the interaction between governance processes in the global cocoa chain and national governance structures in Ghana. It is apparent that as long as Ghana continues with the production of high volumes of good quality cocoa, international buyers will continue to support the partially liberalised system. This is a system of 'joint governance' with an active role for the Ghanaian state and a passive role for international cocoa processors and chocolate manufacturers. Despite the reforms, Ghana is the main intervener in the sector, focusing mainly at maintaining high quality production standards and at increasing volumes of conventional cocoa production. These state-driven interventions are generally top-down and large-scale; they reach the majority of farmers but not always equally. Sharing an interest in maintaining the system in Ghana, international buyers do not exercise significant pressure for further reforms. However, the conditions for cocoa production are changing. The increasing risks for supplier failure urge international processors and manufacturers to exercise more control on the supply side of the chain. The trend towards sustainable cocoa sourcing practices also demands a shift in focusing purely on economic sustainability of cocoa towards paying more attention to social and environmental issues. So far, the Ghanaian government has the tendency to block or slow-down activities in this field and obstructs more direct relations between international buyers and suppliers.

I also analysed the interaction between national governance structures and social structures. Although the Ghanaian state still coordinates the cocoa supply chain, there are visible sign of the gradual introduction of reforms. The role of the government is reduced in for example the provision of extension services and distribution of (subsidised) inputs; also Cocobod privatised the internal marketing of cocoa. New, mainly private, players have entered the scene. The impact of these changing governing structures is felt differently by different farmers. Farm leaders, farm-owners and farmers living in areas where cocoa production is more concentrated tend to benefit more from these changes than other farmers. In most cases the interaction reinforces existing power structures, for example traditional structures, where chiefs play an important leading role. The state-interventions are not directed towards empowering the farmers, but place a higher priority on maintaining their own power. Looking at the long-term perspective this can become problematic as unorganised farmers have difficulty in applying for different services, such as credit and training; some level of organisation is also necessary to get access to alternative markets (such as niche or mainstream certified cocoa).

8.6 A framework for more inclusive upgrading

In this study I proposed a framework for analysing inclusive upgrading which builds on the understanding of different governance levels and how they interact. But a framework for inclusive upgrading needs to include additional dimensions. First of all, it is important to look not only at the number of farmers reached but also at who exactly is targeted. From a development perspective this helps to identify strategies that reach more vulnerable groups. From a private sector perspective this helps to identify the more successful farmers who can serve as role model for others. Second, it is important to look at impact of upgrading and to make a distinction between different types of impact. 'Inclusive upgrading strategies' are strategies that have an impact beyond competitiveness and added value, and create benefits in terms of remunerative farmer income and empowerment. Reaching these impacts is crucial for a sustainable cocoa economy. Understanding opportunities and constraints for inclusive upgrading strategies also requires insight in the trade-offs and constraints faced by different types of farmers.

The strength of this framework for inclusive upgrading is that it contributes to the understanding of longer-term trends towards social in- or exclusion and helps to identify what kind of interventions are needed to steer upgrading into the direction of a more sustainable cocoa economy. Because upgrading is a selective process, it is important to know who will be in and who will be out, and to identify ways of influencing this process. Because in the long-term a large group of farmers is expected to move out of the cocoa sector, upgrading strategies that empower farmers to diversify their risk (diversification) are becoming increasingly important.

Final reflections

Evaluating the partially liberalised system and the interventions taking place was not an easy task. Not only because of the lack of transparency on state-investments and procedures, but also because criticising a relatively beneficial system that has no good alternative should be done cautiously. Nevertheless, I think it is important to share some of my critiques and suggestions for improvement and thus provide some "food for thought" for policy-makers, practitioners and researchers.

The Ghanaian system is under pressure

Others can learn from the example of Ghana, as the state turns out to be able to mitigate risks involved in cocoa production for producers, as well as for other chain actors. Still there two 'risks of inclusion' involved. First, on the level of the producer, the arrangements are sub-optimal and do not create incentives for farmers to behave as entrepreneurs. Also, farmers do not benefit equally from the arrangements in place. Second, the state is inwards oriented and lacks an adaptive approach. This entails a risk for the sector as a whole.

This situation generates some tension, at different levels, which in turn puts pressure on the current system. For example, due to the slow and gradual pace of reforms, LBCs are locked into a system that is not transparent and offers little incentives for high performance and little financial scope for establishing strong relationships with farmers. The lack of farmer organisation is an important reason why farmers do not take full advantage of the (potential) benefits of liberalisation of internal marketing and have no negotiating power. The quality control system is also under pressure and quality performance may continue to fall in a negative spiral. There is also tension between international processing companies and

Cocobod because of the insufficient number of 'light crop' beans for meeting the expansion of cocoa processing capacities, which precludes the government from fulfilling its promises to all the processors of cocoa.

The increasing tensions, especially when combined with a lack of transparency, can destabilise the status quo of the partially liberalised system in place. The incentives for farmers to continue with cocoa production and for private actors to invest in building relationships with suppliers should become stronger and more visible. The interventions of the Ghanaian government are major reinvestments in the cocoa sector, paid mainly from the cocoa export value ('gross FOB'). It can be argued that the farmers themselves pay for these interventions. But farmers do not always benefit, nor are they adequately informed on how 'their' money is being spent. Cocobod should be more transparent on how it reinvests its money in the cocoa sector and how benefits are distributed. This lack of transparency makes any assessment of the benefits of the partially liberalised system problematic and it also undermines Cocobod's legitimacy.

Risk of inertia
Compared to other cocoa producing countries in the region, Ghana generally comes across as relatively favourable country for source cocoa, even in light of the growing attention paid to malpractices in the Ghanaian cocoa sector. Despite the tensions and the problems that occur, Ghana offers still some unique qualities: there is little evidence of child labour, producer prices are stable, the quality of cocoa is relatively good and Ghanaian farmers make relatively little use of chemicals. Instead of seizing this comparative advantage, Ghana initially did not seem to see the need for a pro-active attitude. For example, Cocobod officials were hesitant to discuss the 'unexistent' problem of child labour and would generally stress that volumes of production are more important than more environmentally cocoa practices. Recently, Ghana is developing a more active attitude towards more sustainable cocoa practices in the country. But, instead of a being leader, Ghana is now more of a follower in the pursuit of the sustainable cocoa economy. For example, pilots with mainstream certification schemes are taking place outside Ghana and Cocobod remains reluctant to open up alternative marketing channels. The interests in maintaining the position of the state as the sole external marketer of cocoa are too high. From international buyers Ghana receives little stimulus for change. It runs the risk of inertia: from a cocoa producing country that is relatively beneficial for many farmers (but unfavourable for some), Ghana might slowly turn into a country that excludes the majority of farmers from participation in the global value chain (due to its failure to meet changing requirements).

Worldwide there is a trend towards sustainable cocoa sourcing and Ghana could still opt to take a leading role in this process. In order to motivate farmers to produce more sustainable cocoa, extension services and credit services have to be improved in order to make cocoa farming a profitable occupation. Also the formation and strengthening of farmer groups should be stimulated. Besides incentives to stimulate change, this would require leadership and capacity building of farmers, the private sector and also of the state itself.

Farmer organisation

Farmer organisation contributes to the empowerment of farmers, but not automatically. It is important to avoid imposed forms of cooperation, but there is no agreement on how to facilitate this process of 'bottom-up' organisation, on who should be involved and what kind of organisation is desirable.

Looking at other cocoa-producing countries, one sees that external donors and other (private) agents play a role in facilitating (strengthening) formal farmer organisations. Nevertheless, also in Côte d'Ivoire the majority of farmers are not organised and the organisations that are in place are often not effective. Recently there were some new attempts at organising farmers, through Farmer Field Schools (FFSs), 'input for credit groups' (Wienco) and soap making cooperatives. While governmental services have a top-down tendency, FFSs, soap and credit groups are generally more farmer-driven alternative extension services, provided by NGOs and public-private partnerships. Farmers appreciate these services, but due to the high costs involved these were mostly small scale initiatives that reached only a limited number of farmers. It is a significant challenge to effectively scale-up these approaches, while improving their cost-effectiveness. It is also a challenge to also include other types of farmers in such organisation schemes (women, caretakers etc.). An interesting area for future research is to examine the impact of different organisational modalities, ones that serve different goals and involve different types of farmers (or other actors in the chain).

Incompatibility of change

Globalisation processes influence global and national governance processes. This constantly fluctuating environment creates opportunities and risks for cocoa farmers. For example, in order for global buyers to be able to source sustainable cocoa well organised farmers are required. I illustrated in this study that Ghanaian cocoa farmers lack this organisation. Another example is that the end consumer concern regarding the use of child labour is not compatible with the perception of the average Ghanaian cocoa farmer. Fair trade or sustainable trade are not concepts which have a strong meaning for cocoa farmers.

This incompatibility of changes requires more investment in translating global trends into local action. In order to realise this, leadership within the cocoa growing communities has to be stimulated and benefits of change have to be made more tangible for farmers. A good strategy is to use the existing social network in which farmers are embedded. 'Putting the farmers first', as promoted in sustainable cocoa sourcing campaigns, also requires involving farmers in decision-making processes and management processes. If these crucial steps are not taken, the efforts invested in developing a sustainable cocoa economy will not bear fruit.

Notes

1. In this respect, 'agency' not only refers to the ability of people to accommodate, transform or resist (definition adapted from Post et al., 2002: 3), but also to the ability of the state and of the main drivers within a value chain to actualise change.
2. IMF = International Monetary Fund.
3. IFAD = International Fund for Agricultural Development.
4. TCC = Tropical Commodity Coalition.
5. Important reasons for this drop in cocoa reports in world's largest cocoa producing country are bad management, diseases and rainfall.
6. Peter van Grinsven (2007), Cocoa Sustainability Field Research Manager Masterfoods B.V. Netherlands, a division of Mars. The Role of Knowledge for Tropical Food Chains – Focus on Cocoa – Multi-Stakeholder Workshop on Improving the position of small holders through knowledge exchange between Tropical Food Chains and Research. Wageningen, April 12th, 2007.
7. http://www.unctad.org/en/docs/itetebmisc3_en.pdf (Access date 13 July 2009).
8. http://www.egfar.org/egfar/old_newsletters/special2003/art13.html. (Access date 18 January 2009) EGFAR = Global Forum on Agricultural Research.
9. http://siteresources.worldbank.org/INTWDR2008/Resources/2795087-1192111580172/WDROver2008-ENG.pdf (Access date 18 January 2009).
10. IDRC = International Development Research Centre.
11. For a definition of agency see note 1.
12. Gereffi (1999) added a fifth type, the upgrading of marketing linkages, which refers to a shift to higher value added chains and lead firms. This last type of upgrading is generally not used.
13. This comment is also made by Gibbon and Ponte (2005: 92).
14. Kaplinsky (2000) noted that, from a developing country' perspective, upgrading can be seen as the only way to avoid further reductions in incomes.
15. This involves also researchers from the "Value Chain Governance and endogenous growth" network, which is part of the Development Policy Research Network. The Value Chain network originated from the CERES research school and the thematic CERES network "Value Chains, social inclusion and local economic development" http://value-chains.global-connections.nl/content/value-chain-governance-and-endogenous-growthDPRN/CERES group (Access date 21 September 2009).
16. DFID = Department for International Development in the United Kingdom.
17. An example is the clean clothes campaign. This campaign calls for the consumer to boycott large discount retailers, such as Walmart and Aldi, as they do not take sufficient measures to ensure that no human rights have been violated in the production of their clothing.http://www.cleanclothes.org/campaigns (Access date 3 July 2009)
18. Recent debates on value chains, however, are moving beyond this distinction between industrial organisation and agro-commodities and aim to develop a common framework, including services as a third area of study (personal communication Peter Knorringa, 2009).
19. In the Netherlands a number of development organisations and knowledge institutes utilise the GVC approach, for example: the Royal Tropical Institute, Icco Partner van Ondernemende Mensen (= Interkerkelijke organisatie voor ontwikkelings-samenwerking), HIVOS (= Humanist Institute for Development Cooperation) and SNV (= Netherlands Development Organisation). Examples of international institutes include the FAO (= Food and Agriculture Organisation), USAID (= United States Agency for International Development) and IFAD (= International Fund for Agricultural Development).
20. In later work Gereffi, Humphrey and Sturgeon (2004) distinguish five types of 'coordination': market, modular, relational, captive and hierarchy.
21. UNIDO = United Nations Industrial Development Organisation.
22. UNCTAD = United Nations Conference on Trade and Development.
23. http://www.unctad.org/en/docs/itetebmisc3_en.pdf (Access date 13 July 2009).
24. The most fundamental insight on clustering comes from Marshall's 'Principles of Economics' (1920), which showed how clustering could help enterprises (especially small ones) remain competative. Such advantages included: 'a pool of specialised workers, easy access to suppliers of specialised inputs and services and the quick dissemination of new knowledge'. The external economies of Marshall did not cover joint action as a more 'deliberate force at work' (Schmitz and Nadvi, 1999: 1504).
25. http://www.boci.wur.nl/NR/rdonlyres/98CCE2E3-0FA2-4274-BCA0-20713CA1E125/40798/SandPPaperno1_publicrolesfinal1.pdf http://www.boci.wur.nl/NR/rdonlyres/98CCE23E0FA2-4274-BCA0-20713CA1E125/40798/

SandPPaperno1_publicrolesfinal1.pdf (Access date 17 December 2007).

26. The case studies in their edited volume, based on experiences with clustering in both developing and developed countries, show the limitations. Despite the existence of opportunities to learn and upgrade from both domestic and foreign organisations and in spite of opportunities to connect to domestic or global value chains, clusters are not equally able to take advantage of those external linkages.

27. Local Clusters in Global Value Chains: Exploring Dynamic Linkages Between Germany and Pakistan, can be downloaded at: http://www.ntd.co.uk/idsbookshop/details.asp?id=686 (Access date 18 May 2009).

28. Some examples of other approaches that link different governance levels: Approaches that link GVC to livelihood (cf. Barrientos et al. 2003; Bolwig et al, 2008; Helmsing and Cartwright, 2009). The livelihood approach helps understand economic activities and their impact on poverty and is useful to understand internal power structures and the composition of households. Approaches that link GVC to convention theory and work on improving Africa's marginalisation in the global economy (cf. Gibbon and Ponte, 2005). Convention theory can provide a better understanding of the normative dimensions of governance and its consumption related aspects. The link with political science and conventional trade theory will contribute to generating an understanding of this marginalisation and the type of integration faced by African countries, farms and firms. Recently there are also scholars linking the GVC approach to the Business System model (cf. Andriesse et al, 2009; Oosterveer and Hoi, 2009). The Business System perspective (cf. Whitley, 1998, 2001, 2005) focuses on economic processes as embedded in country-specific social and political institutions and identifies dynamics mostly as endogenous, evolutionary, and national.

29. This definition was adapted from Boschma and Kloosterman (2005: 400).

30. Boschma and Kloosterman (2005: 399-400) use this argument to analyse interactive learning and innovation processes, proposing a more open attitude when assessing the relevance of spatial scales. The scales should be an integral part of the study instead of basing them on presuppositions.

31. There are high levels of concentrations among cocoa processors and chocolate manufacturers. The cocoa sector is dominated by only a small number of large multinationals (see Chapter 4 for more details).

32. The Round Table for a Sustainable Cocoa Economy in 2007 is referred to as RSCE1. The second Round Table meeting is referred to as RSCE2.

33. Without exception all representatives of public agencies that I interviewed in 2003 had obtained a higher position by 2005.

34. MMYE = Ministry of Manpower, Youth and Employment

35. If we introduced a topic which was very interesting for the farmers, a joint response ('aha!') was often given. Or if a question had made them angry different farmers would stand up and protest loudly. This gave us quite a good idea regarding the relevance and importance of the topic for the farmers.

36. See Bittersweet. A short film that I made on the way cocoa farmers in Ghana perceive changes in their environment. http://www.geobrief.tv/movies/376002a.html. (Access date 1 December 2009).

37. IFPRI-ODI = International Food Policy Research Institute – Overseas Development Institute

38. More information on presentations and published articles on this theme can be found at: http://gssp.wordpress.com/2007/12/21/ifpri-odi-cocoa-workshop-proceedings (Access date 19 May 2009).

39. STCP = Sustainable Tree Crop Programme

40. In processing my quantitative data for my statistical analysis of correlations, I have used the data gathered in 2005. I mainly use ordinal and nominal variables. There are some exceptions (for example yield and age); in this case I divided the variables into different categories. I used 'Gamma' and Cramer's V to present the strength (and in case of Gamma also the direction) of relations between variables. I use 'Gamma' to present the correlations between ordinal variables, including bipartite nominal variables, namely: gender (female = 0, male = 1), working together (not working together = 0, working together = 1) and migrant (migrants = 0, non-migrants= 1). To indicate the relationships between ordinal and nominal variables, which is only the case for 'Region', I used Cramer's V. The level of significance is labelled with an asterisk (*) as follows: Approx Sig. 0,05 – 0,1 = *; Approx. Sig. 0,01 – 0,05 = **; Approx. Sig. < 0,01 = ***.

41. I left out 'investigator triangulation', which involves using several different investigators in an evaluation project.

42. Of course my inability to communicate with farmers in their local language was primarily a handicap not an asset.

43. For example, during the period when I conducted the interviews in 2005, many of the farmers had their cocoa rejected by the buying companies because of problems with 'purple beans' (Chapter 5 and 6). Regardless of the topics that I tried to raise in the discussion, the farmers always came 'back to the beans', as this disease threatened them directly at that moment in time.

44. ISSER = Institute of Statistical Social and Economic Research.

45 While they remained largely silent during the plenary session, the farmers were very active in the working groups where a lively discussion developed.
46 Picture available at http://www.fao.org/docrep/X5598E/Ghana.GIF (Access date 21 September 2009).
47 For 2003-04 it is estimated that between 120,000 and 150,000 tonnes was smuggled from Cote d'Ivoire into Ghana (Ruf, 2007b; Brooks et al, 2007 in World Bank, 2007b). Therefore increase in production is often estimated higher.
48 I interviewed 173 farmers in 2003, of which 103 were traced back in 2005. In 2005 I interviewed in total 210 farmers, of which 107 'new'.
49 Production in the Volta region, another cocoa growing region is marginal (913 tonnes in 2002/03, 1,909 tonnes in 2003/04 (Anim-Kwapong and Frimpong, 2004).
50 Source: Financial Times, London, cited by S. Wallace, 2003. http://www.omanhene.com/slavery.php (Access date 4th April 2005).
51 For the cocoa processing industry (cocoa grinders) it is estimated that by the mid-1990s the top ten corporations conducted 70 per cent of all cocoa grindings, with three corporate giants accounting for 50 per cent (Archer Daniels Midland (ADM), Cargill Inc. and Barry Callebaut [http://www.eftafairtrade.org]). (Access date 21 September 2009).
52 Rabobank (2004) Power Point presentation on dynamics in cocoa production, grinding and chocolate confectionary.
53 Schenkel (2006) 'Cacaomarkt draait om Amsterdam' in NRC Handelsblad (28 February 2006: 11).
54 Besides environmental and social criteria also health issues became increasingly important in the cocoa sector, including positive health effects (improving hart-conditions) and negative health effects (cf. obesity).
55 Response to open question in the industry survey.
56 PSOM = Programma Samenwerking Opkomende Markten.
57 WHO = World Health Organisation
58 http://www.waldenassetmgmt.com/social/action/library/01071h.html (Access date 22 February 2006).
59 It is known as the 'voluntary' Harkin-Engel Protocol, signed on the first of October 2001.
60 http://www.treecrops.org.
61 It is estimated that together these programmes will reach around 300,000 cocoa farmers, of the 2 million West African farmers in total (around 14 per cent). There are also some initiatives planned by WCF and STCP; WCF is expected to support 150,000 farmers by 2010 and the STCP another 150,000 by 2012 via the 'Gates Foundation'. So, the projections are that together individual and sector initiatives will reach around one third of all cocoa farmers in West Africa. These training programmes are expected to increase production levels by 25 per cent and to produce an additional 232,000 tonnes per year (TCC, 2009). Projects such as these that are still in planning stagest are not included in the analysis.
62 http://www.cadbury.com/media/press/Pages/cdmfairtrade.aspx (Access date 4 November 2009).
63 http://www.environmentalleader.com/2009/04/13/mars-targets-sustainable-cocoa-sourcing/ (Access date 5 October 2009).
64 Formerly it was the Ghana Standard Board, but at present the international Biscuit, Cocoa, Chocolate and Confectionary Alliance is the main responsible authority for analysing and testing residue levels in cocoa. Following EU regulation, cocoa beans with excess level of residues are rejected on the world market.
65 Barry Callebaut cooperates with the Fair Trade Labelling Organizations International (FLO) and offers a range of fair trade certified products. http://www.barry-callebaut.com/2080 (Access date 16 July 2007).
66 http://www.environmentalleader.com/2009/04/13/mars-targets-sustainable-cocoa-sourcing/ (Access date 5 October 2009)
67 Press release on internet, not an official record. Available at http://www.unctad.org/TEMPLATES/webflyer.asp?docid=2628&intItemID=2022&lang=1 (Access date 17 July 2007).
68 http://www.pressreleasepoint.com/adm-opens-cocoa-plant-kumasi-ghana (Access date 5 November 2009).
69 Global buyers are also involved in community development projects. For example, Cadbury Ghana Limited donated a total of forty wells per year to farmer communities where the Kuapa Kokoo Union is active. Also, as a consequence of signing the Harkin-Engels protocol, in Ghana manufacturers have become involved in educational programmes on farm and labour safety issues, in cooperation with Ghana Cocoa Research Institute (CRIG). Although these kinds of projects are appreciated by a small group of direct beneficiaries, they are not examined in great depth in my study as they have a limited scope.
70 http://www.iscom.nl/upcocoa/ (Access date 13 February 2008).
71 The project is financed by a subsidy from the Cocoa Buffer Fund of the Dutch Ministry of Agriculture (LNV) and by in-kind contributions of ADM Cocoa B.V., Masterfoods B.V./Mars Inc. and IITA/STCP. http://www.iscom.nl/upcocoa/rapporten/Factsheet%20UPCOCOA%20English-2008.pdf (Access date 1 December 2009).
72 ISCOM = Institute for Sustainable Commodities.
73 RIAS = Rabo International Advisory Services.
74 www.iscom.nl/upcocoa (Access date 1 December 2009).

75 http://www.fao.org/sd/dim_kn2/kn2_040401a1_en.htm (Access date 4 February 2008).
76 http://www.boci.wur.nl/NR/rdonlyres/98CCE2E3-0FA2-4274-BCA0-20713CA1E125/40798/SandPPaperno1_publicrolesfinal1.pdf (Access date 17 December 2007).
77 http://www.imf.org/external/pubs/ft/wp/2005/wp0521.pdf (Access date 4 February 2008).
78 ICCO = International Cocoa Organisation.
79 IITA = Agricultural Research for Development in Africa.
80 http://www.icco.org/questions/ liberalisation.htm (Access date 9 August 2004).
81 The National Cocoa and Coffee Board (NCCB), represented by Hope Sona Ebai, presented a paper at a workshop on the liberalisation of the cocoa trade (Lomé 1998), with the title: "Liberalisation on the cocoa and coffee trades in Cameroon".
82 COPAL = Cocoa Producers' Alliance
83 Gockowski started a Cameroon Pilot Project in 2000. http://www.treecrops.org/country/cacaoproduction.asp (Access date 21 September 2009)
84 In Nigeria only 5 per cent of the total volume of loans from formal sources is allocated to smallholders. http://www.nipc-nigeria.org (Access date 09 August 2004)
85 Internet sources: Dow Jones News Wires 2007 and TradeNet Nigeria, 2007. Available at http://www.flex-news-food.com/pages/12986/Africa/Cocoa/nigeria_07_08_cocoa_output_seen_220000_230000_tons___official_dj.html and at http://www.metrobeat.com/documents/?typ=news&news=100002554&lang=en&i=233111&g=allprices (Access date 12 February 2008).
86 The British company 'Gill and Duffud' built this factory.
87 In the early 1980s there were approximately 120,000 people employed by Cocobod; in 2006 this number had dwindled down to only 11,000 (Zeitlin, 2006). Many former employers of Cocobod are still involved in the cocoa sector, as private consultants, providers of extension services and/or as cocoa buyers.
88 Personal communication Vigneri (2007)
89 Personal communication Vigneri (2007). Based on data from Cocobod (2005).
90 During field work in 2005, I personally checked a small number of scales, simply by weighing myself. Without exception all of the scales were manipulated.
91 A recent study by Zeitlin (2006: 7) confirms these findings for a number of other communities. According to his study the average number of LBCs in a village was 3.2 for the cocoa season 2003-04 in the Ashanti Region, Brong Ahafo and the Central Region.
92 Interview CMC (2005).
93 Interviews with Transroyal Commodities, Cocoa Merchants in 2003. and Fedco in 2005.
94 Interview CMC (2005).
95 Almost 58 per cent of the respondents answered that their family taught them how to produce good quality cocoa. Another 8 per cent mentioned family as one of the sources of knowledge (FS 2005).
96 http://www.cocobod.gh/corp_div.cfm?BrandsID=20 (Access date 14 February 2008).
97 CODAPEC = National Cocoa Diseases and Pests Control.
98 Berry (1997), quoted in Lecture 22 (Economics 172) on Issues in African Economic Development (2007) of Profesor Ted Miguel. Department of Economics, University of California, Berkely. Accesible at http://emlab.berkeley.edu/users/webfac/emiguel/e172_s07/lecture22.pdf (Access date 21 September 2009)
99 Annually banks charge between 28 and 36 per cent interest (Ministry of Finance, 1999).
100 http://www.divinechocolate.com http://www.divinechocolate.com (Access date 21 September 2009).
101 Literally speaking, nnoboa means 'mutual assistance in weeding, group action and mutual aid'. It is based on social, ethnic and family actors who are often connected to the area and its traditional land system. It is an expression of solidarity to the members of the traditional society; it is voluntary and performed without written rules or much formality. It tends to be temporary as the group dissolve after the completion of the task.' (Department of Cooperatives, 1990: 147).
102 Based on personal communication Cocobod, (2007); Africa News, 4 October 2007.
103 In ODI Background note. 'The cocoa sector. Expansion, or green and double green revolution?' December 2007. http://www.odi.org.uk/publications/background-notes/0712-cocoa-sector-expansion.pdf (Access date 12 February 2008).
104 Accessible at http://www.treecrops.org/crops/cocoaprodtech.pdf (Access date 12 February 2008).
105 TradeNet Nigeria. www.tradenet.biz/nigeria Available at: http://www.metrobeat.com/documents/?typ=news&news=100002554&lang=en&i=233111&g=allprices (Access date 12 February 2008)
106 Daily Graphic 16/11/05, personal communication LBCs and group discussions with farmers.
107 The impact of some global interventions on a producer-level will be discussed in more detail in Chapter 7.
108 Interview Kenneth Brew 2003, at that time responsible for monitoring and since 2005 promoted as Cocobod's director of seed programme.
109 See also http://www.id21.org/id21ext/s7cmv1g1.html (Access date 15 August 2009).
110 Saurabh Mehra, WCF West Africa Committee and Stephan Weise, Program Manager, IITA/STCP stated this during their

presentation on their West Africa Strategy during the WCF Partnership Meeting Amsterdam, 23 May 2007.
111 This report of Francis Teal and Andrew Zeitlin (Centre for the Study of African Economies, University of Oxford) together with Haruna Mamamah (ECAM Consultancy, Ltd. Accra), published on 11th of March 2006, is based on data gathered by Marcella Vinegri, 2002. The report is available from http://www.gprg.org/pubs/reports/pdfs/2006-04-teal-zeitlin-maamah.pdf (Access date 15 August 2007).
112 For more information on Polly Hill see section 6.2.1.
113 For more information on these different regions check http://www.countryexpertise.nl/ghana/regions/bar.html (Access date 5 May 2009).
114 Whether or not these farmers became farm owners is positively related with the number of years they lived in the community where they currently work. Gamma 0,429*** (FS 2005).
115 When contracted parties are related the sharing arrangement can be different (division in two) (Takane, 2000: 383).
116 Cramer's V 0,439*** (FS 2005).
117 There are different types of arrangements, varying on what is to be shared (the trees or the harvest) and varying on how the land is shared when the establishment of the farm is completed (share the land or return the land to the farm-owner) (Takane, 2000: 384).
118 Cramer's V 0,338*** (FS 2005).
119 The contract under which a caretaker works or whether the farmer is a migrant is not significantly related to yield.
120 WCF Newsletter, August 2007, Issue 31.
121 Mean is 16 and standard deviation is 25.
122 Mean 11 and standard deviation is 7.
123 Mean 19 and standard deviation is 32.
124 Mean 20 and standard deviation is 18.
125 Mean 12 and standard deviation is 12.
126 Gamma 0,564***.
127 The data for cocoa season 2003/04 (based on FS 2005) is less precise as farmers were asked about the number of bags and not the amount of kilograms they produced. In the farmer surveys the most trustworthy indicator for production capacity is the number of bags a farmer produce in a season (as they write down the number of bags sold to a buying company in their pass books), therefore this variable is used.
128 One bag of cocoa contains 62.5 kg of cocoa. One acre = 0.405 ha.
129 A negative consequence is that this development is likely to have negative effects on the environment (Van der Geest, K. et al, forthcoming; Gockowski, 2007).
130 The number of respondents obtaining a very strong position in the community or chain is relatively high. What has contributed to this is that as a researcher and visitor of a community you cannot bypass farmers' with such high status.
131 High status farmers produced significantly higher yields. It is possible that this is a simple correlation as their high status could be due to the fact that they already produced high yields.
132 http://people.tamu.edu/~yarak/gsc97geest.html (Access date 21 September 2009).
133 In a recent study of Ruf (2007b: 19) a trend is observed that a 'new generation in their 20s or 30s is coming in'.
134 http://www.roundtablecocoa.org/showpage.asp?ExpertGroups (Access date 21 September 2009) .
135 Gamma is 0,504***.
136 For male respondents this percentage was considerably lower, namely 60 per cent.
137 Gamma 0,478***.
138 Farmer profiles 2005.
139 Gamma 0,650***.
140 GCMA was established in 1928 and was a mayor buyer of cocoa for a long time. However, due to financial and managerial problems, since 1984 GCMA is no longer operational (GCMA, 2005) .
141 For an overview go to Appendix 6.1
142 In 2003/04, almost 66 per cent of farm-owners received these services, versus 43 per cent of the caretakers (FS 2005).
143 The type of share contracts called Abunu (or yemayenkye) (do and let us share) in exchange for their tasks they receive half share of the cocoa harvest; under a Abusa (also known s Nhwesoo contract) tenants receive only one third of the harvest.
144 This relationship between type of contract and access to quality assistance is highly significant in the Western region Cramer's V 0,540***(FS 2005).
145 Gamma 0,384***.
146 For example, certification of cocoa can result in a higher price paid for this product. But getting a certificate is a costly procedure and the costs may outweigh the benefits.
147 For processors of cocoa price-differentiation does exist. Processing companies that process cocoa within Ghana can buy light-crop beans at a 20 per cent discount (Ministry of Finance, 1999).
148 Isaac Osei, Chief Executive Ghana Cocoa Board, 28 June 2007. Power point presentation. "Sustainable Practices in the Global Cocoa Economy, a Producer's Perspective" Presented at the 4th Indonesia International Cocoa Conference & Dinner 2007. Available at http://www.worldcocoafoundation.org/info-center/document-research-center/documents/3.COCOBOD.pdf (Access date 29 February 2008).
149 http://www.otal.com/images/TOTAL%20Services/CommodityReport/Commodity%20May%202008.pdf (Access date 2 January 2008).
150 WCF Annual Meeting, 2007.
151 High levels of purple beans are problematic for buyers, as these low-quality beans with a

purple colour produce cocoa liquor with less flavour and higher acidity.
152 According to a former Chief Entomologist at CRIG Dr. E. Owusu-Manu. Source Daily Graphic, (15 December 2005).
153 Accessible at http://www.peacefmonline.com/index.php?option=com_content&task=view&id=3158&Itemid=30 (Access date 29 February 2008).
154 For season 2003/04, a quarter of the respondents indicated that they observed a decline in quality (FS 2005).
155 http://www.cocoasustainability.mars.com/Sustainability/Cocoa_Farmer.html (Access date 8 May 2007).
156 Source Daily Graphic 14-01-05 (Ghanian quality daily news paper).
157 I have no further data on other incentives provided by the farmer union.
158 Resigha processes inferior cocoa and cocoa waste into cocoa butter, mainly for the cosmetic industry. Accessible at http://www.janschoemaker.com/index.php?t=2 (Access date 29 February 2008).
159 Source: Ghana Cocoa Annual Report 2005, Global Agriculture Information Network. See website http://www.fas.usda.gov/gainfiles/200510/146131245.doc (accss dat 1 May 2006).
160 Interview QCD (12 May 2003).
161 An important indication of this consistency and reliability is the relatively low level of testing (of bean quality and weights) of Ghana cocoa shipments by the buyers (which itself represents cost savings for international buyers) (Ministry of Finance, 1999).
162 Although recently this has dropped to as low as €15 per metric tonne.
163 Information obtained in interview with QCD (2003). Unfortunately I did not have access to sources that could verify these figures.
164 This was one of the conclusions of a workshop I organised in Ghana, with the title: 'Towards a Sustainable Cocoa Chain, a Ghanaian Perspective' (see also Laven, 2005a).
165 In Chapter 6 I explain the differences in position and how I determined these categories.
166 Gamma 0,279**.
167 Cramer's V= 0,318***.
168 http://findarticles.com/p/articles/mi_m1052/is_/ai_6440781 (Access date 18 May 2008).
169 According to Anim-Kwapong and Frimpong (2004: 5-6) the short-term price-elasticity of supply is estimated at 0.3 per cent and the price elasticity of productivity per 5 year and 10 years is respectively 0.9 and 1.8.
170 Takane argued that the role of price incentives in cocoa production also needs to be reconsidered and placed in wider incentive structures that are embedded in local institutions. Another reason why farmers invest in their farms has to do with land rights; investments in trees is a way of claiming one's land right (different authors in Takane, 2002: 391). An important conclusion of Takane's study is that the farmer's investment in trees 'needs to be understood in terms of both short-term incentives to increase yields and long-term investments to strengthen land–rights'.
171 Environmentalists have opposed the use of this ingredient and recommend Cabamult as an environmentally friendly alternative. However, Cabamult, which is used by forty-five per cent of farmers (of whom less than twenty per cent used protective clothing) is from a human health perspective far more disastrous. According to one of the experts in this field, from the International Pesticide Application Research Centre in the UK, Lindane is more effective and relatively safe to handle (personal communication Bateman, 2007).
172 See also the short documentary "Bittersweet" by Anna Laven. http://www.geobrief.tv (Access date 2 October 2009).
173 The survey indicated that farmers who do not hold a special position in the community make significantly less use of protective clothes than farmers who hold such a position. Gamma 0,462***.
174 Farmers told me that when they first heard about the bonus system they were very excited. They went to the collection points with large jute bags, expecting a lot of cedis. They were very disappointed to find out that the bonus per bag was only marginal (Group discussions 2005).
175 http://www.ghana.gov.gh/ghana/speech_deputy_finance_minister_prof_gyan_baffour.jsp (Access date 2 January 2008)
176 Gamma 0,235*.
177 http://www.ghanaweb.com/GhanaHomePage/economy/artikel.php?ID=48779 (Access date 5 June 2005).
178 Mass spraying is not new. The first successful experiments with mass spraying date back to 1962/63 (ended in 1966), the second attempt was organised in 1973, and the third in 1975. Because of political instability there was no mass spraying for many years (interview CODAPEC, 2005).
179 http://www.csae.ox.ac.uk/output/briefingpapers/pdfs/CSAE-briefingpaper-01-Ghanacocoa.pdf (Access date 10 October 2009).
180 I should stress that the expansion of cocoa farms in the Western Region led to a great loss in biodiversity: forests disappear, mono cropping (without shade) is promoted, and soils are being depleted This is due to the fact that cocoa production is concentrated in this area, which attracts buyers and services. Consequently, the man expansion of cocoa production also takes place in this region.
181 A recent publication of the Tropical Commodity Coalition for sustainable Tea Coffee Cocoa reveals that for 2012 there are two planned initiatives by individual companies (by Cadbury and by Cargill) in

182 GCFS = Ghana Cocoa Farmers Survey.
183 Cramer's V 0,343*** (FS 2005).
184 The mean number of visits to Brong Ahafo was 2.4, while for the Western Region this was 1.7 and for Ashanti region 2.0 (Teal et al., 2006: 15).
185 Gamma 0,191*.
186 http://www.businessghana.com/portal/news/index.php?op=getNews&id=64712 (Access date 11 June 07).
187 http://www.businessghana.com/portal/news/index.php?op=getNews&id=64712 (Access date 11 June 07).
188 In Ghana there were some examples of contract farming on palm oil plantations in the past. This turned out to be difficult as the state had to expropriate land owned by village communities and farm families. Traditional leaders have (chiefs and influential families, and the lawyers representing them have vehemently resisted this process (Daddieh, 1994: 197). Despite these problems the company involved (Unilever) finally managed to expropriate a sufficient area for farming.
189 http://www.tradinorganic.com/ (Access date 5 November 2009).
190 Personal communication with AgroEco in 2009.
191 http://www.chocolonely.nl/ (Accesdate 5 November 2009).
192 http://cecoeco.catie.ac.cr/descargas/Market_of_organic.pdf (Access date 5 November 2009).
193 Personal communication: Rabobank Foundation, 2006.
194 'To facilitate the work of the Union President and the work at the office of the President as a whole, the Trust provided the Union with a brand new Toyota Land Cruiser Prado.' (Kuapa Kokoo Annual Report 2004: 10).
195 Kuapa Kokoo originally retained one third of the business and was closely involved in its development, while Body Shop International owned 14% of the shares at the time. When L'Oreal purchased the Body Shop in 2006 it donated its share to Kuapa Kokoo (http://www.divinechocolate.com/kuapa.htm).
196 Personal meeting with Managing Director, 2008.
197 http://www.nationalcocoashell.com/index2.php (Access date 5 November 2009).
198 During a visit to the Cocoa Processing Company in 2005 I observed that the other ingredients for making chocolate were imported fromFrance (sugar), the Netherlands (milk) and Germany (machinery).
199 In 2006 the installed processing capacity was 65,000 tonne per year.
200 In 2009 ADM opened its factory with a processing facility of 65,000 tonne per year.
201 http://www.cargillcocoachocolate.com/News%20Centre/Cargill%20opens%20cocoa%20processing%20facility%20in%20Ghana.pdf (Access date 5 November 2009).
202 http://www.confectionerynews.com/The-Big-Picture/ADM-opens-Ghana-cocoa-plant (Access date, 5 November 2009).
203 On the other hand farmers in Brong Ahafo had generally better access to inputs (Chapter 6).
204 'Value Chain Governance and endogenous growth' network http://value-chains.global-connections.nl/content/value-chain-governance-and-endogenous-growthDPRN/CERES group.

REFERENCES

Abbott, P. C., J. M. D. Wilcox & W. Muir, A. 2005. *Corporate Social Responsibility in International Cocoa Trade*. Selected Paper prepared for presentation at the 15th Annual World Food and Agribusiness Forum, Sympousyium and Case Conference, Chicago, Iilinois, June 25-28, 2005. West Lafayette, Indiana: Department of Agricultural Economics, Purdue University.

Adusei, E.O. 1993. *Cocoa Production Practices in Ghana*. Report Submitted to The American Cocoa Reseacrh Institute. School of Business and Economics, Long Wood College, August, 1993.

Altenburg, T. 2006. Governance Patterns in Value Chains and their Development Impact. In: *The European Journal of Development Research*, 18: 498-522.

Akiyama, T., J. Baffes, D. Larson, and P. Varangis. 2001. Market Reforms: Lessons from Country and Commodity Experiences. In: *Commodity Market Reforms, Lessons of Two Decades*, eds. T. Akiyama, J. Baffes, D. Larson, and P. Varangis: 5-35. Washington, D.C.: The International Bank for Reconstruction and development/The World Bank.

Amezah, K.A. 2004. DRAFT: *The Impact of Reforms (Privatization of Cococoa Purchases and CSD/MOFA Merger) on Cocoa Extension Delivery*. Accra: Ministry of Food and Agriculture, Directorate of Agric Extension Services.

Amoah, J.E.K. 1998. *Marketing of Ghana Cocoa 1885-1992*. Cocoa Outline Series, No. 2 Accra: Jemre Enterprises Limited

Andriesse, E., N. Beerepoot, B. v. Helvoirt, and G. v. Westen. 2009. DRAFT *Business Systems and Inclusive/Exclusive Development*. Development Policy Review Network (DPRN) 2009/2010.[204]

Anim-Kwapong, G.J., and E.B. Frimpong. 2004. *Vulnerability and Adaptation Assessment under the Netherlands Climate Change Studies Assitance Programme Phase2 (NCCSAP2). Vulnerability of agriculture to climate change-impact of climate change on cocoa production*. CRIG, New Tafo Akim.

Anin, T. E. 2003. *An Economic Blueprint for Ghana*. Accra: Woeli Publishing Services.

Asenso-Okeyre, W. K. 1996. Socio-cultural and Political Aspects of Development in Ghana 21st Century. In: *Workshop on Governance and Development: Perspectives for 21st Century West Africa*, eds. A.E. Ikpi & J. Olayemi, K.: 58-68. Accra, Ghana: Winrock International Institute for Agricultural development and African Rural Social Sciences Research Networks.

Ayenor, G. K., N.G. Röling, P. Padi, A. v. Huis, D. Obeng-Ofori & P. B. Atengdem. 2004. Converging Farmers' and Scientists' Perspectives on Researchable Constraints on Organic Cocoa Production in Ghana: Results of a Diagnostic Study. In: *Wageningen Journal of Life Sciences (NJAS)*, 52: 261-84.

Ayenor, G.K. 2006. Capsid Control for Organic Cocoa in Ghana: results of participatory learning and action research. Wageningen: Wageningen University and Research Centre. Tropical Resource Management Papers, 27.

Barrientos, S., S. McClenaghan & L. Orton. 2001. Stakeholder Participation, Gender and Codes of Conduct in South Africa. In: *Development in Practice*, 11: 575-86.

Barrientos, S., C. Dolan, and A. Tallontire. 2003. A Gendered Value Chain Approach to Codes of Conduct in African Horticulture. In *World Development*, 31 (9): 1511-26.

Bass, H.H. 2006. Structural Problems of West African Cocoa Exports and Options for Improvements. In Successful Cases of Diversification, Upgrading and Redirection of Economies: Strategies to Escape the Commodity Dependence, eds. K. Wohlmuth and M. Nureldin Hussain: 245-63. Münster: Lit Verlag Berlin.

Bebbington, A., and J. Thompson. 2004. *Use of Civil Society Organizations to Raise the Voice of the Poor in Agricultural Policy*. Working Paper. DFID, London, UK.

Beckman, B. 1976. *Organising the Farmers Cocoa politics and National Development in Ghana*. Stockholm: Scandinavian Institute of African Studies, Uppsala.

Bell, M., and M. Albu. 1999. Knowledge Systems and Clusters in Developing Countries. In: *World Development*, 27(9): 1715-34.

Berdegué, J.A., E. Biénabe, and L. Peppelenbos. 2008. *Keys to Inclusion of Small-scale Producers in Dynamic Markets – Innovative practice in connecting small-scale producers with dynamic markets*. International Institute for Environment and Development (IIED), London, UK.

Beuningen, R. v. 2005. *Changes in the Cocoa Market: The role of Cooperatives in the mainstream Cocoa Trade in Côte d'Ivoire*. MA thesis. Faculty of Economics, Groningen: Rijksuniversiteit Groningen.

Blowfield, M. 2003. Ethical Supply Chains in the Cocoa, Coffee and Tea Industries. In: *Greener Management International. The Journal of Corporate Environmental Strategy and Practice*: 15-24.

Bolwig, S., S. Ponte, A. Du Toit, L. Riisgaard, and N. Halberg. 2008. *Integrating Poverty, Gender and Environmental Concerns Into Value Chain Analysis: A Conceptual Framework and Lessons for Action Research*. DIIS Working Paper 2008/16. Copenhagen: Danish Institute for International Studies.

Boomsma, M. 2008. Practices and Challenges for Businesses and Support Agencies. In: *Bulletin 385 – Sustainable Procurement from Developing Countries*. KIT Publishers.

Boschma, R.A., and R.C. Kloosterman. 2005. Learning from Clusters. A Critical Assessment from an Economic-Geographical Perspective. In: *The GeoJournal Library*, 427. Dordrecht: Springer.

Ceglie, G., and M. Dini. 1999. SME Cluster and Network Development in Developing Countries: the Experience of UNIDO. In: *Building a Modern and Effective Development Service Industry for Small Enterprises organized by the Committee of Donor agencies for Small Enterprise Development*. Rio de Janeiro: UNIDO.

Census, Ghana Statistical Service. 2002. 2000 Population & Housing Census. Republic of Ghana, Ghana Statistical Service.

Cocobod. 2000a. *Ghana Cocoa Board Handbook*. Accra: Tha Ghana Cocoa Board.

Cocobod. 2000b. *Regulations and Guidelines for External Marketing of Cocoa*. Ghana Cocoa Board.

Cocobod. 2003. http://www.bog.gov.gh/Private Content/File/Research/Sector%20Studies/ FINANCIAL_%20IMPLICATIONS_COCOA.pdf.

Conservation International Ghana. No date. *Conservation Cocoa in Ghana, Purpose and Progress*. Accra: Conservation International Ghana.

COPAL. 1998. 62nd General Assembly Lome, Republic of Togo, 20th September, 1999. In: *Report on the workshop on the liberalisation of the cocoa trade Lome, (Togo), from 26th to 28th October, 1998*. Lome (Togo): Cocoa Producers' Alliance.

Cortright, J. 2006. *Making Sense of Clusters: Regional Competitiveness and Economic Development*. A discussion paper prepared for the Brookings Institutions Metropolitan Policy Program, March 2006.

Cramer, C. 1999. Can Africa Industrialize by Processing Primary Commodities? The Case of Mozambican Cashew Nuts. In: *World Development*, 27(7): 1247-66.

CREM. 2002. *Analysis of Environmental, Social and Economic Issues Related to Cocoa Products Consumed in the Netherlands. Field Research in Ghana*. Amsterdam: Consultancy and Research in Environmental Management.

Cumbers, A., and D. MacKinnon. 2007. Introduction: Clusters in Urban and Regional Development. In: *Urban studies*, 41(5/6): 959-69, May 2004. Carfax Publishing. Taylor & Francis Group.

Daddier, C.K. 1994. Contract Farming and Palm Oil Production in Côte d'Ivoire and Ghana. In: *Living Under Contract. Contract Farming and Agrarian Transformation in Sub-Saharan Africa*, eds. P.D. Little, and M.J. Watts: 188-215. Madison and London: The University of Wisconsin.

Daviron, B., and S. Ponte, 2005. *The Coffee Paradox: Global markets, Commodity Trade and the Elusive Promise of Development*. London: Zed Books.

Department of Cooperatives. 1990. *History of Ghana Co-operatives 1928-1985*, Accra, Ghana: Department of Cooperatives.

Doherty, B., and S. Tranchell. 2005. New Thinking in International Trade? A Case Study of the Day chocolate Company. In: *Sustainable Development*, 13: 166-76. John Wiley & Sons, Ltd and ERP Environment. Published online in Wiley InterScience (www.interscience.wiley.com). DOI: 10.1002/sd.263.

Duursen, John v., and D. Norde. 2003. *Identifying and Assessing Convergences of Interest for Stakeholders in the Ghanaian Cocoa Chain, in Order to Attain Increased Farm Income and Wider Sustainable Production*. EPCEM internship at AGIDS, University of Amsterdam: Amsterdam.

Dyllick, T., and K. Hockerts. 2002. Beyond the Business Case for Corporate Sustainability. In: *Business Strategy and the Environment*, 11: 130-41.

FAO/WHO 1999. *Food Standards (Codex Alimentarius)*. Available from: URL: http:// www.codex alimentarius.net/

FAO 2005. Statistical Database. Available from http://faostat.fao.org.

Fernández Jilberto, A.E., and B. Hogenboom. 2007. The New Expansion of Conglomerates and Economic Groups. An Introduction to Global Neoliberalisation and Local Power Shifts. In: *Big Business and Economic Development. Conglomerates and Economic Groups in Developing Countries and Transition Economies under Globalisation*, eds. A. Fernández Jilberto, and B. Hogenboom: 1-28. London: Routledge.

Fold, N. 2001. Restructuring of the European chocolate industry and its impact on cocoa production in West Africa. In: *Journal of Economic Geography*, 1: 405-20.

Fold, N. 2002. Lead Firms and Competition in 'Bi-polar' Commodity Chains: Grinders and Branders in the Global Cocoa-chocolate Industry. In: *Journal of Agrarian Change*, 2, 228-247.

Fold, N. 2004. Spilling the Beans on a Tough Nut: liberalization and local supply system changes in Ghana's cocoa and shea chains. In: *Geographies of Commodity Chains*, eds. A.H. Hughes, and S. Reimer: 63-80. New York: Routledge.

GAIN Report. 2005. *Ghana Cocoa Annual Report 2005*. USDA Foreign Agricultural Service. Global Agriculture Information Network.

GCMA. 2005. *3-Year Strategic Plan 2003-2005. Ghana Co-operative Marketing Association Limited "first farmer based organization in Ghana" 2003-2005*, Accra, Ghana.

Gereffi, G. 1999. International Trade and Industrial Upgrading in the Apparel Commodity Chain. In: *Journal of International Economies*, 48: 37-70.

Gereffi, G., M. Korzeniewicz, and R. Korzeniewicz. 1994. Introduction: Global Commodity Chains. In: *Commodity Chains and Global Capitalism*, eds. G. Gereffi, and M. Korzeniewicz: 1-14, Westport: Praeger.

Gereffi, G., J. Humphrey, and T. Sturgeon. 2005. The Governance of Global Value Chains. In: *Review of International Political Economy*, 12: 78-104.

Gibbon, P. 2001. Upgrading Primary Production: A Global Commodity Chain Approach. In: *World Development*, 29: 345-63.

Gibbon, P. 2003. *Commodities, Donors, Value Chains Analysis and Upgrading*. Copenhagen: Danish Institute for International Studies.

Gibbon, P., and S. Ponte. 2005. *Trading Down. Africa, Value Chains and the Global Economy*.

Philadelphia: Temple University Press.

Gilbert, C.L. 2000. Commodity Production and Marketing in a Competitive World. In: *UNCTAD X Presentation*, Bangkok.

Gilbert, C.L., and P. Varangis. 2003. Globalization and International Commodity Trade with Specific Reference to the West African Cocoa Producers. In: *NBER Working Paper Series*, ed. W.P. 9668. Cambridge: National Bureau of Economic Research.

Gilsing, V. 2000. *Preliminary Draft: Cluster Governance, how clusters can adapt and renew over time*. Paper prepared for the DRUID-PhD Conference, Copenhagen, 2000. Available from http://www.druid.dk/conferences/winter2000/gilsing.pdf)

Giuliani, E., R. Rabelotto, and M.P. van Dijk (eds.). 2005. *Clusters Facing Competition: The Importance of External Linkages*. Ashgate Publishing Limited, Aldershot: England.

Global Witness. 2007. *Hot Chocolate: How cocoa fuelled the conflict in Côte d'Ivoire*. A report by Global Witness, June 2009. Available from http://www.globalwitness.org.

Gockowski, J. 2007. *Cocoa Production Strategies and the Conservation of Globally Significant Rainforest Remnants in Ghana. Policy and Impact Analyst*. Presentation at Workshop: Production, markets and the future of smallholders: the role of cocoa in Ghana. November 19, 2007 at Conference Centre, Council of Scientific and Industry Research, Sustainable Tree Crops Program/International Institute of Tropical Agriculture Accra, Ghana.

Goodman. 1997. Agrarian Questions: Global Appetite, Local Metabolism: Nature, Culture, and Industry in *Fin-de-Siècle* Agro-Food Systems. In: *Globalising Food, Agrarian Questions and Global Restructuring*, eds. David Goodman, and M.J. Watts: 1-34. London: Routledge.

Granovetter, M. 1985). Economic Action and Social Structure: the Problem of Embeddedness. In: *The American Journal of Sociology*, 91(3): 481-510.

Graue, E. 1950. Is Cocoa Being Valorized? In: *The Review of Economics and Statistics*, 32: 258-63.

Griffiths, A., and R.F. Zammuto. 2005. Institutional governance systems and variations in national competitive advantage: An integrative framework. In: *Academy of Management Review*, 30(4): 823-42.

Grinsven, P. 2007. *The Role of Knowledge for Tropical Food Chains – Focus on Cocoa*. In Multi-stakeholder workshop on Improving the position of small holders through knowledge exchange between Tropical food Chains and Research. Wageningen.

Guerrieri, P., S. Lammarino, and C. Pietrobelli. 2001. *The Global Challenge to Industrial Districts Small and Medium-sized Enterprises in Italy and Taiwan*, Edward Elgar, Cheltenham.

Giuliani, E., R. Rabellotti, and M.P. van Dijk (eds.). 2005. *Clusters Facing Competition: The Importance of External Linkages*, Ashgate; Hampshire.

Guion, L.A. 2002. *Triangulation: Establishing the Validity of Qualitative Studies*.

Halder, G. 2002. *How Does Globalisation Affect Local Production and Knowledge Systems? The Surgical Instrument Cluster of Tuttlingen*. Germany. Duisberg: INEF.

Hall, R., and D. Soskice. (eds). 2001. *Carieties of Capitalism: The Institutional Foundations of Comparative Advantage*. Oxford: Oxford University Press.

Hamdok, A. 2003. Governance and Policy in Africa: Recent Experiences. In: *Reforming Africa's Institutions: Ownership, Incentives, and Capabilities*, ed. S. Kayizzi-Mugerwa: 15-29. New York: UN University Press.

Haque, I.u. 2004. *Commodities under Neoliberalism: The Case of Cocoa*. In United Nations Conference on Trade and Development. United Nations.

Harris, N. 1986. *The End of the Third World. Newly Industrializing Countries and the Decline of an Ideology*. London: I.B. Tauris.

Helferich, J. (Vice President for Research and Development M&M/Mars). 1999. *An Industry Perspective on Sustainable Cocoa Development in Africa: Conditions Needed for Effective Public/Private Partnerships to Work*, Paper presented at STCP Forum, 1999.

Helmsing, B. 2002. Partnerships, Meso-Institutions and Learning: New Local and Regional Economic Development Initiatives in Latin America. In: *Realigning Actors in an Urbanizing World, Governance and Institutions from a Development Perspective*, eds. I.S.A. Baud, J. Post, L. de Haan, and T. Dietz: 79-100. Department of Geography and Planning, University of Amsterdam: Ashgate, England.

Helmsing, A.H.J., and Cartwright, A. 2009. Enterprise, livelihood and local economic development, with a case study of the Wool Chain Initiative, Eastern Cape, South Africa. In: *Journal of Development Alternatives and Area Studies*, 28 (March).

Heslin, P.A., and J.D. Ochoa. 2008. Understanding and Developing Strategic Corporate Social Responsibility. In: *Organizational Dynamics*, 37: 125-44, 2008. Available at SSRN: http://ssrn.com/abstract=1149001

Hill, P. 1963. *The Migrant Cocoa-Farmers of Southern Ghana*. Cambridge: University Press.

Hopkins, T.K., and I. Wallerstein. 1986. Commodity Chains in the World Economy Prior to 1800. In: *Review*, 10(1): 157–70.

Hopkins, T.K., and I. Wallerstein. 1994. Commodity Chains: Construct and Research. In: *Commodity Chains and Global Capitalism*, eds. G. Gereffi, and M. Korzeniewicz, Praeger.

Hospes, O. and J. Clancy. 2009. *DRAFT Conceptualizing Social Inclusion and Values in Biofuel Chains*. "Value Chain Governance and endogenous growth" network http://value-chains.global-connections.nl/content/value-chain-governance-and-endogenous-growth DPRN/CERES group.

Humphrey, J. 2004. *Commodities, Diversification and Poverty Reduction*. Rome: Institute of Development Studies.

Humphrey, J., and J. Schmitz. 1995. *Principles for Promoting Networks of SMEs, Small and Medium Enterprise Program*, UNIDO, Vienna, Discussion Paper No. 1.

Humphrey, J., and H. Schmitz. 2000. Governance and Upgrading: Linking Industrial Cluster and Global Value Chain Research. In: *IDS Working Paper 120*. Institute of Development Studies: University of Sussex and Institute for Development and Peace: University of Duisburg

Humphrey, J., and H. Schmitz. 2001. Governance in Global Value Chains. In: *IDS Bulletin*, 32.

Humphrey, J., and H. Schmitz. 2002. *Developing Country Firms in the World Economy: Governance and Upgrading in Global Value Chains*. Duisburg: Institüt für Entwicklung und Frieden der Gerhard-Mercator-Universität Duisberg.

IFAD. 2001. *Rural Poverty Report 2001 – The Challenge of Ending Rural Poverty*. http://www.ifad.org/poverty/chapter2.pdf, Access date 08-10-03.

ICCO. 2005. *Facts and Figures on Fair-trade Cocoa*. Consultative Board on the World Cocoa Economy, Fifth meeting, London 6th June 2005. CB/5/CRP.1. 31 May 2005.

ICCO. 2006. *Review of the Market Situation Executive Committee*. One hundred and thirty-first meeting. London: EBRD offices.

ICCO. 2007. *ICCO Annual Report 2006/2007*.

IDRC. 2006. *Globalization, growth, and Poverty Program initiative. Description of the Program for 2006-11*, IDRC.

IFPRI. 2002. *Empowering Women and Fighting Poverty: cocoa and land rights in West Africa*. IFPRI.

IMF. 2009. *Impact of the Global Financial Crisis on Sub-Saharan Africa*. International Monetary Fund, African Department. Accessible at http://www.imf.org/external/pubs/ft/books/2009/afrglobfin/ssaglobalfin.pdf

Jaeger, P. 1999. *The Market for Cocoa and its Relevance to African Production*. USAID/STCP.

Jong, A., de, and A. Harts-Broekhuis. 1999. Cocoa Production and Marketing in Cameroon and Ghana. The Effects of Structural Adjustment and Liberalization. In: *Agricultural Marketing in Tropical Africa. Contributions from the Netherlands*, eds. H.L. v. d. Laan, T. Dijkstra, and A. v. Tilburg: 87-108. Leiden: Ashgate.

Joosten, F., and D. Eaton. 2007. Public Sector Roles in Agri-food Chains. Regulatory Strategies and Functions in Food Safety, Corporate Social Responsibility and Seed Sector Development. In: *Markets, Chains and Sustainable Development. Strategy and Policy paper 1*. Wageningen: Wageningen UR.

Kalb, D., W. Pansters, and H. Siebers. 2004. Conflictive Domains of Globalization and Development. In: *Globalization and Development. Themes and Concepts in Current Research*, eds. D. Kalb, W. Pansters, and H. Siebers: 1-8. Dordrecht: Kluwer Academic Publishers.

Kalb, D. 2004. Time and Contention in "the great globalization debate". In: *Globalization and Development. Themes and Concepts in Current Research*, eds. D. Kalb, W. Pansters, and H. Siebers: 9-48. Dordrecht: Kluwer Academic Publishers.

Kaplinsky, R. 2000. Spreading the Gains from Globalisation: What can be Learned from Value Chain Analysis. In: *IDS Working Paper 110*. Institute of Development Studies.

Kaplinsky, R. 2001. *Globalisation and Unequalisation: What Can Be Learned from Value Chain Analysis*. Institute of Development Studies University of Sussex, and Centre for Research in Innovation Management, University of Brighton.

Kaplinsky, R. 2004. *Competitions Policy and the Global Coffee and Cocoa Value Chains*. Institute of Development Studies University of Sussex, and Centre for Research in Innovation Management, University of Brighton.

Kaplinsky, R., and M. Morris. 2003. *Governance Matters in Value Chains, Developing Alternatives*, 11-18.

Keesing, D., and S. Lall 1992. Marketing Manufactured Exports from Developing Countries: Learning Sequences and Public Support. In: *Trade Policy, Industrialisation and Development*, ed G. Helleiner: 176-93, Oxford University Press.

Kennedy, Loraine. 1999. Cooperating for Survival: Tannery Pollution and Joint Action in the Palar Valley (India). In: *World Development*, 27 (9): 1673-91.

KIT, Faida MaLi, and IIRR. 2006. *Chain Empowerment: Supporting African farmers to Develop Markets*. Royal Tropical Institute, Amsterdam; Faida Market Link, Arusha; and International Institute of Rural reconstruction, Nairobi.

KIT, AgroEco & Tradin. Forthcoming. *Feasibility Study on Organic Cocoa Production and Marketing In West Africa*. CFC/ICCO.

Koning, M. de, and B. de Steenhuijsen Piters. 2009. *Farmers as Shareholders. A close look at recent experience*. Amsterdam: KIT Publishers.

Kooiman, J. 1993. Findings, Speculations and Recommendations. In: *Modern Governance, New Government-Society Interactions*, ed. J. Kooiman: 249-62. London: Sage Publications.

Knorringa, P. 1999. Agra. An Old Cluster Facing the New Competition. In: *World Development*, 27(9): 1587-1604.

Knorringa, P. 2002. Urban Cluster Trajectories in Developing Countries: Beyond the Industrial District Model. In: Realigning Actors in an Urbanizing World, Governance and Institutions from a Development Perspective, eds. I.S.A. Baud, and J. Post: 63-78. Department of Geography and Planning, University of Amsterdam, Ashgate, England.

Industrial District Model. In: *Realigning Actors in an Urbanizing World, Governance and Institutions from a development Perspective*, eds. I.S.A. Baud, J. Post, L. de Haan and T. Dietz: 63-78. Department of Geography and Planning, University of Amsterdam, Ashgate, England.

Knorringa, P. and L. Pegler. 2006 *Globalisation, Firm Upgrading and Impacts on Labour*. The Hague: Institute of Social Studies.

Kuapa Kokoo. 2002. *2001-2002 Annual Report*. Kumasi: Kuapa Kokoo Limited.

Kuapa Kokoo. 2004. *2004 Annual Report*. Kumasi: Kuapa Kokoo Limited.

Laan, T. 2006. *Upgrading by Producing for Niche Markets: The Case of Small Scale Cocoa Farmers in Talamanca, Costa Rica*. M.Sc. International Development Studies Thesis, ISHSS/University of Amsterdam.

Lall, S. 2005. Rethinking Industrial Strategy: The Role of the State in the Face of Globalization. In *Putting Development First. The Importance of Policy Space in the WTO and International Financial Institutions*, ed K. Gallagher: 33-68, London: Zed Books.

Lambooy, J. 2002. Firms, Regions and Resources in a Globalizing Economy: A Relational view. In: *Realigning Actors in an Urbanizing World, Governance and Institutions from a development Perspective*, eds. I.S.A. Baud, J. Post, L. de Haan, and T. Dietz: 25-42, Department of Geography and Planning, University of Amsterdam, Ashgate, England.

Larsen, M.N. 2003 *Quality Standard Setting in the Global Cotton Chain and Cotton Sector Reform in Sub-Saharan Africa*. DIIS Working Paper 03.07. Copenhagen: Danish Institute for International Studies.

Laven, A. 2005. *Towards a Sustainable Cocoa Chain, a Ghanaian Perspective*. Report of a workshop held at the British Council on the 31st of March 2005, in collaboration with professor V.K. Nyanteng and SNV, Netherlands Development Organisation. British Council, Accra, Ghana.

Laven, A. 2007. Who is interested in good quality cocoa from Ghana? In: *Tropical Food Chains. Governance Regimes for Quality Management*, eds. R. Ruben, M. van Boekel, A. van Tilburg, and J. Trienekens: 189-210. Wageningen: Wageningen Academic Publishers.

Little, P.D. 1994. Contract Farming and the Development Question. In: *Living Under Contract. Contract Farming and Agraraian Transformation in Sub-Saharan Africa*, eds. P. Little, and M.J. Watts: 216:47. Madison: The University of Wisconsin Press.

Little, P.D., and M.J. Watts (eds.) 1994. *Living Under Contract. Contract Farming and Agraraian Transformation in Sub-Saharan Africa*. Madison: The University of Wisconsin Press.

LMC International & University of Ghana School of Administration. 2000. *Liberalisation of External Marketing of Cocoa in Ghana*. 72. Cocoa Sector Reform Secretariat, Accra.

Long, N. 2001. *Development Sociology. Actors Perspectives*. London and New York: Routledge. Taylor and Francis Group.

Losch, B. 2002. Global Restructuring and Liberalization: Côte d'Ivoire and the End of the International Cocoa Market? In: *Journal of Agrarian Change*, 2: 206-27.

Masdar, L. 1998. *Cocoa Board Ghana, Socio-economic Study, Final Report*. Berks: The Cocoa Board Ghana.

Maskell, P. and A. Malmberg. 1999. Localised Learning and Industrial Competitiveness. In: *Cambridge Journal of Economics*, 23: 167-85, Cambridge Political Economy Society, 1999.

Mayoux, L. no date. *Case Study: Kuapa Kokoo, Ghana, Enterprise Development Impact Assessment Information System (EDIAIS)*. Institute for Development Policy and Management (IDPM) of Manchester University; Women. Sustainable Enterprise Development Ltd (WISE Development) and DFID.

McCormick, D. 1999. African Enterprise Clusters and Industrialization: Theory and Reality. In: *World Development*, 27(9) 1531-51.

McGahan, A., and M.E. Porter. 1997. How Much Does Industry Matter, Really? In: *Strategic Management Journal*, Summer Special Issue 18:15-30.

Mehra, S., and S. Weise. 2007. *West Africa Strategy*. Presentation on WCF Partnership Meeting WCF West Africa Committee and IITA/STCP, Amsterdam, May 23, 2007.

Meyer-Stamer, J. 2002. *Clustering and the Creation of an Innovation-Oriented Environment for Industrial Competitiveness: Beware of Overly Optimistic Expectations*. Revised Version of Paper prepared for International High-Level Seminar on Technological Innovation, sponsored by the Ministry of Science and Technology of China and United Nations University, Beijing, September 5-7, 2000, Duisburg, February 2002.

Milburn, J. 1970. The 1938 Cold Coast Cocoa Crisis: British Business and the Colonial Office. In: *African Historical Studies*, 3: 57-74.

Ministry of Finance. 1999. *Ghana Cocoa Sector Development Strategy*. Accra: Ministry of Finance.

Ministry of Finance. 2007. http://www.ghana.gov.gh/ghana/speech_deputy_finance_minister_prof_gyan_baffour.jsp

MMYE. 2007. *Labour Practices in Cocoa Production in Ghana*. Pilot Survey. National Programme for the Elimination of the Worst Forms of Child Labour in Cocoa.

Nadvi, K. 1995. *Industrial Clusters, and Networks: Case Studies of SME Growth and Innovation*, Working Paper: UNIDO Discussion Paper, pp 1-78, UNIDO small medium program, Vienna. Available at http://www.unido.org/userfiles/PuffK/Nadvi.pdf.

Nadvi, K. 1999. Collective Efficiency and Collective Failure: the Response of the Sialkot Surgical Instrument Cluster to Global Quality Pressures. In: *World Development*, 27 (9):1605-26.

National Cocoa and Coffee Board [NCCB]. 2008. *Liberalisation on the Cocoa and Coffee Trades in Cameroon*. Paper presented at a workshop on the liberalisation of the cocoa trade (Lomé 1998) by Hope Sona Ebai.

Nadvi and Barrientos, 2004. *Industrial Clusters and Poverty Reduction. Towards a methodology for poverty and social impact assessment of cluster development initiatives*. UNIDO. Economy Environment Employment. Funded by the Swiss Agency for development Cooperation.

Nadvi, K., and G. Halder. 2002. Local Clusters in Global Value Chains: Exploring Dynamic

Linkages in Germany and Pakistan. In: *IDS Working Paper 152*. Brighton: Institute of Development Studies.

Nadvi, K., and H. Schmitz. 1994. *Industrial Clusters in Less Developed Countries: Review of Experiences and Research Agenda*. Institute of Development Studies, Discussion Paper, No. 339, Institute of Development Studies, University of Sussex, Brighton.

Neven, D., and C.L.M. Dröge, no date. *A Diamond for the Poor? Assessing Porter's Diamond Model for the Analysis of Agro-Food Clusters in the Developing Countries*. Department of agricultural Economics, Department of marketing and Supply Chain Management. Michigan State University.

North, D.C. 1981. *Structure and Change in Economic History*, WW Norton, New York.

Nuijten, M. 2004. Governance in Action. Some theoretical and practical reflections on a key concept. In: *Globalization & Development. Themes and Concepts in Current Research*, eds. D. Kalb, W. Pantsers & H. Siebers: 103-30. Dordrecht/Boston/London: Kluwer Academic Publishers.

Ogunleye, K.Y.,and J.O. Oladeji. 2007. Choice of Cocoa Market Channels among Cocoa Farmers in Ila Local Government Area of Osun State, Nigeria. In: *Middle-East Journal of Science and Research*, 1(1):14-20.

Oosterveer, P., and Hoi, Pham, v. 2009. DRAFT *Governance and the Greening of Agro-food Chains; the cases of Vietnam and Thailand*. DPRN 2009/2010.

Oxfam. 2004. *The Commodity Challenge. Towards an EU Action Plan*. A Submission by Oxfam to the European Commission, Oxfam.

Palmer, R. 2004. *For Credit Come Tomorrow: Financing of Rural Micro-Entrepreneurs: Evidence from Nhawie-Kuma, Atwima District, Ghana*. Edingburgh: Centre for African Studies Edingburgh University.

Palpacuer, F. 2000. *Preparatory Note, Characterizing Governance in Value Chain Analysis*, IDS/Rockefeller Foundation Meeting on Global Value Chains, Bellagio, Italy, September 25-October 2, 2000. Montpellier: CREGO, University of Montpellier II, France.

Pietrobelli, C., and R. Rabelloti 2005. Upgrading in Global Value Chains: Lessons from Latin American Clusters. In: *Clusters Facing Competition: The Importance of External Linkages*, eds. E. Guiliani, R. Rabelloti, and M.P. van Dijk: 13-38. Hampshire: Ashgate Publishing Limited. Burlington: Ashgate Publishing Company.

Piore, M.J., and C.F. Sabel 1984. *The Second Industrial Divide: Possibilities for Prosperity*. Basic Books, Inc., New York.

Ponte, S. 2002. *The 'Latte Revolution'? Winners and Losers in the Restructuring of the Global Coffee Marketing Chain*. Working Paper Sub-series on Globalisation and Economic Restructuring in Africa no. xiii CDR Working Paper 01.3 June 2001. Centre For Development Research, Copenhagen.

Ponte, S., and B. Daviron. 2005. *The Coffee Paradox: Global Markets, Commodity Trade and the Elusive Promise of Development*. London: Zed Books.

Porter, M.E. 1990. *The Competitive Advantage of Nations*. The Free Press. New York.

Porter, M.E. 1998. *On Competition*. Harvard Business Review Books.

Porter, M.E. 2008. *Clusters, Innovation, and Competitiveness: New Findings and Implications for Policy*. Institute for Strategy and Competitiveness, Harvard Business School, 22 January 2008.

Post, J., I.S.A. Baud, L. de Haan, and T. Dietz. 2002. Introduction. In: *Realigning Actors in an Urbanizing World, Governance and Institutions from a Development Perspective*, eds. I.S.A. Baud, J. Post, L. de Haan and T. Dietz: 1-22. Department of Geography and Planning, University of Amsterdam: Ashgate, England.

PSOM. 2004. *Establishing a Quality Chain for Vietnamese Cocoa*. Qualification proposal. PSOM04/VN/3. May 2004. Gerkens Cacao BV, Cargill Asia Pacicic Limited, Masterfoods Veghel BV. EVD.

Raikes, P., M.F. Jensen, and S. Ponte. 2000. *Global Commodity Chain Analysis and the French Filière Approach: Comparison and Critique*. Working Paper Subseries on Globalisation and Economic Restructuring in Africa no. ix. 26. Copenhagen: Centre for Development Research.

Republic of Côte d'Ivoire. 1998. *Liberalisation of Cocoa and Coffee marketing in Côte d'Ivoire*. Paper presented on workshop. Lomé: Ministry of Agriculture and Animal Resources, Ministry for the Promotion of External Trade.

Rhodes, R. 2000. Foreword. In: *The new politics of British local governance*, ed. G. Stoker, xi-xv. London: MacMillan Press.

Ruf, F. 2007a. *The Cocoa Sector. Expansion, or green and double green revolutions*. ODI Background Note.

Ruf, F. 2007b. *The new Ghana Cocoa Boom in the 2000s. From Forest Clearing to Green Revolution*. Report prepared for the Ministry of Finance and Economic Planning, Ghana. CIRAD/University of Ghana.

Schmitz, H. 1989. *Flexible Specialisation: a new paradigm of small-scale industrialization?* Brighton: Institute of Development Studies.

Schmitz, H. 1995. Small Shoemakers and Fordist Giants; Tale of a supercluster. In: *World Development*, 23(1):139-50.

Schmitz, H. 1999. Global Competition and Local Cooperation: Success and Failure in the Sinos Valley, Brazil. In: *World Development*, 27(9): 1627-50.

Schmitz, H., and P. Knorringa 1999. *Learning from Global Buyers*. IDS Working Paper 100. Brighton: Institute of Development Studies.

Schmitz, H., and K. Nadvi. 1999. Clustering and Industrialization: Introduction. In: *World Development*, 27: 1503-16.

Sewell, W.H., Jr. 1992. A Theory of Structure: Duality, Agency, and Transformation. In:

The American Joruanl of Sociology, 98: 1-29. The University of Chicago Press

Singh, A. 2002. Competition and Competition Policy in Emerging Markets: International and Developmental Dimensions. In: *G24 Technical Group*, 56. Beirut: University of Cambridge.

Smakman, F. 2003. *Local Industry in Global Networks Changing Competitiveness, Corporate Strategies and Pathways of development in Dinagpore and Malaysia's Garment Industry*. Faculteit Geowetenschappen. Utrecht: Universiteit Utrecht.

STCP. 2005. Income and Collective Action: Groups Sales and Farmer field Schools in the Cocoa Sectors of Cameroon and Côte d'Ivoire. In: *Impact Brief, Sustainable Tree Crops Program*. Published by the International Institute of Tropical Agriculture.

STCP, and IITA. 2002. *Child Labor in the Cocoa Sector of West Africa. A synthesis of findings in Cameroon, Côte d'Ivoire, Ghana, and Nigeria*, Under the auspices of USAID/USDOL/ILO.

Stiglitz, J. 2002. *Globalization and its discontents*. London: Pengion group.

Takane, T. 2000. Incentives embedded in institutions: The case of share contracts Ghanaian cocoa production. In: *The Developing Economies*, XXXVIII-3 (September 2000): 374–97.

Takane, T. 2002. *The Cocoa Farmers of Southern Ghana, Incentives, Institutions, and Change in Rural West Africa*. Chiba: Institute of Developing Economies, Japan External Trade Organization.

Teal, F., and M. Vigneri. 2004. *Production Changes in Ghana Cocoa Farming Households under Market Reforms*. The Centre for the Study of African Economies Working Paper Series. Centre for the Study of African Economies.

Teal, F., A. Zeitlin, and H. Maamah 2006. *Ghana Cocoa Farmers Survey 2004: Report to Ghana Cocoa Board*, Accra, March 11, 2006.

Tiffen, P. no date. The Creation of Kuapa Kokoo. Available from http://www.divinechocolate.com

Tiffen, P., H. Maamah, B. Lewin, K. Bong-Jin, and E. Bryla 2002. *Ghana: Cocoa Price Risk Management*. Phase II Report, February 2002.

Ton, G., G. Hagelaar, A. Laven, and S. Vellema. 2008. *Chain Governance, Sector Policies and Economic Sustainability in Cocoa*. Markets, Chains and Sustainable Development Strategy and Policy paper 12, Wageningen UR.

Tropical Commodity Coalition. 2008. *Sweetness follows. A rough guide towards a sustainable cocoa sector*. Den Haag: Tropical Commodity Coalition.

Tropical Commodity Coalition. 2009. *Cocoa Barometer 2009*. The Hague: Tropical Commodity Coalition.

UNDP. 2003. *Human Development Report 2003*. United Nations Development Programme.

UNCTAD. 2001a. *Improving the Competitiveness of SMEs in Developing Countries. The role of finance to enhance enterprise development*. United Nations. New York and Geneva. http://www.unctad.org/en/docs/itetebmisc3_en.pdf (Access date 13 July 2009).

UNCTAD. 2001b. *International Cocoa Agreement*. United Nations Cocoa Conference.

UNCTAD/WTO. 2001. *Cocoa, A guide to trade practices*. Geneva: International Trade Centre UNCTAD/ WTO.

UNIDO. 2004. *Industrial Clusters and Poverty Reduction, Towards a methodology for poverty and social impact assessment of cluster development initiatives*. This Report was prepared by Khalid Nadvi and Stephanie Barrientos, Institute of Development Studies, University of Sussex, United Kingdom, in cooperation with staff of UNIDO's Small and Medium Enterprises Branch.

Van der Geest, K., A. Vrieling, and T. Dietz. Forthcoming. Migration and Environment in Ghana: A cross-district analysis of mobilty and vegetation dynamics. Submitted to *Environment and Urbanization*.

Van der Geest, S. no date. *GSC Notes and Queries elderly people in Ghana: ongoing anthropological research*, University of Amsterdam.

Vargas, M.A. 2001. *Forms of Governance, Learning Mechanisms and Upgrading Strategies in the Tobacco Cluster in Rio Pardo Valley – Brazil*. Sussex: IDS – University of Sussex.

Vermeulen, S., J. Woodhill, F.J. Proctor, and R. Delnoye 2008. *Chain-wide Learning for Inclusive Agrifood Market Development: a guide to multi-stakeholder processes for linking small-scale producers to modern markets*. International Institute for Environment and Development, London, UK and Wageningen University and Research Centre, Netherlands, the Netherlands.

Verrest, H. 2007. *Home-Based Economic Activities and Caribbean Urban Livelihoods. Vulnerability, Ambition and Impact in Paramaribo and Port of Spain*. PhD dissertation, University of Amsterdam. Amsterdam: Amsterdam University Press.

Visser, E. 2004. *A Chilean Wine Cluster? Governance and Upgrading in the phase of internationalization*. Division of Production, Productivity and Management. ECLAC/GTZ Project on "Natural Resource-based Strategies Development" (GER 99/128). Santiago de Chile, September 2004. ECLAC, UN.

Vuure, R.P. 2007. *The Interaction of Civic Conventions aimed at Improving Gender Equality for Women Working in the Ghanaian Cocoa Sector with their Respective Livelihood Strategies*. M.Sc. International Development Studies Thesis, ISHSS/University of Amsterdam.

Walker, L. 2006. Maintaining a Delicate Balance of Supply. In: *The Public Ledger. World Commodities Weekly*, 12.

Watts, M., and D. Goodman. 1997. Fin-de-siècle Agro-Food systems. In: *Globalising Food. Agrarian questions and global restructuring*, eds. D. Goodman, and M. Watts. London: Routlegde.

Wayo Seini, A. 2002. *Agricultural Growth and Competitiveness Under Policy Reforms in Ghana*. Accra: Institute od Statistical, Social&Economic Research University of Ghana, Legon.

WCF 2007. *World Cocoa Foundation Newsletter*, August 2007, Issue 31.

Wennink, B., S. Nederlof, and W. Heemskerk (eds.) 2007. *Access of the Poor to Agricultural Services. The role of farmer organizations in social inclusion.* Bulletin 376. Development Policy & Practice. Royal Tropical Institute. KIT Publishers: Amsterdam.

Westen, G. v. 2002. Embeddedness: The role of local factors in economic development. In: *Re-aligning Government, Civil Society and the Market: New Challenges in urban and regional development,* eds. I.S.A. Baud, J. Post, L. de Haan, and T. Dietz. Essays in honour of G.A. de Bruijne, Amsterdam Institute for Global Issues and Development Studies (AGIDS), University of Amsterdam.

Whitley, R. 1998. Internationalization and Varieties of Capitalism: the limited effects of cross-national coordination of economic activities on the nature of business systems. In: *Review of International Political Economy,* 5(3): 445-81.

Whitley, R. 2001. Developing Capitalisms: The Comparative Analysis of Emerging Business Systems in the South. In: *Understanding Business Systems in Developing Countries,* eds. G. Jakobsen, and J. Torp: 25-41. New Delhi, Thousand Oaks and London Sage.

Whitley, R. 2005. How National are Business Systems? The Role of States and Complementary Institutions in Standardizing Systems of Economic Coordination and Control at the National Level. In: *Changing Capitalisms? Internationalization, Institutional Change, and Systems of Economic Organization,* eds, G. Morgan, R. Whitley, and E. Moen: 190-231. Oxford: Oxford University Press.

World Bank. 2002. *From Action to Impact – The Africa Region's Rural Strategy,* Washington D.C.

World Bank. 2007a. *World Development Report 2008. Agriculture for Development. Overview.* Washington: The International Bank for Reconstruction and Development/The World Bank.

World Bank. 2007b. *Ghana. Meeting the Challenge of Accelerated and Shared Growth.* Country Economic Memorandum. (In Three Volumes) Volume III: Background Papers. Report No. 40934-GH. November 28, 2007. PREM 4. Africa Region. Document of the World Bank.

Yeung, H.W. 2003. Practicing New Economic Geographies: A Methodological Examination. In: *Annals of the Association of American Geographers,* 93(2), 2003: 442–62. Association of American Geographers. Oxford: Blackwell Publishing.

Zeitlin, A. 2006. *Market Institutions and productivity: Microeconomic Evidence from Ghanaian Cocoa.* Oxford: University of Oxford.

Summary

1   Local producers in a global economy

Proponents of globalisation, consisting of advocates of neo-liberal free markets and liberal civil society, have long argued that free trade will lead to economic growth and improvements in the livelihoods of all, including poor farmers in developing countries. But it has become clear that gains from globalisation are not distributed equally. Small-scale farmers in developing countries must improve or upgrade their businesses, if they are to cope with the challenges of globalisation, increased competition and price fluctuations.

There are different opinions on which upgrading strategies small-scale farmers should follow. There are also different ideas on which actors should facilitate and/or support the poor in this process. International institutions, such as the World Bank, emphasise that farmers themselves are responsible for change. In the process of adding value to their product or production process, farmers can learn from 'lead firms' higher up in the value chain (the value chain perspective) or can realise change through joint action and collective efficiency (clustering perspective). Others argue that in less developed countries, where the majority of raw material production is in the hands of small producers, capturing higher margins for unprocessed commodities requires public action.

It is increasingly emphasised that in order to understand the relationship between power structures, poverty and upgrading it is important to look at the interaction between different governance levels. In this study the exceptional case of cocoa in Ghana is used to analyse how governance in the international cocoa chains – driven by a combination of concentration, sustainability standards and uncertainty in global sourcing – combines or contrasts with interventions and regulations by the state at a national level, and social organization and performance at local, farmers' level. The study focuses on the capacity of the state to solve problems and shows how a national state stands between local producers and global buyers of cocoa.

Cocoa production is concentrated in West Africa and employs around 14 million workers worldwide. It is estimated that about 3 million smallholders account for 90 per cent of production. Cocoa is a primary commodity mainly produced for export, with little added value. In addition to being the world's second largest producer, Ghana has some particularities that make it an interesting case for assessing the interaction between private and public policies. First, Ghana is the only cocoa producing country in the region that has only *partly* liberalised its marketing and pricing system; the government still plays a governing role in the sector. Second, it is the only country that provides traceable cocoa and the high quality cocoa it produces fetches an additional premium.

A number of important insights will arise from the Ghanaian case. First, it contributes to understanding of how producers of agricultural export commodities benefit from being inserted in a global value chain, one which is increasingly driven by multinational cocoa processors and chocolate manufacturers. Second, it contributes to the recent discussion on hybrid governance structures, where both public and private actors play a governing role. Ghana is unique because of the strong role of the state. Lastly, looking at differences among cocoa farmers contributes to understanding processes of in- and exclusion of particular farmers, and hereby the role of the state.

2 Upgrading for development

The discussions on globalisation and its impact on development emphasise the challenges faced by entrepreneurs in developing countries, i.e. intensified competition and price-fluctuations. Since the late nineties, the focus has been especially on small and medium enterprises (SMEs). Recently, with agriculture back on the development agenda as the 'engine for economic growth', the emphasis has shifted from SMEs to small-scale agricultural producers. At the same time, these small-scale producers are increasingly viewed as independent entrepreneurs. In order to support sustainable growth and reduce poverty, these agricultural 'firms' have to improve their competitiveness. Competitiveness is not only based on better pricing and improved quality of products but it is also important whether there is sustainable production within the chain.

In the literature on competitiveness, the concept of upgrading highlights the options available to producers for obtaining better returns. This concept, which finds its origin in political economy and industrial economics, has been applied in various bodies of literature, from the value chain approach to cluster studies. The dominant view is that upgrading is the outcome of organisational learning and inter-firm networking. Essentially the value chain approach proposes that the inclusion of entrepreneurs in value chains offers the possibility to engage in learning processes and to acquire new knowledge from external buyers. This approach concentrates on 'global chain governance', which is defined as 'authority and power relationships that determine how financial, material, and human resources are allocated and flow within the chain'. A value chain may be characterized by different forms of coordination in various segments of the chain.

It has been argued that the buyers' risks for supplier failure determine the type of relationships between local producers and external buyers. But it is not clear how the specifics of the value chain ('tight or loose' organisation) are linked to a particular outcome for producers. This implies that there are power relations in a chain that have to be unravelled, as poor producers run up against them. Although the value chain approach captures different categories of governance and coordination, the focus is on inter-firm relationships between buyers and suppliers. The focus on vertical networks does not capture more local inter-firm relations, for example joint actions among producers, and interactions with local institutions. Alternatively, cluster studies focus on local level governance structures, which are viewed as the main facilitators of upgrading and innovation. Having relationships in a cluster facilitate the creation of new products and services. The cluster literature uses the concept of 'cluster governance' to refer to the intended, collective actions of cluster actors aimed at upgrading a cluster. Advantages from clusters usually derive from an 'optimal mix between cooperation and competition among its members'. When this balance is disturbed, clustering can jeopardize competition. Understanding the conditions under which such an optimal balance can occur requires insight in the level of 'embeddedness' of economic activity. Many of the social relationships are geographically localised. People are not simply workers or managers; they are also consumers, citizens, church-goers, kin, and community members.

Several authors have highlighted the limitations of both approaches and the necessity to combine the 'horizontal networks' with the 'vertical networks'. But, by combining these two approaches some of their limitations, which are similar, are not yet removed. Both approaches assume the presence of an open market system. But this condition is not always 'fully' present. In some large sectors of several commodity producing countries, there are no truly 'free markets'. To understand how governance is linked to processes of in- and exclusion I brought back the role of the state into the discussion. Accordingly, a new theoretical orientation needs to be introduced. I chose to use a dynamic comparative framework for labelling public-private interaction in a global value chain to explain variations in national industrial competitiveness, which draws on both the strategic management and political economy literature. The framework points out four types of interaction in a value chain: 'state governance' (a situation where transactions are coordinated through state involvement and the value chain is coordinated through market forces), 'joint governance' (a situation where transactions are coordinated through state involvement and the value chain is coordinated through chain integration), 'market governance' (a situation where transactions and the value chain are both coordinated through market forces) and 'corporate governance' (a situation where transactions are coordinated through state involvement and the value chain is coordinated through chain integration).

The multi-level entry point makes it possible to discuss the capacity of the (national) state to solve problems and to elaborate on how the state's performance depends on governance at international level. In order to value the state's performance and understand how performance affects the way particular farmers are in-or excluded requires bringing in again governance structures at a local level. Combining global, national and local governance structures will show how a national state stands between local producers and global players and offers the opportunity to discuss more precisely how state functions can or cannot help to address the challenges involved in being inserted in an increasingly trader-driven value chain. The case of cocoa in Ghana is particularly interesting because the sector is gradually and only partially liberalised. It is an example of a 'hybrid' case where the state is still keeping a rent in monopolising export trade in a situation where the international market is increasingly driven international buyers.

3 Research questions, methods and respondents

The central question in this study is how different governance processes interact in creating opportunities and constraints for more inclusive upgrading among small-scale cocoa farmers in Ghana. I distinguish between three levels of governance: first, the global chain governance (referring to power relations in the global cocoa chain); second, state governance (referring to the level of state involvement in the Ghanaian cocoa sector); and third, the social structures (in which cocoa farmers are embedded locally). The different dimensions contained in this question demand for a combination of different research tools and concepts. The value chain approach is used as a tool to identify upgrading opportunities and constraints for cocoa farmers in Ghana by considering the existing power relations in the global chain and by looking at changes in these relations. At the national level, the introduction of reforms in Ghana is taken as a 'key-turning point' to understand local upgrading opportunities and constraints, and how these have changed overtime. At the local level, I seek to explain the different impact that shifts in governance structures and upgrading opportunities (along with the constraints that result from these changes) have on farmers, resulting in unequal benefits. Central concepts of the cluster literature, such as 'embeddedness' and 'joint action', are used to identify social structures which constrain or facilitate upgrading strategies of individual cocoa farmers and thus affect the way they benefit from these strategies.

To answer the research question, I combined qualitative with quantitative data. I held in-depth interviews with actors involved in the cocoa sector in Ghana, including the farmers, local and international buyers, governmental bodies, farmer organisations, NGOs, banks and research institutes. In addition I gathered information through a number of informal discussions with the world's major cocoa buyers. In addition, I organised two multi-stakeholder workshops, one in the Netherlands (2003) and one in Ghana (2005), with key representatives of industry and other public and private actors. I administered three surveys: 1) a survey among a small number of cocoa processors, chocolate manufacturers and some of the institutions that represent their interests; 2) a survey among 173 farmers, held in 2003; and 3) a second farmer survey, held in 2005, among 103 farmers that participated in the 2003 survey and 107 additional cocoa farmers. The farmer surveys were conducted in 34 communities in 17 districts in 4 cocoa-growing regions of Ghana (Western region, Brong Ahafo, Ashanti and Central region), using a stratified sampling procedure. In addition, I held group discussions in around one third of the communities that I visited.

4 The risky business of cocoa

In this chapter the emphasis is on understanding how the main interests of global actors who currently govern the cocoa chain are being manifested locally, both through their involvement in local upgrading strategies in Ghana as elsewhere, and through their establishment of more direct relations with cocoa suppliers and the formation of new public-private partnerships.

For getting this understanding I have taken a global value chain perspective. The global cocoa chain is increasingly driven by international buyers (traders) of cocoa. From the mid 1950s until the 1980s, cocoa chains were first driven by associations of producers and later by the state, with significant variations between countries. In West Africa, where cocoa production is concentrated, Anglophone countries produced under marketing board systems, while Francophone countries used stabilisation funds. Now all are under the control of international buyers, with the exception of Ghana. There are different reasons for this shift in governance. For example, global buyers have become stronger actors in these chains due to takeovers and an increase in the scale of their operations. But the increased governing role of global cocoa processors and manufacturers can also be explained by the increased risks for supplier failure, playing at different levels. At the global level the risks for supplier failure increased due to changes in demand that favoured sustainable cocoa production methods, put on the agenda by both advocacy movements and public-private institutions. As a result, international traders and chocolate manufacturers have become more dependent on the local suppliers operating at the bottom of a chain. This also entails greater responsibility, in particular to provide producers with the information as well as the new technologies they need to comply with new production and process standards. At the national level marketing reforms in cocoa producing countries had quite an impact on the organisation of the cocoa chain. Prior to reforms the marketing boards (or stabilisation funds) governed the supply chain. The reforms stipulated a reduction of state involvement in the provision of marketing channels and services for cocoa, in order to open these markets to competition. While reforms are evaluated positively in abolishing inefficient marketing boards and initially increasing the producer price, their negative impacts in terms of quality of the produce, farmer income and conditions under which cocoa producers operate gave reasons for concern. Also, as a result of reforms, tracing the cocoa back to the cocoa buyer became (even) more problematic. There are also local and regional factors that form a threat to global buyers. For example, the concentration of cocoa production in West Africa is perceived as a risk, especially with the recent political crisis in Côte d'Ivoire. Also heavy rains (or in some cases water shortages), adversely affected the volume of cocoa production. Particularly damaging are the outbreaks of pests and diseases, such as Witches Broom, Black Pod and the Swollen Shoot Virus Disease. Other local risks have to do with the average high age of farmers and their tree stock.

Global buyers responded in different ways to these risks. First of all, global buyers started looking for ways to spread their risks by decreasing their dependence on specific countries (for example by looking for new suppliers in other regions), sectors (for example processors shift attention to the processing of other commodities), and quality of the cocoa (for example by searching for technological innovations that compensate for variations in bean quality without compromising customer demand for intermediate goods with specific properties). Another consequence is that global buyers actively sought new alliances with current local suppliers, and started offering them assistance in optimising their operations.

Working together with other actors is also seen as a way of spreading risks and public-private partnerships have been launched in this context.

An interesting observation is that while it has become strategically important to make on-farm investments, at the same time the location of the farm and the exact owner seem to become less important. In the practice of global sourcing of cocoa, a consistent supply is foreseen to become a major problem for the industry; how to govern the cocoa chain in order to cope with this problem is an important question. In order to guarantee future supply and demand for cocoa, 'working directly with farmer groups', 'trading with cooperatives' and 'strengthening relations with suppliers' are perceived by international buyers as the main opportunities. Manufacturers often engage with farmers in a more indirect way (through membership of organisations such as the World Cocoa Foundation or participation in public-private partnerships such as the Sustainable Tree Crop Programme) while processors seemed to look for direct ways of interaction (buying directly from farmer cooperatives). However, this holds not true for every country. In Ghana, for example, the strong role of the state obstructs direct (trading) relations. How this relative autonomy of the state in Ghana helps to address the problem of insecure quantity and links to processes of in- and exclusion is the focus of the next section.

5 The role of the state in a liberalised cocoa sector

In this chapter the focus is on the changing role of the state in the Ghanaian cocoa sector and how these changes affect the conditions under which cocoa farmers operate and in turn define their opportunities and constraints for upgrading.

The concept of 'state governance' is used to assess the level of state involvement in contrast to coordination through market mechanisms. Like many other sectors in developing countries, the cocoa sector in Ghana is in transition, with marketing and institutional reforms being gradually introduced. As a result the role of the state is changing and new actors have entered the sector. In contrast to other cocoa-producing countries in the region, the Ghanaian government has remained the main coordinator of the cocoa supply chain.

Ghana is world's second largest producer of cocoa. Around 30 per cent of Ghana's total earning derives from cocoa exports and almost on third of its population depends on cocoa for its livelihood. The Ghanaian government has always been actively involved in the development of its cocoa sector. During colonial times, public involvement was initially combined with private efforts aimed at stimulating cocoa production and improving quality. From the late 1940s onward, a system of state control was put in place, which was further consolidated during the early years of independence. With the introduction of structural adjustment programs in the late 1980s, the control over the cocoa sector in other cocoa producing countries shifted from the state to multinational buyers of cocoa. In Ghana, however, the state continued to play a major role. In order to avoid the generally negative experiences of cocoa producing countries that liberalised over-night (such as Nigeria), the

Ghanaian government opted for the gradual introduction of reforms in the cocoa sector. Just like in Côte d'Ivoire, also a country with a high stake in cocoa, the Ghanaian government did not want to leave its strategic position in the sector.

The Ghanaian government still controls external marketing and regulates internal marketing, pricing systems, processing activities, research, quality control and the provision of services. In short, it retained its role as the coordinator of the cocoa supply chain. Nevertheless, with the reforms the government abandoned some of its former duties, which were taken over by other public, private and civil actors. Extension services were merged with the services of the Ministry of Food and Agriculture, the input distribution system was privatised and internal marketing was liberalised. The opportunities and incentives for the actors to assume their new roles were sometimes still limited by the state, in some cases resulting in serious drawbacks. The state had difficulty in successfully managing the cocoa sector. For example, the quality control system was pressured by the increased volumes of cocoa. The new unified extension services appeared problematic and was heavily criticised, mainly for its lack of adequate personnel and expertise. Farmers were made particularly vulnerable through their lack of effective organization, which resulted in a lack of bargaining power. As a result farmers were unable to fully benefit from the reforms. The reforms were also not optimal for private licensed buying companies. The reforms made it possible for them to enter domestic marketing but they were not allowed to play a role in external marketing of cocoa. Another weakness of the partially liberalised system is that the export margins received by the state are still high; Cameroon and Nigeria have lower government margins. Although part of this money is reinvested in the cocoa sector, Cocobod officials do not know the exact allocation mechanism behind these reinvestments. My fieldwork indicated three different types of reinvestments. First, Cocobod reinvests part of its marketing margin back into the cocoa sector through small bonuses, giving farmers incentives to remain involved in cocoa production and to increase their volume of production. A second type of reinvestments is through a public spraying programme in order to combat two major diseases threatening cocoa production in Ghana. A third reinvestment is through offering processing companies a 20 per cent discount on light-crop beans, which actively stimulated foreign processors to outsource part of their processing facilities to Ghana. Although indirectly all cocoa farmers pay for these investments individually they do not always benefit. The reinvestments do not go without problems and furthermore are not transparent, thus undermining the credibility of a partial liberalised system.

Despite the tensions and weaknesses of the Ghanaian system, which is an example of a government retaining a key steering role together with the private sector ('joint governance'), it proved as quite favourable for cocoa producers and for the other actors involved in the sector. Due to the reliable marketing system, Ghana enjoys a high reputation for honouring its contract and offering relatively high quality produce. Other benefits of the partially liberalised systems include the intact price stabilisation, the gradual price increases for farmers, tax decreases and increased volumes of production. Also the services provided to farmers are generally better than in fully liberalised countries.

So, partial liberalisation may indeed be a viable alternative model to full liberalisation. Currently the system can count on the support of global buyers and international donor organisations. However, it is not unthinkable that changes in preferences may eventually bring another wave of liberalisation. In order to be prepared for changes in demand in global markets the Ghanaian government should invest more in the capacity of other actors, especially the farmers and the private sector, empowering them to contribute more to building a strong cocoa sector, which would also survive in a changing environment.

6 Who are the cocoa farmers?

Upgrading can be enabled or hindered by powerful players in a chain, but also by governments and social structures. In this chapter the focus is on differences among cocoa producers and the social structures in which they are embedded that might influence the impact of shifts in governance and the success of upgrading strategies. Building on lessons drawn from the GVC literature and cluster literature, I questioned farmers on several issues related to land-ownership, volume of production, gender, age and social networks, amongst others. This knowledge clarifies the extent to which social relations, economic features and spatial characteristics influence the respondents' decision-making in economic choices, their responses to interventions and the extent to which they benefit from interventions.

The analysis made clear that there are significant differences between the respondents. For example, the farmers in Brong Ahafo, a more remote region with lower population density, had less favourable opportunities for cocoa production. There are few extension officers that travel to this region where farms are spread out far apart. Also the number of buying agents in the villages is low. It seems that farmers producing in the Western region are better off; the concentration of cocoa production in the Western region has attracted buyers and service providers. Besides region, the analysis made it clear that the context played a major role producing different outcomes for different groups of farmers. For example, the matrilineal system of inheritance stimulates land fragmentation. This makes it difficult for farmers to make cocoa farming a profitable business. This system hit families of small-scale farmers and caretakers the hardest. An important observation was that both 'exclusion' and 'inclusion under unfavourable conditions' are not natural processes, but an outcome of social structures, land and shareholding systems and interventions.

Another part of the analysis was looking for significant correlations between the different variables. It turned out that among the respondents differences between the farmers were inter-related. For example, land is not equally accessible to all farmers. This is a problem, not only because land is needed for the production of cocoa, but also because land is often requested as collateral for obtaining credit at a financial institution. Without land the participation in farmer groups is also restricted. This has consequences on yet another level because participation in farmer groups affects farmer's access to training, for example in the farmer field schools. Consequently the majority of farmers who receive training are farm owners.

Caretakers, many women, migrants and younger farmers have more chance of being left out. This is naturally problematic, but even more problematic as without training and options to raise their productivity, cocoa production can become less attractive to young farmers, which threatens the future of this important economic sector for Ghana.

7 The risks of inclusion

In this chapter I looked at the interaction between different governance levels (global, national and local) by zooming in on some of the interventions taking place in the cocoa sector in Ghana. The main questions I try to answer in this chapter is: how do upgrading strategies and the resulting interventions benefit different groups of cocoa farmers, and to what extent do these groups participate actively in the activities and initiate their 'own' strategies? I assess the inclusiveness of interventions by looking at the mechanism behind the intervention, their scope, expected impact, constraints and unexpected trade-offs.

There are multiple interventions leading to upgrading, which interact with each other and are executed by different actors involved in the cocoa chain. In order to make an overview, I identified a large number of interventions that affect Ghanaian cocoa producers and structured these around sub-strategies. These are in turn linked to three main upgrading strategies for small-scale farmers in developing countries identified in the literature. Sub-strategies for capturing higher margins for unprocessed cocoa *(upgrading strategy 1)* are contributing to producing better quality cocoa, increasing productivity and the production of higher volumes of cocoa, and producing under more remunerative contracts. This first group of sub-strategies is dominated by large-scale public interventions. The Ghanaian government is the main intervener in safeguarding quality standards and increasing volumes of production. International buyers share the agenda of the government but play a rather passive role. Control and standard setting affect all farmers, but interventions that provide services are not easily accessible to everyone. For my respondents, the main determining factors that determined access to such services were ownership, social position in the community and location (region). The farmers themselves are responsible for producing high quality cocoa, but toil under diminished incentives. In terms of volumes of production and productivity, the farmers are actively involved and have developed different ways of increasing their volume of production, for example through effective pest management, planting new varieties of cocoa and working together in labour exchange groups.

Sub-strategies for producing new forms of existing commodities *(upgrading strategy 2)* are divided into producing for specialty/niche markets, development of non-traditional uses of cocoa and diversification into non-traditional products, and other (non-farm) income-generating activities. This second group of substrategies concerns mainly multi-stakeholder initiatives and interventions of NGOs, which are generally small-scale and exclusive. Among the respondents ownership and region played a decisive role in access to this type of partnerships.

Sub-strategies for localising commodity processing and marketing *(upgrading strategy 3)* are processing cocoa waste, processing cocoa beans and the marketing of cocoa beans. This third group is exclusive or does not reach farmers at all. The multinational buyers are the main interveners; they share an interest with the government in outsourcing part of their processing capacity to Ghana. The interventions aimed at marketing reach farmers indirectly.

I made a selection of four interventions which are discussed in-depth in this chapter. I analysed two large-scale public interventions, both falling under upgrading strategy 1; one aimed at the production of high quality cocoa, the quality control system, and the other at increasing the volumes of cocoa production, the public mass-spraying programme. These interventions differ both in the type of impact and type of farmers they reach. Indirectly, farmers themselves pay for these public re-investments in the cocoa sector. A problem is a lack of transparency on costs and benefits and how these are distributed. A third intervention I looked at, that falls under upgrading strategy 2, was a medium-scale multi-stakeholder initiative (which includes public, private and civil actors), namely the only formal farmer union, the Kuapa Kokoo Farmer Union (KKFU). This Farmer Union, which encompasses around 50,000 farmers and their families, produces a small share of its beans for the fair trade market. In addition to opening up an alternative marketing channel, membership in the union also empowers farmers. A fourth intervention that I discussed was an intervention by international processing companies, which outsourced part of their processing capacity to Ghana. This intervention, which comes under upgrading strategy 3, has no direct impact on farmers but does contribute to the long-term demand for Ghanaian cocoa by consolidating relations between Cocobod and international processing companies.

The aim of the analysis was to obtain insights in how the interests of the different players in the cocoa chain are manifested locally, who dominates the upgrading agenda and which upgrading issues are prioritised. Furthermore, I wanted to highlight the strengths and weaknesses of the interventions. In my analysis I illustrated the impact of each selected intervention on the farmers' position in a chain *(individual level)* by making use of an existing matrix, called the 'empowerment matrix'. I also developed the scenario matrix to reflect on the cocoa sector in its totality *(collective level)*. This 'scenario matrix' is built around two dimensions: changes in demand, moving from 'product' to 'process' requirements, and the level of liberalisation. It provides an enhanced understanding of the vulnerability of the current system by looking at changes in context. This contributes to the identification of more inclusive upgrading strategies that are (also) effective on a longer term.

The Ghanaian case, often presented as best practice, embodies two important dimensions: first, it is unique due to its partially liberalised economy; and second, it is exceptional for its production of large quantities of premium quality cocoa. The partially liberalised system reflects partly the strong role of the Ghanaian state and partly the global buyers' interest to maintain or only slightly modify the Ghanaian system. The production of premium quality cocoa reflects both the capacity of the national government to coordinate the supply chain as well as the existing high demand for premium quality cocoa.

The upgrading strategies taking place in Ghana, which focus on quality and volume of production, reflect these dimensions. But the conditions underlying these dimensions are not fixed. First, there is a trend in the global cocoa chain that product requirements become less important. Second, it is not sure if a partially liberalised system is the end-stage of the reforms. Understanding the position of Ghanaian cocoa farmers in the chain and the kind of upgrading strategies that are beneficial for farmers require a dynamic perspective, not only by drawing lessons from the past and making comparisons with experiences in fully liberalised countries, but also by taking into account possible future scenarios. Ghanaian farmers are better off now, but what if the main pillars that underpin their strong position disappear?

It turns out that Ghana is not well-prepared for change. Farmers and the private sector are particularly vulnerable. Looking at experiences in other cocoa growing countries in the region that fully liberalised their cocoa sector it has become clear that weak farmer organisations and a weak private sector are severe bottlenecks for farmers to benefit from further reforms. Nevertheless, the Ghanaian government is not investing in capacity development of private buyers of cocoa and farmer organisations. The lack of investment in farmer organisation also makes it increasingly difficult to meet (changes in) demand, for which being organised becomes more and more a prerequisite. Moreover, this has contributed to a lack of agency among farmers to change their position and to benefit more from the current partially liberalised system. More inclusive upgrading requires more emphasis to be placed on empowering farmers and local private actors, it also requires more awareness (beforehand) of whom interventions are likely to include and whether they intensify unequal social structures in the Ghanaian society or contribute to transforming them.

8 Conclusions

The objective of this study was to develop a thorough understanding of the opportunities and constraints that producers of primary commodities face in their effort to improve their position in the global value chain. This research is particularly relevant now, as farmers are increasingly expected to behave more like 'firms'. However, as my results show, agents higher up in the chain, and the institutional environment that surround farmers, still largely determine scope for change and the direction of 'progress' available to farmers. The case of Ghana, which is partially liberalised, shows an example of a value chain in which the role of the state is strong; the state still intervenes directly in sectoral development and stands between local producers and global buyers of cocoa.

In Ghana the state functions as 'balancer' as well as 'bottleneck'. As a balancer it mainly protects cocoa farmers from price-fluctuations on the world market, providing them with a stable income and reinvesting part of its income back into the cocoa sector. It also guarantees international buyers the supply of premium quality beans. So, the risks for suppliers as well as for international buyers are mediated by the state, assuring smooth trade relations. In a partially liberalised system the state can also function as 'bottleneck', preventing other public, private and civil actors from taking a more active role and contributing to accomplishing development goals. The state as hinderer of development is linked to the lack of transparency on how the state calculates and distributes the costs, benefits and risks involved in cocoa production and marketing.

Others can learn from the example of Ghana, as the state turns out to be able to mitigate risks involved in cocoa production for producers, as well as for other chain actors. Still there two 'risks of inclusion' involved. First, on the level of the producer, the arrangements are sub-optimal and do not create incentives for farmers to behave as entrepreneurs. Also, farmers do not benefit equally from the arrangements in place. Second, the state is inwards oriented and lacks an adaptive approach. This entails a risk for the sector as a whole.

My case study showed that a partially liberalised system has benefits but also poses risks. To overcome these risks it is important to 'put farmers first', which is not only about increasing their benefits but also about empowering them to become active agents in the chain.

Nederlandse samenvatting

1 Lokale producenten in een mondiale economie

Voorstanders van globalisering hebben lang verkondigd dat vrijhandel zal leiden tot economische groei en verbeteringen in de leefomstandigheden van iedereen, ook die van arme boeren in ontwikkelingslanden. Maar het is duidelijk geworden dat niet iedereen in gelijke mate profiteert van de voordelen die globalisering biedt. Globalisering biedt ook uitdagingen, zoals toenemende concurrentie en prijsschommelingen waar kleinschalige boeren niet zo maar tegen opgewassen zijn. Om dergelijke uitdagingen het hoofd te bieden is het nodig dat juist deze kwetsbare ondernemers hun bedrijfsvoering verbeteren.

Er zijn verschillende ideeën over het type verbeteringen dat kleinschalige boeren zouden moeten nastreven. Er zijn ook verschillende ideeën over wie hen daarbij zou moeten helpen. Internationale instellingen zoals de Wereldbank, benadrukken dat de boeren zelf verantwoordelijk zijn voor verandering. In het proces van het toevoegen van waarde aan hun product of het verbeteren van hun productieproces *(upgrading)*, kunnen boeren leren van 'leidende' ondernemingen die zich hoger in de keten bevinden (het ketenperspectief) of kunnen boeren veranderingen tot stand brengen door middel van gezamenlijke actie en collectieve efficiëntie (het clusterperspectief). Anderen zijn van mening dat in ontwikkelingslanden, waar de meerderheid van de productie van grondstoffen in handen is van kleine producenten, het vastleggen van hogere marges voor onbewerkte grondstoffen publieke inmenging vereist.

Om de relatie tussen machtstructuren, armoede en *upgrading* te begrijpen is het belangrijk te kijken naar de interactie tussen de verschillende niveaus van bestuur. In deze studie kijk ik naar een unieke casus, namelijk de cacaosector in Ghana. Deze casus laat zien hoe het bestuur *(governance)* in de internationale cacaoketens – aangedreven door een combinatie van concentratie, verduurzaming en onzekerheid over het toekomstige aanbod van cacao – combineert of juist contrasteert met staatsinterventies en nationale regelgeving, en met de sociale structuren waarin boeren zijn ingebed. Dit proefschrift richt zich op de capaciteit van de staat om problemen op te lossen en laat zien hoe een nationale overheid een positie in kan nemen tussen lokale producenten en mondiale opkopers van cacao.

In de cacaosector werken wereldwijd ongeveer 14 miljoen mensen. Geschat wordt dat ongeveer 3 miljoen kleine boeren goed zijn voor 90 procent van de productie. De productie concentreert zich voornamelijk in West-Afrika, met Ghana als de op één na grootste cacaoproducent van de wereld, na Ivoorkust. Ghana heeft een aantal specifieke kenmerken die het een interessante casus maken om te kijken naar de interactie tussen privaat en publiek beleid. Ten eerste is Ghana het enige cacaoproducerende land in de regio waar de cacaosector slechts gedeeltelijk geliberaliseerd is; de overheid speelt nog steeds een controlerende rol. Ten tweede is Ghana het enige cacaoproducerende land waar de cacaoboon traceerbaar is. Bovendien is Ghanese cacao uniek vanwege haar hoge kwaliteit, waarvoor de overheid een extra premie ontvangt. Bestudering van de Ghanese casus draagt in de eerste plaats bij aan een beter begrip van de manier waarop boeren die grondstoffen produceren voor de exportmarkt kunnen profiteren van deelname aan internationale ketens (in dit geval een keten die in toenemende mate wordt gestuurd door multinationale cacaoverwerkers en chocolademakers). In de tweede plaats draagt de casus bij aan de recente discussie over hybride *governance* structuren, waarin zowel publieke als private actoren een rol spelen. Ten slotte, door te kijken naar verschillen tussen cacaoboeren, streef ik om inzicht te krijgen in de processen van in- en uitsluiting van bepaalde groepen boeren, en de rol die de overheid daarbij speelt.

2 Upgrading ten behoeve van ontwikkeling

In discussies over globalisering en de impact die dit proces heeft op ontwikkeling worden de uitdagingen voor ondernemers in ontwikkelingslanden benadrukt. Sinds eind jaren negentig, is de focus vooral gericht op het midden- en kleinbedrijf (MKB). Met het terugplaatsen van landbouw op de ontwikkelingsagenda (als de 'motor voor economische groei'), is de nadruk verschoven van het MKB naar kleinschalige agrarische producenten. Tegelijkertijd worden deze kleinschalige boeren steeds meer gezien als zelfstandige ondernemers. Om duurzame groei en armoedevermindering mogelijk te maken moeten deze agrarische 'bedrijven' hun concurrentiepositie verbeteren. Het vermogen om te concurreren is niet alleen gebaseerd op betere prijsstelling en verbeterde productkwaliteit, maar het is daarnaast ook steeds belangrijker dat er sprake is van duurzame productie in de keten.

In de literatuur over competitiviteit en *upgrading* worden de mogelijkheden benadrukt voor producenten om een beter rendement te krijgen. Het concept *upgrading* wordt breed toegepast, van de ketenbenadering tot de clusterbenadering. De heersende opvatting is dat *upgrading* het resultaat is van organisatorisch leren en netwerkvorming tussen bedrijven. De ketenbenadering concentreert zich op globale sturing in de keten, waarin samenwerking binnen de keten een concurrentievoordeel kan opleveren. Centraal in de ketenbenadering staat dat producenten onderaan de keten kunnen leren van actoren hoger op in de keten. Het concept *global chain governance* is een belangrijke dimensie in de ketenbenadering, waarbij het gaat om machtsrelaties die bepalen hoe financiële, materiële en personele middelen stromen binnen de keten. Afzonderlijke segmenten in de keten kunnen verschillend gecoördineerd worden.

Type relaties die ontstaan tussen internationale opkopers en lokale producenten worden deels bepaald door het risico dat opkopers lopen dat lokale producenten niet de gevraagde producten leveren. Maar het is niet duidelijk hoe specifieke kenmerken van een keten ('strak' of 'los' georganiseerd) gekoppeld zijn aan een bepaalde uitkomst voor betrokken producenten. Kleinschalige producenten kunnen tegen machtsverhoudingen in de keten aanlopen. Deze relaties moeten ontrafeld worden. In de ketenbenadering ligt de nadruk op relaties tussen actoren in de keten, vooral tussen internationale opkopers en lokale producenten. De focus is dus op de verticale netwerken en veel minder op de relaties tussen bedrijven op lokaal niveau, zoals gezamenlijke acties van boeren en interacties met lokale instellingen. De clusterbenadering richt zich juist wel op deze horizontale netwerken. Binnen deze benadering worden lokale relaties tussen bedrijven en de interactie met lokale instituten beschouwd als een van de belangrijkste krachten achter modernisering en innovatie. In de clusterliteratuur staat het begrip *cluster governance* centraal, wat verwijst naar collectieve acties van actoren binnen een cluster die gericht zijn op het verbeteren van de cluster. Voordelen van clusters komen meestal voort uit een optimale mix tussen samenwerking en concurrentie tussen haar leden. Inzicht in de voorwaarden waaronder deze optimale balans tot stand komt vereist op haar beurt inzicht in het niveau van inbedding van de economische activiteit in sociale relaties *(embeddedness)*. Veel sociale relaties zijn geografisch bepaald, en mensen zijn niet alleen werknemers of managers, ze zijn ook consumenten, burgers, kerkgangers, en leden van de lokale gemeenschap.

Verschillende auteurs hebben gewezen op de beperkingen van beide benaderingen en de noodzaak om 'horizontale netwerken' met 'verticale netwerken' te combineren. Maar een combinatie van de twee benaderingen is niet genoeg om de beperkingen van de afzonderlijke benaderingen, die in zekere mate vergelijkbaar zijn, te niet te doen. Beide benaderingen veronderstellen de aanwezigheid van een open markt systeem. Maar aan deze voorwaarde wordt niet altijd voldaan. Bovendien is het belangrijk om naar de rol van de staat te kijken om te begrijpen wie er profiteert van *upgrading* en wie er mogelijk wordt buitengesloten. Om de rol van de staat niet uit het oog te verliezen is dus een andere benadering nodig. Hierbij heb ik gebruik gemaakt van een dynamisch vergelijkend raamwerk dat het mogelijk maakt om publiekprivate interactie in een keten te plaatsen. Dit raamwerk is gebaseerd op de literatuur afkomstig uit zowel strategisch management als uit politieke economie en het helpt variaties te verklaren in het vermogen van de industrie om op nationaal niveau te concurreren. Het raamwerk geeft de volgende vier mogelijke types van interactie in een keten aan: *state governance* (een situatie waarin transacties worden gecoördineerd door de staat en de keten wordt gecoördineerd door marktkrachten); *joint governance* (een situatie waarin transacties worden gecoördineerd door de staat en de keten wordt gecoördineerd door middel van ketenintegratie); *market governance* (een situatie waarin transacties en de keten beide worden gecoördineerd door middel van marktwerking); en *corporate governance* (een situatie waarin transacties worden gecoördineerd door de staat en keten wordt gecoördineerd door middel van ketenintegratie). Deze benadering maakt het mogelijk om de capaciteit van de (nationale) staat te bespreken wat betreft haar vermogen om problemen op te lossen en om de relatie te laten zien met de manier waarop de keten internationaal georganiseerd is. Om te begrijpen hoe de prestaties van de staat uitwerken op processen van in- en uitsluiting voor bepaalde groepen boeren moet het lokale *governance* niveau weer worden ingebracht. De combinatie van globale, nationale en lokale *governance* structuren zal laten zien hoe een nationale staat zich bevindt tussen lokale producenten en internationale spelers en biedt de gelegenheid om nauwkeuriger te bespreken in hoeverre de staat lokale producenten kan helpen om het hoofd te bieden aan de uitdagingen van globalisering. De casus van cacao in Ghana is bijzonder interessant omdat de sector geleidelijk en slechts gedeeltelijk is geliberaliseerd. Het is een voorbeeld van een 'hybride' casus, waar de staat nog steeds de nationale sector coördineert in een keten die in toenemende mate gestuurd wordt door internationale opkopers.

3 Onderzoeksvragen, de methoden en de respondenten

De centrale vraag in dit proefschrift is hoe verschillende *governance* processen op elkaar inwerken in het creëren van kansen en beperkingen voor inclusieve *upgrading* voor kleinschalige cacaoboeren in Ghana. Ik maak onderscheid tussen drie niveaus van *governance*: ten eerste, bestuur in een internationale keten (verwijzend naar de machtsverhoudingen in de internationale cacaoketen), ten tweede de rol van de staat in de ontwikkeling van een landbouw sector (verwijzend naar het niveau van de overheidsbemoeienis in de Ghanese cacaosector), en ten derde, de sociale structuren (refererend aan bestaande structuren op lokaal niveau waar cacaoboeren onderdeel van uitmaken). De verschillende niveaus vragen om een combinatie van verschillende onderzoeksmethoden en concepten. De ketenbenadering wordt gebruikt als een instrument om de kansen en beperkingen te analyseren die het bestuur in een keten, en de veranderingen die hierin plaats vinden, bieden voor cacaoboeren in Ghana. Op nationaal niveau wordt de invoering van hervormingen in Ghana genomen als een belangrijk keerpunt. Dit wordt gebruikt om de verbeteringsmogelijkheden voor cacaoboeren in Ghana te begrijpen. Op lokaal niveau probeer ik inzicht te krijgen in de impact van verschuivingen in bestuur en upgradingstrategieën op boeren, en de manier waarop mogelijke voordelen en risico's worden verdeeld. Ik gebruik centrale begrippen uit de clusterliteratuur, zoals 'inbedding' en 'gezamenlijke actie', om sociale structuren te identificeren die van invloed zijn op het proces van *upgrading* en de manier waarop boeren van upgrading profiteren.

Voor de beantwoording van de onderzoeksvraag, combineer ik kwalitatieve met kwantitatieve data. Ik heb diepte-interviews gehouden met actoren die op verschillende manieren betrokken zijn bij de cacaosector in Ghana, waaronder cacaoboeren, lokale en internationale opkopers en verwerkers van cacao, de overheid, vertegenwoordigers van boerenorganisaties, maatschappelijke organisaties, banken en onderzoeksinstellingen. Daarnaast heb ik data verzameld op basis van een aantal informele gesprekken met 's werelds grootste cacao-opkopers en heb ik twee multi-stakeholder workshops georganiseerd, één in Nederland (2003) en één in Ghana (2005), met de belangrijkste vertegenwoordigers van het bedrijfsleven en andere actoren. Ik heb tijdens mijn veldwerk drie surveys gehouden: 1) onder een klein aantal van de cacaoverwerkers en chocolademakers, en enkele van de instellingen die hun belangen behartigen; 2) onder 173 cacaoboeren in 2003, en 3) onder 210 boeren in 2005, waaronder 103 boeren die eerder deelnamen aan de survey van 2003 en 107 nog niet eerder geïnterviewde cacaoboeren. De surveys werden uitgevoerd in 34 gemeenschappen in 17 districten in 4 cacaoproducerende regio's van Ghana (Western Region, Brong Ahafo Region, Ashanti Region en Central Region), met behulp van een gestratificeerde steekproef. Daarnaast organiseerde ik groepsdiscussies in ongeveer een derde van de gemeenschappen die ik heb bezocht.

4 Cacao, een riskante zaak

In dit hoofdstuk laat ik zien hoe belangen van multinationale spelers, die momenteel de internationale cacaoketen besturen, zich lokaal manifesteren. Dit doe ik door te kijken naar hun betrokkenheid bij lokale upgradingstrategieën, het ontstaan van meer directe relaties met cacaoproducenten en nieuwe publiekprivate samenwerkingsverbanden.

In dit hoofdstuk maak ik gebruik van het ketenperspectief. De internationale cacaoketen wordt in toenemende mate gestuurd door internationale opkopers van cacao (handelaars, verwerkers en chocolademakers). Vanaf midden jaren 50 tot de jaren 80, werden cacaoketens in eerste instantie gestuurd door producentenverenigingen en later door de staat, met aanzienlijke verschillen tussen de cacaoproducerende landen. In West-Afrika, waar cacaoproductie is geconcentreerd, produceerden Engelstalige landen onder zogenaamde *marketing board* systemen, terwijl de Franstalige landen gebruik maakten van stabilisatiefondsen. Tegenwoordig zijn alle ketens onder de controle van internationale opkopers, met uitzondering van Ghana. Er zijn verschillende redenen voor deze verschuiving. Zo zijn internationale opkopers van cacao sterkere ketenspelers geworden als gevolg van overnames en schaalvergroting. Een andere reden is dat voor opkopers van cacao de risico's zijn toegenomen dat het aanbod en de kwaliteit van cacao onvoldoende wordt. Op internationaal niveau is dit verhoogde risico vooral een gevolg van veranderingen in de vraag, waarbij de voorkeur in toenemende mate uitgaat naar duurzaam geproduceerde producten. Als gevolg hiervan is de afhankelijkheid tussen internationale handelaren en fabrikanten van chocolade enerzijds en lokale leveranciers van cacao anderzijds toegenomen. Dit veronderstelt ook een grotere verantwoordelijkheid voor opkopers om cacaoproducenten te voorzien van adequate informatie en hen toegang te geven tot (kennis over) nieuwe technologieën die nodig zijn om te voldoen aan de nieuwe voorwaarden die worden gesteld aan cacaoproductie. Op nationaal niveau hebben de introductie van markt- en prijshervormingen in de cacaosector een behoorlijke invloed gehad op de wijze waarop de sector georganiseerd is. Voorafgaand aan de hervormingen stuurde de overheid de sector. De hervormingen bedongen een terugtrekkende rol van de staat en stelden de markten open voor concurrentie. Hoewel hervormingen op sommige punten doorgaans positief worden beoordeeld, zoals de afschaffing van inefficiënte *marketing boards* en de aanvankelijke verhoging van de producentenprijs, geven de negatieve effecten van de hervormingen op het gebied van kwaliteit, het inkomen van de boer en de gebrekkige traceerbaarheid van de cacao reden tot bezorgdheid. Er zijn ook lokale en regionale factoren die een bedreiging vormen voor de internationale cacaoverwerkers en chocolademakers. Zo wordt de concentratie van cacaoproductie in West-Afrika beschouwd als een risico, zeker met de recente politieke crisis in Ivoorkust. Ook zware regenval (of in sommige gevallen watertekorten), bedreigt de omvang van de cacaoproductie. Bijzonder schadelijk zijn de uitbraken van ziekten en plagen. Andere lokale risico's hebben te maken met de gemiddeld hoge leeftijd van de boeren en hun bomenbestand.

Internationale opkopers van cacao reageren verschillend op deze risico's. Allereerst, vindt risicospreiding plaats door het verminderen van hun afhankelijkheid van bepaalde landen (bijvoorbeeld door op zoek te gaan naar nieuwe leveranciers van cacao in andere regio's), sectoren (bijvoorbeeld door het verwerken van andere grondstoffen), en kwaliteit van de cacao (bijvoorbeeld door technologische innovaties die variaties in kwaliteit bonen compenseren). Een andere reactie is dat de internationale verwerkers en fabrikanten actief samenwerking zoeken met lokale leveranciers, en bijvoorbeeld zijn begonnen met het ondersteunen van producentenorganisaties in het optimaliseren van hun activiteiten. Samenwerken met andere actoren in de keten en het aangaan van publiekprivate samenwerkingsverbanden wordt ook gezien als een manier om risico's te spreiden.

Een interessante constatering is dat terwijl het nu van strategisch belang lijkt om te investeren in de cacaosector en in het boerenbedrijf, op hetzelfde moment de locatie en type boer minder belangrijk lijken te worden.

In de praktijk wordt voor de industrie de aanvoer van een consistent aanbod van cacao als een van de grootste problemen gezien. De vraag is hoe met deze dreiging in het achterhoofd de cacaoketen het beste gestuurd kan worden. Om een consistent aanbod te garanderen word directe samenwerking met groepen boeren, handel met boerencoöperaties and het versterken van relaties met boeren als belangrijkste mogelijkheden genoemd. Chocolademakers ondersteunen boeren vooral indirect (bijvoorbeeld via lidmaatschap van organisaties zoals de *World Cocoa Foundation*, of door participatie in publiekprivate samenwerkingsverbanden zoals het *Sustainable Tree Crop Programme*) terwijl cacaoverwerkers vooral inzetten op rechtstreekse interactie (direct opkopen van boerencoöperaties). Echter, dit geldt niet voor elk land. In Ghana bijvoorbeeld, verhindert de sterke rol van de staat directe (handels)relaties. Hoe in Ghana deze relatieve autonomie van de staat helpt om risico's te verminderen en hoe dit zich vertaalt in processen van in- en uitsluiting van bepaalde groepen boeren is de focus van de het volgende hoofdstuk.

5 De rol van de staat in een geliberaliseerde cacaosector

In dit hoofdstuk gebruik ik het concept van *state governance* gebruik om te aan te geven hoe de rol van de staat zich verhoudt tot de rol van de markt in het coördineren van de cacaoketen. Net als veel andere sectoren in ontwikkelingslanden, neemt in Ghana de rol van de overheid in de cacaosector af. Door het geleidelijk invoeren van marketing en institutionele hervormingen verandert de rol van de Ghanese staat en nemen nieuwe spelers een deel van de taken van de overheid over. In tegenstelling tot andere cacaoproducerende landen in de regio, is de Ghanese overheid nog steeds de belangrijkste coördinator van de nationale cacaoketen.

Ghana is een belangrijke cacaoproducent en cacao is een belangrijk export product voor Ghana. Bijna een derde van de bevolking is afhankelijk van cacao als belangrijke bron van inkomsten. De Ghanese regering heeft altijd een actieve rol gespeeld in de cacao-sector. Tijdens de koloniale tijd speelden zowel de overheid als de private sector een rol in het stimuleren van de cacaoproductie en een verbetering van de kwaliteit. Vanaf eind jaren 40 werd een systeem van overheidscontrole geïnitieerd. Dit systeem werd geconsolideerd en versterkt gedurende de eerste jaren van Ghana's onafhankelijkheid. Met de invoering van de structurele aanpassingsprogramma's eind jaren 80 door de Wereldbank, begonnen internationale opkopers van cacao de keten in toenemende mate te controleren. In Ghana, speelt de overheid echter nog steeds een grote rol. Een belangrijke reden voor de Ghanese regering om te kiezen voor de introductie van geleidelijke hervormingen was de negatieve ervaring van cacaoproducerende landen die hals over kop hervormingen doorvoerden. Net als in Ivoorkust, waar de cacao-sector ook een zeer belangrijke sector is, wil Ghana haar strategische positie in de sector niet verlaten.

De Ghanese regering controleert nog steeds de verhandeling van cacao voor binnenlandse verwerking en export *(external marketing)* en coördineert de binnenlandse handel *(internal marketing)*. De Ghanese *cocoa marketing board* (Cocobod) beheert het systeem van prijsstabilisatie, verwerkt een deel van de cacao voor de regionale markt, doet cacao-onderzoek, is verantwoordelijk voor de kwaliteitscontrole en verleent informatie aan boeren. De overheid is, ondanks de hervormingen, nog steeds de coördinator van de nationale keten, maar een aantal van haar taken zijn overgenomen door andere actoren. De cacaovoorlichtingsdienst van Cocobod werd samengevoegd met de algemene landbouwvoorlichtingsdienst van het Ministerie van Voedsel en Landbouw, het distributiesysteem van *inputs* werd geprivatiseerd en binnenlandse handel in cacao werd geliberaliseerd. Niet alle spelers waren voorbereid op hun nieuwe rol, en in sommige gevallen werden zij ernstig beperkt door de overheid. Maar ook de staat kreeg het langzamerhand moeilijker om de sector succesvol te besturen. Zo kwam bijvoorbeeld het systeem voor kwaliteitscontrole onder druk te staan door een toename in volume van cacaoproductie. Ook verliep het samenvoegen van de voorlichtingsdiensten problematisch en het nieuwe systeem werd zwaar bekritiseerd, voornamelijk vanwege het ontbreken van voldoende personeel en cacao-expertise. In het gedeeltelijk geliberaliseerde systeem vormen vooral de cacaoboeren een kwetsbare schakel, een belangrijke reden hiervoor is hun gebrek aan effectieve organisatie, wat op haar beurt resulteert in een zwakke onderhandelingspositie van boeren. Mede hierdoor zijn boeren niet in staat om volledig te profiteren van de hervormingen. De hervormingen zijn ook niet optimaal voor private opkopers van cacao. Hoewel de hervormingen het mogelijk maakten om als private opkoper van cacao de binnenlandse markt te betreden, werd het hun niet toegestaan om (een gedeelte van) de cacao te exporteren. Het gedeeltelijk geliberaliseerde systeem heeft nog een ander zwak punt, namelijk dat de overheid een aanzienlijke marge ontvangt over de export prijs. Over deze 'verborgen' marge wordt niet openlijk gecommuniceerd. Hoewel een deel van deze extra opbrengsten wordt geïnvesteerd in de cacao-sector, is het niet duidelijk wat het exacte allocatiemechanisme is. Uit mijn veldwerk komen drie typen herinvesteringen naar voren. In de eerste plaats herinvesteert Cocobod een deel van haar marge terug in de cacao-sector door middel van het geven van kleine bonussen aan boeren. Hierdoor ontvangen boeren een stimulans om cacao te produceren en hun productie volumes te verhogen. Een tweede herinvestering is een *public spraying programme* gericht op het bestrijden van belangrijke ziekten die de cacaoproductie in Ghana bedreigen. Een derde herinvestering is middels een korting van 20 procent op *light-crop* bonen die aangeboden wordt aan de verwerkende industrie. Deze prijssubsidie stimuleert buitenlandse cacaoverwerkers een deel van hun verwerkingscapaciteit te verplaatsen naar Ghana. De investeringen gaan niet zonder problemen en het is niet altijd duidelijk hoe het geld precies besteed wordt, en wie hier vervolgens van profiteert. Dit gebrek aan transparantie doet afbreuk aan het vertrouwen in het huidige systeem.

Ondanks deze problemen is het heersende systeem in Ghana, waar de overheid en internationale private spelers een alliantie vormen *(joint governance)*, relatief gunstig voor zowel cacaoboeren als actoren verder op in de cacaoketen. Door haar betrouwbare handelssysteem heeft Ghana een goede reputatie opgebouwd als het aankomt op het naleven van contracten en het aanbod van relatief hoogwaardige cacao. Andere voordelen van het gedeeltelijk geliberaliseerde systeem zijn onder andere het intact gebleven systeem van prijsstabilisering, de geleidelijke prijsverhogingen voor boeren, de belastingvermindering en de toegenomen omvang in productie van cacao. Ook biedt Ghana nog steeds betere diensten aan haar boeren dan in volledig geliberaliseerde buurlanden. Wat dat betreft biedt een systeem van gedeeltelijke liberalisering wel degelijk een alternatief model voor volledige liberalisering.

Vooralsnog kan het Ghanese systeem rekenen op de steun van internationale opkopers en internationale donororganisaties. Het is echter niet ondenkbaar dat hun voorkeuren veranderen. Een dergelijke verandering kan uiteindelijk een nieuwe golf van liberalisering te weeg brengen. Om voorbereid te zijn op veranderingen in wereldmarkt zou de Ghanese overheid meer moeten investeren in de capaciteit van andere actoren, in het bijzonder de boeren en de private sector. Zodat zij beter in staat worden gesteld om bij te dragen aan een sterke cacao-sector, die ook zou overleven in een veranderende context.

6 Wie zijn de cacaoboeren?

Upgrading kan mogelijk gemaakt of gehinderd worden door machtige spelers in een keten, maar ook door overheden en sociale structuren. In dit hoofdstuk ligt de nadruk op de verschillen tussen de cacaoproducenten en de sociale structuren waar zij onderdeel van uitmaken en die mogelijkerwijs het succes van upgradingstrategieën en de impact van verschuivingen in *governance* beïnvloeden. Voortbouwend op de lessen uit de keten¶en clusterliteratuur, heb ik boeren onder andere gevraagd naar hun grondbezit, het volume van productie, *gender*-relaties, hun leeftijd en deelname aan sociale netwerken. Hierdoor heb ik meer inzicht gekregen in de mate waarin sociale relaties, economische kenmerken en locatie van invloed zijn op de manier waarop boeren reageren op interventies en in hoeverre zij in staat zijn om van interventies te profiteren. De analyse maakte duidelijk dat er aanzienlijke verschillen zijn tussen de respondenten. Bijvoorbeeld, voor boeren in Brong Ahafo, een meer afgelegen gebied met een lagere bevolkingsdichtheid, is het relatief moeilijker om hun productie te verhogen. Er zijn maar weinig voorlichters die met regelmaat een bezoek brengen aan boeren in dit gebied. Ook zijn er relatief minder lokale opkopers in dorpen in deze regio aanwezig. Het lijkt erop dat cacaoboeren die in de westelijke regio wonen beter af zijn. In de *Western region* is cacaoproductie sterk geconcentreerd, hierdoor is het vestigingsklimaat voor lokale opkopers van cacao relatief aantrekkelijker in deze regio en dienstverleners kunnen boeren makkelijker bereiken. De analyse maakte duidelijk dat naast regio, context een belangrijke rol speelt in het genereren van verschillende uitkomsten voor verschillende groepen boeren. Bijvoorbeeld, het matrilineaire systeem van overerving van land stimuleert landversnippering. Dit maakt het in toenemende mate moeilijk voor boeren om hun boerenbedrijf winstgevend te maken. Dit systeem treft families van kleine boeren en boeren die zelf geen land bezitten het hardst. Een belangrijke constatering is dat zowel 'uitsluiting' als 'insluiting onder gunstige omstandigheden' geen natuurlijke processen zijn, maar een resultaat van maatschappelijke structuren, landrechten en interventies.

Een ander deel van de analyse was gericht op het vinden van correlaties tussen de verschillende variabelen die gebruikt zijn om aan te duiden op welke manier boeren van elkaar verschillen. Een uitkomst was dat er onder de respondenten verschillen zijn die significant met elkaar samenhangen. Bijvoorbeeld, land is niet voor iedereen even toegankelijk. Dit is een probleem, niet alleen omdat land nodig is voor de productie van cacao, maar ook omdat land vaak wordt gevraagd als onderpand voor het verkrijgen van krediet bij een financiële instelling. Zonder grond is ook mogelijke deelname aan boerenorganisaties beperkt. Dit heeft weer gevolgen voor de kans die je als boer hebt op het verkrijgen van toegang tot bepaalde trainingsprogramma's. De meerderheid van de boeren die toegang heeft tot opleidingen is landeigenaar. Boeren zonder eigen land, veel vrouwelijke boeren, boeren die migrant zijn en jongere boeren hebben minder kans om te profiteren van dit soort programma's. Dit is vanzelfsprekend geen wenselijke situatie, en wordt problematisch als hierdoor de cacao-sector minder aantrekkelijk wordt voor jonge boeren, wat een bedreiging zou vormen voor de toekomst van deze belangrijke economische sector voor Ghana.

7 Het risico van insluiting

In dit hoofdstuk heb ik gekeken naar de interactie tussen de verschillende *governance* niveaus (mondiaal, nationaal en lokaal) door in te zoomen op een aantal van de interventies die plaatsvinden in de cacao-sector in Ghana. De belangrijkste vragen die ik in dit hoofdstuk probeer te beantwoorden zijn: hoe profiteren verschillende groepen boeren van upgradingstrategieën en de daaruit voortvloeiende interventies, en in welke mate nemen deze groepen actief deel aan de interventies en initiëren zij eigen strategieën? Op basis van de manier waarop de interventie wordt geïntroduceerd (stimulerend of dwingend), het bereik van de interventie, de verwachte impact, mogelijke beperkingen en onverwachte *trade-offs* geef ik aan in hoeverre de interventie inclusief kan worden genoemd.

Er zijn meerdere interventies die leiden tot *upgrading*. Deze interventies, die worden geïnitieerd door verschillende actoren in de keten, werken op elkaar in. Om een overzicht te maken, heb ik een groot aantal interventies geïdentificeerd die van invloed zijn op cacaoboeren. Deze interventies heb ik vervolgens gestructureerd rond een aantal substrategieën. Deze heb ik op hun beurt gekoppeld aan drie belangrijke upgradingstrategieën voor kleinschalige boeren in ontwikkelingslanden, die in de literatuur worden aangegeven. Substrategieën voor het creëren van hogere marges voor onbewerkte cacao *(upgradingstrategie 1)* kunnen bijdragen aan het produceren van betere kwaliteit cacao, het verhogen van de productiviteit en de productie van grotere volumes van cacao, en de productie van cacao onder betere contracten. Deze eerste groep van substrategieën wordt gedomineerd door grootschalige publieke interventies. De Ghanese overheid is de belangrijkste speler als het gaat om de bescherming van de kwaliteitsnormen en het investeren in een toename van cacaoproductie. Internationale opkopers delen dit streven van de regering, maar spelen zelf een tamelijk passieve rol. Kwaliteitscontrole en het zetten van standaarden hebben betrekking op alle boeren, maar interventies die bestaan uit vormen van dienstverlening bereiken lang niet alle boeren even gemakkelijk. Voor de respondenten uit de survey kwamen als belangrijkste factoren die (mede de) toegang tot dergelijke diensten bepalen naar voren: landeigendom, sociale positie in de gemeenschap en locatie (regio). De boeren zijn zelf verantwoordelijk voor de productie van hoge kwaliteit cacao, maar ontvangen steeds minder prikkels om de traditioneel hoge kwaliteit te waarborgen. Boeren dragen actief bij aan het verhogen van hun productie en streven naar het verbeteren van hun productiviteit. Boeren maken hierbij gebruik van verschillende strategieën, bijvoorbeeld door een effectieve bestrijding van plagen, de aanplant van nieuwe variëteiten bomen en door samen te werken met andere boeren.

Substrategieën die bijdragen aan de productie van nieuwe vormen van bestaande producten *(upgradingstrategie 2)* zijn onderverdeeld in de productie voor niche markten, de ontwikkeling van niet-traditionele toepassingen van cacao en diversificatie naar niet-traditionele producten, en andere (niet-agrarische) inkomstengenererende activiteiten. Deze tweede groep substrategieën bestaat voornamelijk uit multi-stakeholder initiatieven en interventies van maatschappelijke organisaties, die over het algemeen kleinschalig en uitsluitend zijn. Onder de respondenten speelden landeigendom en de regio een doorslaggevende rol in het bepalen van toegang tot dit soort activiteiten.

Substrategieën die bijdragen aan het lokaal verwerken en verhandelen van grondstoffen *(upgradingstrategie 3)* zijn bijvoorbeeld te vinden in de verwerking van cacaoresten, de verwerking van cacaobonen en de verkoop van cacaobonen. Deze derde groep bereikt de overgrote meerderheid van boeren slechts indirect. De internationale opkopers zijn de belangrijkste spelers en delen de interesse van de Ghanese overheid in het verplaatsen van een deel van hun cacao verwerkingscapaciteit naar Ghana.

Ik heb een selectie gemaakt van vier interventies die in dit hoofdstuk in detail worden besproken. Ik bespreek twee grootschalige publieke interventies (beiden vallend onder strategie 1); één gericht op de productie van hoge kwaliteit cacao, het systeem voor kwaliteitscontrole, en de andere op het verhogen van de cacaoproductie, het publieke *spraying* programma. Deze publieke interventies verschillen zowel in de impact die ze hebben op boeren als in de groep boeren die ze bereikt. Indirect, betalen boeren zelf voor deze publieke interventies. Een probleem is een gebrek aan transparantie over de kosten en baten van deze investeringen en de wijze waarop deze zijn verdeeld. Een derde interventie betreft een kleiner *multi-stakeholder* initiatief (vallend onder strategie 2) gericht op de enige formele boerenorganisatie in Ghana, de Kuapa Kokoo Farmer Unie (KKFU). Deze boerenunie telt ongeveer 50.000 leden. Boeren die lid zijn van Kuapa Kokoo produceren een klein deel van hun bonen voor de *fair trade* markt. Lidmaatschap van Kuapa Kokoo biedt toegang tot een alternatief verhandelingkanaal en draagt verder bij aan de versterking van boeren. Een vierde interventie die ik toelicht gaat over de lokale verwerking van cacao. Deze interventie (vallend onder strategie 3), die gepleegd wordt door internationale cacaoverwerkers, heeft geen directe impact op de boeren, maar draagt door het consolideren van de betrekkingen tussen Cocobod en de verwerkende industrie wel bij aan een continue vraag naar Ghanese cacao.

Het doel van de analyse was om inzicht te krijgen in hoe de belangen van de verschillende spelers in de cacaoketen zich lokaal manifesteren, wie de upgradingagenda bepaalt en welke upgradingstrategieën de voorkeur hebben. Bovendien wilde ik de sterke en zwakke punten van de verschillende interventies markeren. In mijn analyse heb ik de impact van interventies op de boer (individueel niveau) geïllustreerd door gebruik te maken van een bestaande matrix, de *'empowerment matrix'*. Daarnaast heb ik een 'scenario matrix' ontwikkelt die me in staat stelt om te reflecteren op de cacao-sector in zijn geheel (collectief niveau). Deze tweede matrix bestaat uit twee dimensies: veranderingen in de vraag (van eisen aan productkwaliteit naar eisen aan het productieproces), en in de mate van liberalisering (van gedeeltelijk naar volledige liberalisering). Door te kijken naar mogelijke veranderingen in de context heb ik geprobeerd een beter begrip te krijgen van de kwetsbaarheid van het huidige Ghanese systeem. Dit draagt bij tot het identificeren van upgradingstrategieën die (ook) effectief zijn op de langere termijn.

De Ghanese casus wordt vaak gepresenteerd als voorbeeld en belichaamt twee belangrijke dimensies: ten eerste, het is een unieke casus door de gedeeltelijk geliberaliseerde economie, en ten tweede is Ghana uitzonderlijk voor de productie van grote hoeveelheden hoge kwaliteit cacao. Het Ghanese systeem reflecteert deels de sterke rol van de Ghanese staat en deels geeft het de belangen van internationale opkopers van cacao aan in het handhaven of slechts lichtjes wijzigen van het Ghanese systeem. De productie van hoge kwaliteit cacao weerspiegelt zowel de capaciteit van de nationale regering om de sector te coördineren, als ook de bestaande vraag naar hoogwaardige cacao.

De upgradingstrategieën die plaatsvinden in Ghana weerspiegelen deze dimensies, en richten zich vooral kwaliteit en volume van productie. Maar de omstandigheden waaraan deze dimensies ten grondslag liggen, staan niet vast. Ten eerste is er een trend in de internationale cacaoketen dat producteisen minder belangrijk worden dan eisen die worden gesteld aan productieprocessen. Ten tweede, is het niet zeker of een gedeeltelijk geliberaliseerd systeem een daadwerkelijk eindstadium is. Dit maakt dat inzicht in de positie van Ghanese cacaoboeren in de keten en het soort upgradingstrategieën die gunstig zijn voor boeren een dynamisch perspectief eist, niet alleen door lessen te trekken uit het verleden en het maken van vergelijkingen met andere landen, maar ook door rekening te houden mogelijke toekomstige scenario's. Ghanese boeren zijn momenteel beter af, maar wat als de belangrijkste pijlers die ten grondslag liggen hun sterke positie verdwijnen?

Het blijkt dat Ghana niet goed is voorbereid op verandering. Boeren en de private sector zijn in het bijzonder kwetsbaar. Kijkend naar ervaringen in andere cacaoproducerende landen in de regio, die de sector volledig geliberaliseerd hebben, is het duidelijk geworden dat gebrekkige boerenorganisaties en een zwakke private sector belangrijke knelpunten zijn voor boeren om te profiteren van verdere hervormingen. Toch investeert de Ghanese overheid niet in capaciteitsopbouw van deze actoren. Een zwakke organisatie van boeren maakt het ook steeds moeilijker om te voldoen aan (veranderingen in) de vraag. Bovendien heeft dit bijgedragen aan een het onvermogen van boeren om zelf hun positie te veranderen en om meer te profiteren van huidige systeem. Meer inclusieve upgrading vereist dat er meer nadruk wordt gelegd op de versterking van boeren en lokale private actoren, het vereist ook meer aandacht (vooraf) voor de doelgroep en verwachte impact van interventies.

8 Conclusies

Het doel van deze studie was om goed inzicht te krijgen in de mogelijkheden en beperkingen van boeren in ontwikkelingslanden om hun positie in internationale ketens te verbeteren. Dit onderzoek is relevant nu steeds meer verwacht wordt dat boeren zich als 'bedrijven' gedragen. Echter, mijn resultaten laten zien dat actoren hoger in de keten en de institutionele omgeving die boeren omringd, nog steeds in grote mate bepalend zijn voor de ruimte die beschikbaar is voor boeren om verbeteringen door te voeren en voor de richting van 'vooruitgang' die boeren kunnen inslaan. Het geval van Ghana, een land met een cacao-sector die gedeeltelijk geliberaliseerd is, toont een voorbeeld van een keten waarin de rol van de overheid nog steeds sterk is. De Ghanese overheid neemt een positie in tussen lokale producenten en internationale opkopers van cacao.

In Ghana functioneert de staat zowel als *balancer* en als *bottleneck*. Als *balancer* beschermt de staat cacaoboeren tegen prijsschommelingen op de wereldmarkt, voorziet zij boeren van een stabiel inkomen en investeert zij een deel van de cacaoinkomsten terug in de cacao-sector. Het Ghanese systeem garandeert internationale opkopers het aanbod van hoge kwaliteit bonen. Het huidige systeem vermindert zowel de risico's voor de boeren als voor internationale opkopers, dit draagt bij aan goede handelsbetrekkingen. In een gedeeltelijk geliberaliseerd systeem kan de overheid ook functioneren als *bottleneck*. Zo maakt de Ghanese overheid het voor een aantal andere publieke, private en maatschappelijke actoren moeilijk een meer actieve (en effectieve) rol te vervullen. De overheid gaat in sommige gevallen ontwikkeling tegen, wat gekoppeld is aan het gebrek aan transparantie over de wijze waarop zij kosten, baten en risico's berekent en verdeelt.

Anderen kunnen leren van het voorbeeld van Ghana en de sterke rol van de overheid, aangezien hierdoor de risico's voor zowel boeren als andere keten actoren verminderd worden. Toch zijn er twee risico's die te maken hebben met deelname aan het systeem. In de eerste plaats zijn op het niveau van de producent de arrangementen sub-optimaal en bieden geen stimulansen voor boeren om zich te gedragen als ondernemers. Ten tweede, is de staat te veel naar binnen gericht en mist een adaptieve benadering. Dit brengt een risico voor de sector als geheel.

Deze casus laat zien dat een gedeeltelijk geliberaliseerd systeem voordelen biedt, maar ook risico's met zich meebrengt. Om deze risico's te overbruggen is het belangrijk om 'boeren voorop te stellen'. Dit gaat niet alleen over het realiseren van meer voordelen voor boeren, maar gaat ook over het daadwerkelijk versterken van boeren en hun op deze manier in staat te stellen actieve spelers in de keten te worden.

Appendix 3.1 Overview participants workshops

Workshop 1: participants cocoa workshop, 29 October 2003, Amsterdam

Name	Organisation	Position
Ingrid Aaldijk	Stichting Milieunet / De Derde Kamer	
Amma Asante	Gemeenteraad Amsterdam	Raadslid, Onderzoeker
Isa Baud	AGIDS	Director
Harald Bekkers	ASSR	PhD Candidate
Anita Blom	Min of Social Affairs and Employment	Policy Advisor
Jenny Botter	ICCO, Interchurch Organization for Development	Policy Officer Fair Economic Development
Freek van Breemen	Senior-trader	Commercieel manager West-Africa
Dick de Bruin	Sitos Group	Managing Director
Kees Burger	ESI VU	Head Econ Research Div
Mark Clayton	Common Fund for Commodities (CFC)	
S. Delodder	Rabobank	Research and advisory
Chris Dutilh	Unilever / Stichting Duurzame Voedingsmiddelenketen	
Joost Engelberts	Dahltv	Researcher
Mauk Faber	Rabobank International	Director
Eelco Fortuijn	FairFood	Director
P. van Goor	Theobroma B.V.	Trader
Peter van Grinsven	Masterfoods	Cocoa Sustainability Field Research Manager
Henk Hartoch	IUCN	
Aagje van Heekeren	Kocon/Derdekamer	Consultant/lid
Anouk van Heeren	CREM	Consultant
Jan Hoijtink	NIDO	
Ard Hordijk	Twynstra Gudde Management Consultants	Consultant
Dr.Carel A. van Houten	M.A.T.R.I.X. bv	Senior Consultant
Koert Jansen	Triodos Bank	Investment officer
Gerd Junne	UvA / FMG	Hoogleraar
Jan W. Kips	Daarnhouwer en Co. BV	Director
P.L.M.Koopmans	Continaf B.V.	Senior Trader
Wouter Klootwijk	Dahltv	Researcher
Antoine Legrand	Cargill	
Marlies Lensink	Koffiecoalitie	Campaign-coördinator
Frank van der Linden	FvdL Consultancy	Directeur
Viktor Mattousch	applying at AGIDS	PhD
Karel Menu	Unicom(International)BV	
Milah Wouters	KPMG	
Alan Muller	Erasmus Univ. Rotterdam	Researcher
Annelot Tempelman	SOMO	Researcher
Sjoerd Panhuysen	Koffiecoalitie	
Frans Paul van der Putten	EIBE Nyenrode	CSR Research
Esther Schouten	PWC	
Mrs Viparat Sookkaew	M.A.T.R.I.X. bv	Consultant
Marcel Spaas	FairFood	Campaign manager
Eduard Stomp	De Duurzaamheidsdesk	Director

Taco Terheijden	Cargill	Trader
Johan Verburg	Novib Oxfam Netherlands	CSR advisor
Hugo Verkuijl	KIT Senior Advisor	
Martin Versteeg	Sitos Group	
Anje Wind	Rabobank Foundation	Programme Manager Africa
Fred Zaal	AGIDS	Universitair Docent
Wouter Zant	VU	researcher
Moniek de Zwaan	Stichting DOEN	Programma manager Ontwikkelingssamenwerking

Workshop 2: List of participants, 31 March 2005, Ghana

Group 1   Farmers

Name	Address	Function
1. Kwame Donkoh	Kokoase, Amenfi West, WR	Cocoa farmer
2. Bernice Donkoh	Kokoase, Amenfi West, WR	Cocoa farmer
3. Agnes Donkoh	Kokoase, Amenfi West, WR	Cocoa farmer
4. Francis Asare	Afeaso, Twifo Praso, CR	Cocoa farmer
5. James Otoo	Bobi, Twifo Praso, CR	Cocoa farmer
6. Alex Amoah	Abekankw, Twifo Praso, CR	Cocoa farmer
7. Anthony Arhin	Dunkwa, Upper Denk.Dunkwa, CR	Cocoa farmer
8. Charles Adjei	Ohiamatuo, Amenfi West, WR	Cocoa farmer
9. Patience Oye	Ayamfuri, Upper Denkyise, CR	Cocoa farmer
10. Isaac K. Gyamfi	Kumasi, Ashanti	Sustainable Tree Crop Programme

Facilitator: Eric A. Agyare, SNV.
Recorder: Henry Anim-Somuah, student

Group 2   Private sector

Name	Organisation	Position
1. F. Frimpong	Cocoa Merchants	HR Manager
2. E.G. Asante	Cocoa Research Institute Ghana	Agric. Economist
3. N. Leibel	Cadbury	Manager
4. F. van Breemen	ADM	Manager
5. P. van Grinsven	Masterfoods	Manager
6. Solomon K. Addo	Barry Callebaut	Commodity analyst
7. William Nuamah	Reiss&Co Ltd	Agronomist
8. Marc Kok	Wienco	Agronomist
9. Nana Kwantwi-Barimah	Farmers Alliance Company	Acting Managing Director
10. Ralph Odei-Tettey	Wienco	Agronomist

Facilitator: Lawrence Attipoe, SNV
Recorder: Robert Assan-Donkoh, Student

Group 3	Public Sector	
Name	Organisation	Position
1. Philip Twu	Cocobod	Deputy Director
2. Anim Kwapung	CRIG	Researcher
3. T.G. Essandoh	CPC Ltd	Chief Accountant
4. R. Poku Kyei	Ministry of Finance	Special advisor
5. Ernest Dame	Department Cooperatives	Deputy Registrar
6. Kizito Ballans	WAWDA	DCD
7. Cosmos Marisu	MOFEP	Asst. Econ. Plan Officer

Facilitator: Maureen Odoi, SNV
Recorder: Isaac K. Asare, Student

Group 4	External Agents	
Name	Organisation	Position
1. J.Sinclair	Sitos (Gh)	
2. D. Snoeck	CIRAD	Researcher
3. F. Ruf	CIRAD	Researcher
4. Bob Hensen	Netherlands Embassy	Second Secretary
5. Greg Vaut	USAID/WARP	Public Private Alliances
6. Shaun Robertson	USAID/WARP	Phytosanitary Advisor
7. Ayenor Godwin	COS Project, PhD student	researcher
8. Dr. D.B. Sarpong	University of Ghana	Senior Lecturer
9. A.B. Andani	University of Ghana	Student
10. Mizane Yohannes	Grass roots Africa	Researcher
11. Yaw Osei-Owusu	Conservation International	Manager
12. Okyeame Ampadu-Agyei	Conservation International	Country Director
13. Zakaria Sulemana	Action Aid	CEF Coordinator

Facilitator: Pedro Arens, SNV
Recorder: Eric K. Doe

Appendix 6.1 Relations between farmer characteristics

Gamma	Position on farm	Kind of contract	Position in community	Yield	Gender	Working together	Education	Age	Migrant	Region Cramer's V
Position on farm		-	0,436***	0,375***	0,465**	-	0,391***	0,446***	0,667***	-
Kind of contract	-		-	-	-	0,519***	-	-	-	0,248***
Position in community	0,436***	-		0,257***	-	-	-	0,315***	-	0,338***
Yield	0,375***	-	0,257***		0,478***	-	0,210**	-	-	0,259***
Gender	0,465**	-	-	0,478***		0,650***	0,504***	-	-	0,221**
Working together	-	0,519***	-	-	0,650***		-	-	-	0,272***
Education	0,391***	-	-	0,210**	0,504***	-		-0,321***	-	-
Age	0,446***	-	0,315***	-	-	-	-0,321***		-	-
Migrant	0,667***	-	-	-	-	-	-	-		-
Region Cramer's V	0,272***	0,338***	-	0,259***	0,221**	0,272***	-	-	-	

Appendix 7.1   Analysing upgrading strategies

Strategy 1: Capturing higher margins for unprocessed cocoa	Intervention (identified between 2002 and 2005)	Activity	Mechanism	Farmers reached	Expected impact	Constraints	Trade-offs
Sub-strategy 1.1. Capturing higher margins for unprocessed cocoa by producing better quality cocoa	International institutions FAO/EU	Setting standards	Judicial (repressive)	All farmers (inclusive)	Impact 1: Meeting standards creates access to international market	Some governments pose more stringent standards than others; Some of the standards are subject to debate: (e.g. the application of the chemical Gammalin and its impact on health and environment).	Not identified
	International buyers	Paying premium	Economic incentive (stimulating)	All farmers (inclusive)	Impact 1: Premium adds value to the bean (decline in premium lowers value)	No transparency on distribution of premium; No transparency on distribution of costs involved in meeting 'premium standards'; No price-differentiation (farmers have no choice: no real incentive).	– The cocoa that does not meet the standards can only be sold at fraction of its price.
		Reject beans	Economic disincentive (repressive)	On one occasion (2005) Cocobod responded by sanctioning LBCs, who (temporarily) stopped buying cocoa.	Impact 1: No export of 'inferior' cocoa	Not identified	– Farmers have no storage facilities; this is why rejection of beans hits farmers hardest → farmers are the main risk takers.
	Ghanaian government QCD	Quality control	Control (repressive)	All farmers (inclusive)	Impact 1: Meeting standards creates access to market	Little incentives for local buyers (PCs), who are responsible for first check, to be strict on quality control, which also reduces incentives for farmers to follow traditional fermentation and drying practices; Corruption among quality control officials; Fragmentation of extension services: farmers receive little and sometimes conflicting advice; Quality control system is not directed towards prevention of inferior cocoa.	+ QCD system contributes to traceability of cocoa. QCD system contributes to premium price paid for Ghanaian cocoa; Cost saving for international buyers (low level of additional testing). – Quality control system is expensive. There is no transparency on distribution of costs (and benefits) of system; Farmers carry large share of the costs for delivering only premium cocoa to the market (cf. pre-selection), these are not taken into account.

Appendix 7.1 continued

Strategy 1: Capturing higher margins for unprocessed cocoa

Intervention (identified between 2002 and 2005)	Activity	Mechanism	Farmers reached	Expected impact	Constraints	Trade-offs
MoFA	Extension services	Learning (stimulating)	Large-scale but exclusive: Access to quality extension depends on location, social position, level of ownership, level of education and yield (Box 6.2 and Appendix 6.1). Survey indicates that around 57% of farmers received quality assistance, mainly coming from public extension services (FS, 2005)	Impact 1: Extension services help to meet standards Impact 2: extension services involve dissemination on good agricultural practices, pest management and new technologies. Impact 3: Extension services provides farmers with access to information	Merge of cocoa extension services with MoFA weakened services to farmers (in number and quality); adoption rates of new technologies (farmers lack capital and production costs are high); Public extension services are top-down and not based on farmer-experiences.	Not identified
CRIG	Research/listing recommended practices	Learning (stimulating)	Inclusive: Information is broadcasted on radio and published in newspapers	Impact 1: Meet international standards Impact 3: access to information	Many research outcomes do not reach the farmers; Illiteracy constrains farmers from understanding written information	Not identified
CMC	Sanctioning LBCs	Economic disincentive (repressive)	Large-scale, but only on one occasion. Then, as a result, some LBCs stopped buying cocoa, adversely affecting the farmers who normally sell to them.	Impact 1: contributes to competitiveness of Ghanaian cocoa (attempt to protect image of Ghanaian superior cocoa)	No consensus on causes for quality decline, resulting in conflicting advice to farmers.	− Farmers and LBCs are hit hardest by sanctioning. LBCs temporarily stopped buying cocoa, affecting farmers livelihood (Chapter 5) Farmers indicated that also the possibilities to overcome (temporary) financial constraints has become difficult; 'now creditors and buying companies are [even more] reluctant to give farmers loans because they are afraid their cocoa will be rejected'.
Local private sector	Quality control	Control (repressive)	All farmers (inclusive)	Impact 1: Meeting standards creates access to market	Competition on volume instead of on quality or prices. This provides LBCs and their clerks little incentive to be very strict on quality control.	+ Ghana's position as producer of world's finest cocoa is increasingly at risk.
	Drying cocoa	Applying existing knowledge (agency)	Large scale. Involved LBC and its suppliers (farmers)	Impact 1: PCs take over tasks of farmers, paying them the (same) floor price	Not identified	+ Farmers do not have to spend time and labour on drying cocoa − Disincentive for farmers to deliver quality cocoa

Appendix 7.1 continued

Strategy 1: Capturing higher margins for unprocessed cocoa

Intervention (identified between 2002 and 2005)	Activity	Mechanism	Farmers reached	Expected impact	Constraints	Trade-offs
PBC	Training farmers	Learning (stimulating)	Small-scale, occasionally and exclusive: only farmers selling to PBC. In 2005 this was around 30 per cent of the farmers. I have no information on exactly how many farmers have been trained by PBC.	Impact 1: Meeting standards to create access to markets Impact 3: farmers get access to information	Training was focused on avoiding purple beans entering the market, without agreement on exact causes for decline.	Not identified
Farmer groups KKFU	Purple bean seminars	Learning (stimulating)	Small-scale and exclusive: only members of Kuapa Kokoo (KKFU), which has around 50.000 members. I have no data on the exact number of participants in seminars.	Impact 1: meeting standards to create access to markets Impact 3: farmers get access to information	Training was focused on avoiding purple beans entering the market, without agreement on exact causes for quality decline.	Not identified
KKFU	Small bonus for dried cocoa	Economic incentive (stimulating)	Small-scale, occasionally and exclusive: only members of KKFU. I have no data on how many KKFU members actually received this extra bonus.	Impact 1: meeting standards to create access to markets	Not identified	Not identified
Individual farmers	Traditional fermentation and drying practices	Applying existing knowledge (agency)	Inclusive	Impact 1: meeting standards to create access to markets	Farmers receive conflicting advice on the required number of days for drying and fermentation	- Labour intensive process.
	Pre-selection of good pods/beans	Applying existing knowledge (agency)	Inclusive	Impact 1: Meeting standards to create access to markets		- Costs for pre-selection are borne by the farmers and not taken into account in cost-benefit calculations.
	Pest-management	Applying existing knowledge (agency)	Inclusive	Impact 2: Avoiding loss of premium quality beans		Increase use of chemicals can result in excess levels of pest residues, making the cocoa not suitable for export.

Sub-strategy 1.2: Increase in productivity and higher volumes of production

Intervention	Activity	Mechanism	Farmers reached	Expected impact	Constraints	Trade-offs
International institutions Research institutes (CIRAD)	Research	Learning (stimulating)	Difficult to estimate	Impact 2: research for large part directed towards more effective pest management, new varieties. If the knowledge becomes available for farmers this can contribute to higher yields	Information does not always reach farmers Low adoption rates	Not identified

Appendix 7.1 continued

Strategy 1: Capturing higher margins for unprocessed cocoa

Intervention (identified between 2002 and 2005)	Activity	Mechanism	Farmers reached	Expected impact	Constraints	Trade-offs
International buyers	Research on pests and diseases, integrated pest management, new varieties, etc.	Learning (stimulating)	Large scale but indirect and exclusive. Reaches farmers through extension services. Not all farmers have access to these services.	Impact 2: yields Impact 3: access to new knowledge	Information does not always reach farmers Low adoption rates	Not identified
Ghanaian government	Increase producer-price	Economic incentive (stimulating)	Inclusive: All farmers receive floor price	Impact 1: Farmers receive higher margin	Increase in price does not automatically increase farmer income due to rise in production costs and costs of living. In Ghana producer-price is based on % net FoB price, where Cocobod retains a large share of FoB.	- Margins for other actors reduce. For LBCs this resulted in less motivation/capacity to invest in maintaining good relation with farmers.
	Bonuses (compensation)	Economic incentive (stimulating)	Large-scale but exclusive: Region and social network turned out to be significant in determining access to bonuses. Bonuses have not been handed out every year.	Impact 2: Remunerative income	Bonuses are very small: farmers were disappointed Delays in payments Problems with registration	- Bonus is actually not an extra but a way of compensation for situations where exchange rate and world market price exceed projected value. If farmers don't get bonus this is a cost for them.
CODAPEC	Mass-spraying programme	Economic incentive and learning (stimulating)	Large-scale but exclusive: farmer survey in 2005 indicates that only 6 per cent receives 4 times spraying. Frequency of spraying depends on region and social position (appendix/box)	Impact 2: Higher yields Impact 3: Farmers experience the impact of weeding and application of chemicals	Spraying gangs favour their relations and farmers obtaining a strong position. Part of chemicals is allocated to black market. Logistical problems (fuel, timing, etc.)	+ A condition for receiving spraying was weeding. This in itself contributed to higher yields. - The mass spraying programme is supposed to be 'free'. But costs are paid from difference between net and gross FoB price. The spraying programme does not stimulate entrepreneurial behaviour of farmer. The mass spraying programme obstructs the introduction of more friendly methods of pest management and makes it more difficult to introduce the production of organic cocoa.

252

Appendix 7.1 continued
Strategy 1: Capturing higher margins for unprocessed cocoa

Intervention (identified between 2002 and 2005)	Activity	Mechanism	Farmers reached	Expected impact	Constraints	Trade-offs
CODAPEC	High-tech programme (fertilizer on credit)	Economic incentive and learning (stimulating)	Small-scale and exclusive (pilot).	Impact 2: Significant increase in yields. Impact 3: Farmers experience impact of fertilizer and are responsible for pay-back	Problems with pay-back loans	Not identified
Cocobod	Rehabilitation of (abandoned) cocoa farms	Economic incentive (stimulating?)	Small-scale and exclusive (no of farmers)	Impact 2: increase in volume of cocoa production	Farmers are not eager to return to farms	Not identified
MoFA	Extension services	Learning (stimulating)	Large-scale but exclusive (see earlier)	Impact 2: see earlier Impact 3: see earlier	See earlier	Not identified
CSSVD	Swollen shoot programme	Judicial (repressive)	Large-scale and inclusive: all farmers with affected trees are targeted	Impact 2: in the short-run it affects income but in the long-run it avoids further income losses.	Farmers do object to removal of sick trees.	+ Farmers are compensated
CRIG	Research and development of new varieties	Learning (stimulating)	Exclusive. Reaches farmer through extension services. Not all farmers have access to these services.	Impact 2: yields Impact 3: access to new knowledge	Information does not always reach farmers. Low adoption rates	Not identified
Local private sector Wienco (input provider)	Provision of fertilizer on credit to farmer groups, combined with extension services	Economic incentive and learning (both stimulating)	Small-scale and exclusive (number)	Impact 2: yields Impact 3: empowerment through training and the formation of groups	Costly exercise for intervener → reason for small-scale Core-business for intervener is selling inputs	+ Involvement of private sector in provision of credit turns out to be more successful than public interventions in this field. − Stimulates the use of fertilizer and chemicals, which can have a negative effect on the environment and health
LBCs/PCs	Provision of credit	Economic incentive (stimulating)	Very small-scale and exclusive: in 2005 2% of farmers obtained credit from a PC/LBC	Impact 3: Empowerment (ability to make a choice for investments)	Problems with paying back. Farmers are not loyal to PCs Yields are vulnerable for hazards	Not identified
Banks	Provision of credit	Economic incentive (stimulating)	Very small-scale and exclusive: In 2005, one per cent of farmers obtained a loan from a bank	Impact 3: Empowerment	High-interest rates, demand for collateral, demand for guarantees, lack of savings, small amounts of credit, lack of trust, problems with paying back loans etc. Yields are vulnerable for hazards	In case a harvest gets lost, farmers have to pay back loan, including interests, or run the risk for loosing collateral.

253

Appendix 7.1 continued

Strategy 1: Capturing higher margins for unprocessed cocoa

Intervention (identified between 2002 and 2005)	Activity	Mechanism	Farmers reached	Expected impact	Constraints	Trade-offs
Multi-stakeholder initiatives/PPP STCP	Farmer Field Schools: Farmer-based extension services and training	Learning (stimulating) and applying knowledge (agency)	Small-scale and exclusive: pilot phase only took place in Ashanti region.	Impact 2: schools generate knowledge on more efficient and more responsible (socially and environmentally (IPM)) practices Impact 3: Involvement of farmers in development of knowledge, training of farmers, improving information flow	High costs involved Mass-spraying obstructs the introduction of more environmentally friendly ways of pest management In producing countries the programme offered by STCP focused more on marketing and information systems, this is not necessary in Ghana for now, but might be in the farmers' future interest.	+ Farmers have also been trained on labour issues such as child labour STCP is an attempt to form (informal) farmer groups.
CI, KKFU, MoFA and CRIG	Farmer field schools in conservation areas	Learning (stimulating) and applying knowledge (agency)	Small-scale and exclusive: in total between 120-150 farmers participated in these schools and were trained as trainers, being mainly KKFU members (KKFU excludes shareholders)	Impact 2: Schools generate knowledge on more efficient and environmentally friendly practices (IPM/shade management) Impact 3: Involvement of farmers in development of knowledge, training of farmers, improving information flow	High costs involved (due to expensive experts) Farmers have difficulty applying all the knowledge due to lack of financial capital	+ These schools operated in nature conservation areas. Attention for inter-cropping and shade management. This is a way of diversification (risk-management). After the pilot, the ToT continue to meet other farmers and exchange knowledge and give advice
Farmer groups Informal (nnoboa)	Exchange labour and knowledge	Economic incentive, learning (stimulating) Applying existing knowledge (agency)	Large-scale but exclusive: for example women are far less involved in nnoboa.	Impact 2: Time-efficiency Impact 3: Working together and ex-changing knowledge	Lack of trust Lack of incentives	− In some cases the farmers' services provided on other farms are not reciprocated
KKFU	Credit unions	Economic incentive (stimulating)	Small-scale and exclusive: Only members of KKFU that have enough savings can apply for a small loan (doubling their savings)	Impact 3: Empowerment (ability to make a choice for investments)	Small loans Lack of savings Some credit unions are not operational Problems with pay back Yields are vulnerable for hazards	Not identified
Ad hoc organisation	Get advice/training/ access to products	Agency	Small-scale but including all farmers that are present at the time of meeting	Impact 3: Access to information/ empowerment	Not identified	+ Ad hoc organisation is simple and without too much costs (it is advisable to inform chief in advance) and can be very effective

Appendix 7.1 continued

Strategy 1: Capturing higher margins for unprocessed cocoa

Intervention (identified between 2002 and 2005)	Activity	Mechanism	Farmers reached	Expected impact	Constraints	Trade-offs
Individual farmers	Planting new varieties	Applying existing knowledge (agency)	Inclusive	Impact 2: yields	New varieties cannot be 'cloned'. Farmers often neglect this with as a result that second generation of new varieties is not successful.	Not indentified
	Applying good farm practices	Applying existing knowledge (agency)	Inclusive	Impact 2: yields	High production costs. Lack of capital. Fragmentation of extension services	Not identified
	Pest management	Applying existing knowledge (agency)	Inclusive	Impact 2: yields	Inadequate use of chemicals (related to illiteracy). Black market selling forbidden/ outdated chemicals	Little use of protecting clothes
	Using fallow land	Using existing resources	Only farmers that have fallow land (this excludes most caretakers)	Impact 2: yields	Increase in yield requires also extra investment.	– If no extra investment/labour available this entails the risk of increasing already heavy workload of farmer and family members.
	Hire more labour	On-farm investment	Only farmers that can afford hiring labour	Impact 2: yields	Labour costs are high	
	Savings and apply for credit	On-farm investment	Small-scale and exclusive: Very few farmers have access to (informal) credit (in 2005 a little bit more than 10 per cent of farmers)	Impact 2: yields	No bank-farm relation → lack of trust. Banks are far away. Barriers for getting credit (see earlier)	See earlier
	Participation in training	Learning (agency)	Training involves mainly farm-owners and more pro-active farmers	Impact 3: Empowerment	Farmers lack will and time to participate. Not every training has an impact (not always possible to apply new technologies)	Not identified
International institutions / International banks	Financing forward sales	Economic incentive (stimulating)	Inclusive	Impact 1: Contributes (indirectly) to premium price paid for Ghanaian cocoa	Not identified	Not identified
International buyers	Forward sales premium	Economic incentive (stimulating)	Inclusive	Impact 1: Contributes to premium price paid for Ghanaian cocoa	Not identified	Not identified

Sub-strategy 1.3: Producing under more remunerative contracts

255

Appendix 7.1 continued

Strategy 1: Capturing higher margins for unprocessed cocoa	Intervention (identified between 2002 and 2005)	Activity	Mechanism	Farmers reached	Expected impact	Constraints	Trade-offs
	Ghanaian government	Forward sales	Economic incentive (stimulating)	Inclusive	Impact 1: Contributes to premium price paid for Ghanaian cocoa	Not identified	Not identified
	Local private sector LBCs	Investment in (selection of) PCs	Trust-building	Indirect, but large-scale	Impact 2: contributes to less cheating and better relations between buyers and farmers	Not identified	Not identified
	LBCs/PCs	Prompt payment, provision of services, credit, subsidized inputs etc.	Building social capital	Small-scale and exclusive	Impact 2: Contributes to more remunerative contracts	Lack of loyalty among farmers; No bargaining between farmer (groups) and local buyers; LBCs/PCs are financially constrained to invest on a large scale.	No loyalty No (informal) farmer contracts